Computerized Circuit Analysis with SPICE

Computerized Circuit Analysis with SPICE

A Complete Guide to SPICE with Applications

THOMAS W. THORPE

A Wiley-Interscience Publication

JOHN WILEY & SONS, INC.

New York Chichester Brisbane Toronto Singapore

PSPICE is a registered trademark of MicroSim Corporation.
MicroSim is a registered trademark of MicroSim Corporation.
Texas Instruments is a registered trademark of Texas Instruments Corporation.
Motorola is a registered trademark of Motorola Corporation.

In recognition of the importance of preserving what has been written, it is a policy of John Wiley & Sons, Inc., to have books of enduring value published in the United States printed on acid-free paper, and we exert our best efforts to that end.

Library of Congress Cataloging in Publication Data:

Thorpe, Tom, 1944-
Computerized circuit analysis with SPICE: a complete guide to SPICE with applications / written by Tom Thorpe. – 1st ed.
p. cm.
"A Wiley-Interscience publication."
Includes bibliographical references.
1. Electric circuit analysis – Data processing. 2. SPICE (Computer program) I. Title.
TK454.T47 1991
621.381′548 = 02855369 – dc20 91-23497
ISBN 0-471-55164-3 CIP

Printed in the United States of America

Printed and bound by the Hamilton Printing Company.

10 9 8 7 6 5 4 3 2 1

To

Thomas W. Thorpe Jr.

PREFACE

This text was written to fill the void created by the development and introduction of computerized circuit analysis programs. Until now, it was assumed that the user understood the proper use and implementation of these programs without requiring an explanation of the basic concepts. This occurred as well with the SPICE circuit analysis package. Much has been written and published on how to apply SPICE using advanced modeling techniques. Little, if any, has been written on understanding the basics associated with the computerized circuit analysis concepts. This text provides that basic understanding of SPICE, allowing the reader to work progressively, with each new chapter building on the information presented in the previous chapter.

We begin with a general discussion of SPICE and address the basic circuit elements (resistors, capacitors, and inductors). Next we demonstrate the proper use of these elements by presenting the concept of a nodal analysis and AC analysis. The building process continues with the introduction of semiconductor devices, which lead naturally to the realization of the DC analysis. The convergence of an AC and DC analysis is then shown along with efficient circuit-definition techniques. The transient analysis is discussed with appropriate examples to support that discussion. The final chapter combines the results of all presented capabilities and demonstrates an overall analysis capability.

At the completion of the seven chapters, we have discussed the use and application of all major SPICE commands. To facilitate the understanding of the SPICE capability, examples of actual analyses are presented in each chapter. Each analysis is designed to support the goals of that chapter. Analysis results are also provided to allow a correlation between stimulus (command) and response.

Discussing only a capability fails to address the basic problem of how to best use the capability. The text spends extensive time (1) providing actual circuit examples, (2) providing detailed analysis of the resulting data, and (3) discussing ways in which the data can be analyzed and extracted. The emphasis is on understanding how data is produced rather than how to produce the data.

Thomas W. Thorpe

Monument, Colorado
December 1991

ACKNOWLEDGMENTS

This book is dedicated to my parents Thomas W. Thorpe Jr. and Kathryn M. Thorpe who were wise enough and persistent enough to instill the philosophy that a person's limitations are mainly self - imposed.

A special thanks goes to my wife, Sandra K. Thorpe, for her love, understanding and patience during the entire process of publishing this book, to my son, Thomas W. Thorpe, for his undying faith in me, and my daughter, Tammy L. Thorpe, for her faith and extensive typing and editing support in finalizing the book. Without their help this book would not have been completed.

The cover of this book was computer generated by a close talented friend of many years, Craig A. Lindley, who possesses a rare mix of creative flair and solid technical understanding which enabled him to find an optimal mix of abstract art and aesthetic technical appeal. To Craig I give a big thanks.

I additionally want to thank the gracious contributions of MicroSim Corporation and Texas Instruments Corporation for permitting the use of their SPICE models which enabled the book to take on an additional dimension of worth and usefulness.

T.W.T.

CONTENTS

FIGURES

TABLES

Computerized Circuit Analysis with SPICE

1

INTRODUCTION

The Simulated Program with Integrated Circuit Emphasis (SPICE) circuit analysis capability presented in this text represents the evolutionary development of the original SPICE software developed at the University of California at Berkeley in the early 1970s. Here, we use SPICE 2G.6, which is the basis for all current SPICE packages running today. The differences in SPICE packages lie mostly in the features associated with the specific package. SPICE-based packages may have schematic capture, worst-case/Monte Carlo analysis, optimization analysis, and Computer Aided Design (CAD) interface capabilities. Technology now exists for an engineer or designer to (1) draw a schematic using a CAD package, (2) convert it automatically to a SPICE format, (3) analyze the data, (4) make corrections, and (5) produce a circuit board layout. This integrated capability provides the designer with a very powerful capability to realize the rapid design, analysis, and prototyping of a new product.

1.1 WHAT IS SPICE?

SPICE represents a software circuit analysis package used extensively by both the research and development and the manufacturing segments of the electronics industry. SPICE provides a rapid and highly accurate means of simulating and verifying electronic circuit performance in ways that would prove costly and time consuming using conventional laboratory test and evaluation approaches. SPICE then is a tool which, properly used, can serve as a high-speed and cost-effective mechanism for the verification of any electronic circuit under a variety of adverse environmental conditions. As we proceed, the power and flexibility associated with this tool becomes increasingly obvious.

SPICE is a general-purpose circuit-simulation program for nonlinear DC, non-linear transient, and linear AC analyses. Circuits may contain resistors, capacitors, inductors, mutual inductors (transformers), independent voltage and current sources, four types of dependent sources, transmission lines, switches, and the four most common semiconductor devices: diodes, bipolar junction transistors (BJT), junction field effect transistors (JFET), and metal oxide substrate field effect transistors (MOSFET).

SPICE has models with generic parameters which can be adapted for a variety of popular semiconductor devices. The user need only to specify the pertinent model parameters to realize a highly accurate analysis. The model for the BJT is based on the integral charge model of Gummel and Poon; however, if the Gummel-Poon parameters are not specified, the model reduces to the simpler Ebers-Moll model. In either case, charge storage effects, ohmic resistances, and a current-dependent output conductance may be included. The diode model can be used for either junction diodes or Schottky barrier diodes. The JFET model is based on the FET model of Shichman and Hodges.

1.2 WHY LEARN SPICE?

As alluded to above, SPICE can validate an electronic design in far less time than conventional validation techniques at a fraction of the cost. Traditionally, the electronic industry has, in its design cycle of a product or circuit, performed the following steps, to realize a working design of a circuit:

1. The design engineer initially designs the circuit, based on a set of requirements, using traditional design practices and his or her best engineering judgement.

2. The design next enters a prototype phase in which parts are procured and resources allocated.

3. The design is then built (prototyped) using expensive manpower and capital equipment resources. It often possesses flaws that require design modification or redesign. If flaws do exist, an additional cost is incurred in the ordering of new or additional parts. Additional resources must also be expended to verify the finished stable design.

4. The design is finally verified for proper performance using a laboratory testing environment to simulate the actual intended environment. This testing typically represents a major investment of time and effort on the part of a company.

In any design using this approach, issues of circuit stability due to temperature affects on components, initial component tolerances, or component aging must be considered. The risk of testing the design and missing an obvious problem remains high. This may well place the integrity of the product and the reputation of the manufacturer at risk with the customer.

The use of SPICE, before the prototype phase where the procurement of materials and the allocation of resources must be realized, presents a low-cost and low-risk alternative. The designer, knowing the expected circuit response, can choose to computer simulate the circuit in parts or in its entirety and in doing so can determine the performance characteristics of the design based on component tolerance variations due to environmental influences. Computer time is inexpensive and efficient when compared to the expense of building a new prototype. The use of SPICE, therefore, presents a highly attractive alternative for realizing a successful product development. Additionally, many SPICE packages provide the ability to interface with schematic capture and circuit board layout software packages and to perform analyses tuned to the performance assurances required by some governmental programs.

1.3 CAPABILITIES OF SPICE

The SPICE circuit analysis package provides significant flexibility to the user and is capable of modeling and simulating virtually any linear and digital circuit. The user need only define the circuit and the circuit components and perform the desired analysis. Although circuit analysis packages have existed for years in various forms and have various capabilities, SPICE represents an optimal compromise to currently available packages. Part of the reason for the success of SPICE is the program's ability to accurately simulate many different circuit component parameters without the necessity of generalized linear assumptions or detailed parameter definitions. An example of this lies in a beta curve of a traditional transistor. In properly modeling the beta curve, which looks much like a bell shaped distribution when collector current is plotted against beta, three approaches are possible:

1. A linear beta curve can be assumed that follows a traditional collector-current versus base-current relationship (ic = beta * ib)

2. A beta curve that consists of many data points in the form of beta versus collector current or base current versus collector current.

3. A beta curve has its peak, low end, and high end beta points modeled thus, defining a bell-shaped beta curve. This approach requires the definition of only three points.

Of the three approaches, early circuit analysis software chose to use approach 1 which represents the simplest approach. The problem that occurs with approach 1 is that unrealistic results exist when the transistor is used at very low and very high collector currents. Some second-generation circuit analysis programs chose to use the second approach. Highly accurate analyses were obtained, relative to a specific transistor's parameters, at the expense of excessive modeling time. The creators of the SPICE software chose the third option, which provided good simulation accuracy with a minimal amount of effort on the part of the user.

1.3.1 TYPES OF ANALYSES

The SPICE circuit analysis package provides the user with the ability to perform four basic types of analyses:

1. Operating point
2. DC
3. AC
4. Transient

It is essential to gain an understanding of the four types of analysis before proceeding further in the text. The reason simply is that you must know the type of analysis required to provide the data and results you want. The differences between the four types of analysis are elaborated upon below. The basic premise in describing the differences is that you can picture a laboratory environment in which you have power supplies, sweep frequency generators, oscilloscopes, AC/DC voltmeters, and function/pulse generators. The equipment is assumed to be at your disposal for the sole purpose of making measurements associated with a circuit contained on a circuit card. The circuit might be a collection of resistors and capacitors forming some type of passive filter, or it might be the functional elements in support of a stereo power amplifier consisting of both active and passive components. The focus, in describing the four types of analyses, will be on the circuit card with various pieces of the laboratory equipment attached. Before we connect any measurement equipment, assume that you have attached a power supply(s) and have DC power applied to the circuit card.

1.3.1.1 OPERATING POINT ANALYSIS

The operating point analysis capability of SPICE represents the circuitry on the circuit card as being under power without the application of any input stimulus. Assume that you chose to measure (1) all voltages at all points in the circuit referenced to ground, (2) the current being supplied by the power supply(s) powering the circuit, and (3) the power that the circuit is consuming. The power that the circuit is consuming is simply the output voltage of the power supply(s) multiplied by the current flowing into or out of the power supply.

The SPICE circuit analysis program does just this prior to attempting any type of analysis or simulation. This process is known as obtaining the DC OPERATING POINT of the circuit and is critical to the realization of any successful simulation because the DC operating point determines the bias points within a circuit, which directly affects the gain and linearity characteristics of the circuit. This analysis may prove highly valuable in diagnosing a complex resistor matrix, determining a component selection error, or realizing potential over stress conditions that exist in the circuit. SPICE, in computing current flow, uses the electron flow convention, thus, a current flowing from the positive terminal of a power supply, through the circuit, and returning to the negative terminal will be shown as a negative current and thus represents electron flow.

1.3.1.2 DC ANALYSIS

To understand a DC analysis, one must envision the circuit card discussed above, connected to a power supply(s). The input to the circuit card is from a variable voltage source, which can be adjusted in incremental steps between a start and end voltage. Assume you are monitoring the variable voltage source and an output from the circuit card with separate voltmeters. Next to the card you have constructed a data sheet that contains two vertical columns with the first labeled "Input voltage" and the second labeled "Output voltage." You begin by setting the input voltage source to 0.0 V and observe the output voltage reading. You record the input & output voltage on the data sheet. You then increment the input voltage source by 1.0 V and again measure the output voltage and record your readings. The process continues until such time as the input voltage source is at the maximum voltage of interest relative to its affects on the circuit. The data sheet might resemble Table 1-1.

Table 1-1. Recorded Input versus Output Data

Input voltage	Output voltage
-10.00	-5.45
-9.00	-5.45
-8.00	-5.44
-7.00	-5.42
-6.00	-5.04
-5.00	-5.00
-4.00	-4.00
-3.00	-3.00
-2.00	-2.00
-1.00	-1.00
0.00	0.00
1.00	1.00
2.00	2.00
3.00	3.00
4.00	4.00
5.00	5.00
6.00	5.40
7.00	5.42
8.00	5.44
9.00	5.45
10.00	5.45

You could then choose to store the data away for future reference, or you might choose to plot the data on graph paper. If you were to plot the data, it might be logical to have a vertical scale labeled "Output voltage" and a horizontal scale labeled "Input voltage." You would then have a mechanism to graphically represent the data and the graph would, like the data, provide information which showed the effect on the output as a function of the input. Such a graph might look as shown in Figure 1-1.

This type of information is Input versus Output data and is useful in determining the DC performance characteristics of a circuit. SPICE does exactly this when requested to perform a DC analysis and does it with the same capability that you could realize in the laboratory, but without the element of laboratory measurement error involved.

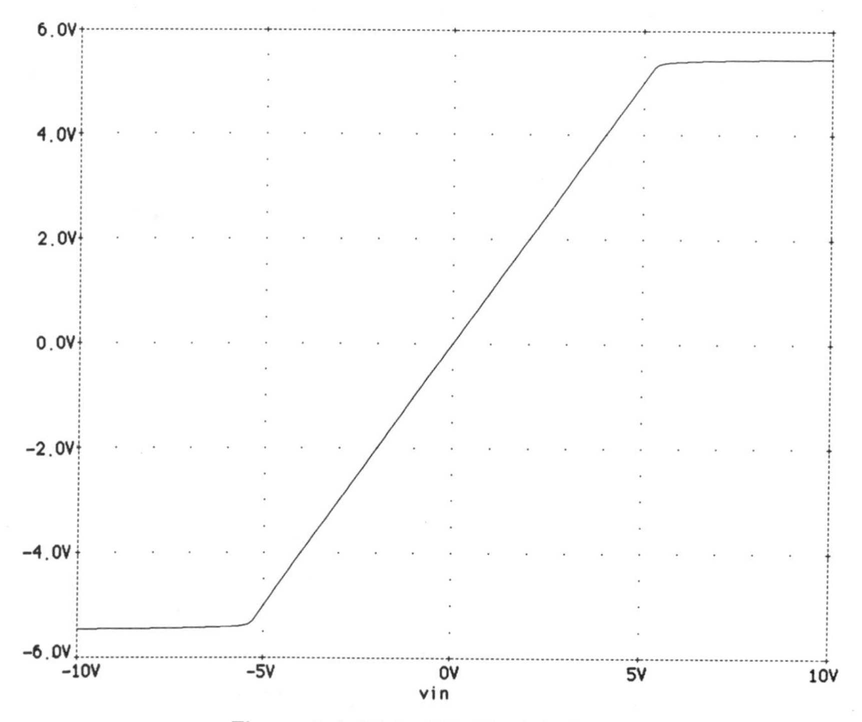

Figure 1-1. Plot of Table 1-1 data

1.3.1.3 AC ANALYSIS

To understand an AC analysis, it is necessary to again envision the circuit card discussed above connected to a power supply(s). The input connected to the card is from a sine wave generator with a fixed output amplitude. The sine wave generator has a fixed-amplitude output; therefore, you need only to monitor the frequency being produced by the generator, which is done with a frequency counter connected in parallel with the sine wave generator. The frequency range of interest has a starting frequency, an ending frequency, and a fixed step size to get from the starting to the ending frequency. Connect an AC voltmeter (broad band) to the output point of the circuit card that you want to monitor and record. Consistent with the procedure followed in the DC analysis approach, construct a data sheet with two vertical columns. The first labeled "Input Frequency in Hz" and the second labeled "Output in DB." Begin by setting the sine wave generator to 100 Hz and observe the output AC voltage reading on the AC voltmeter. Convert the AC voltage reading to DB and record the frequency at the input and at the output on the data sheet. Then increment the input frequency to 200 Hz and again measure the output voltage, recording your readings. The process continues until the input frequency source has been set to all frequencies of interest and its affects on the output of the circuit. Such a data sheet is illustrated in Table 1-2.

You could again choose to store the data away for future reference, or you might choose to plot the data in graph form. If you were to plot the data then it might be logical to have a vertical scale labeled "Output in DB" and a horizontal scale labeled "Input Frequency in Hz." This mechanism to graphically represent the data would, like the data, provide information that showed the effect on the output gain or loss as a function of the input frequency. Such a plot of the data is shown in Figure 1-2.

Table 1-2. Recorded Data of DB Loss

Input Frequency in Hz	Output in DB
10.00	-0.001
100.00	-0.103
1000.00	-5.417
2000.00	-10.760
10000.00	-30.480
100000.00	-69.440

This Frequency versus Output data and is useful in determining the AC performance characteristics of a circuit. SPICE does exactly this when requested to perform an AC analysis. The resulting plot is referred to as a Bode plot. Should you have chosen, in addition to recording the amplitude, to measure the phase shift through the circuit or the DB gain/loss of the circuit, then SPICE would have provided that capability also. The main elements of the AC analysis are as follows:

1. Uses a fixed amplitude sine wave source.

2. Varies the input frequency over a specified frequency range in a

specified step size.

3. Provides printed or plotted data that represents frequency versus some desired output parameter (amplitude, phase, decibels, etc.).

It must be noted that an AC analysis is a frequency domain analysis and has no relationship to any time dependent analysis. The frequency variation of the input is over a specified frequency range and as such has no period measurement associated with the input frequency.

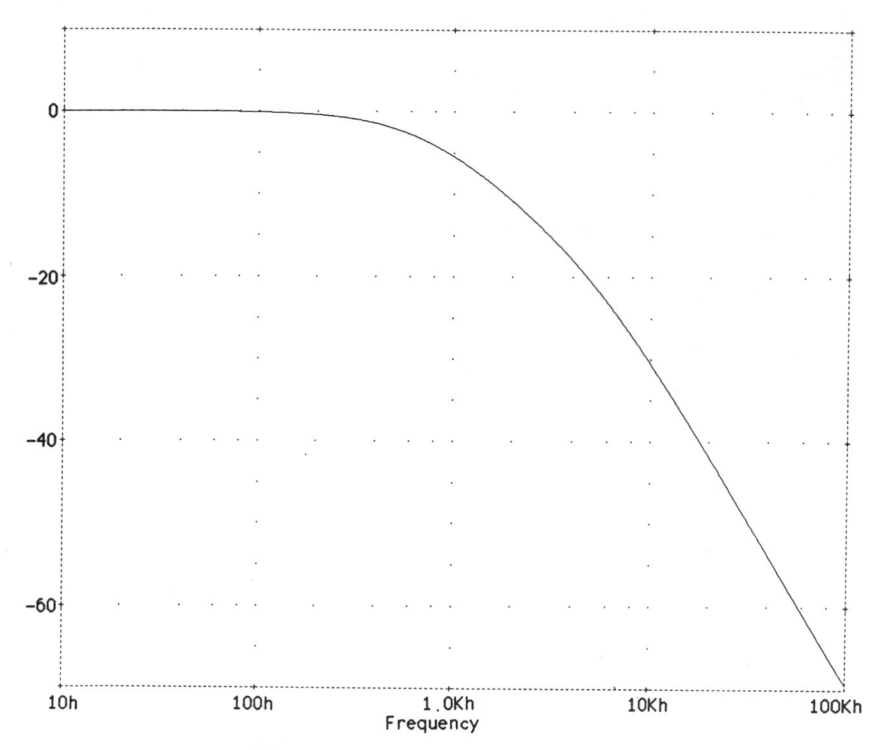

Figure 1-2. Plot of Table 1-2 data

1.3.1.4 TRANSIENT ANALYSIS

To understand a transient analysis, we must again envision the circuit card, discussed above, connected to a power supply(s). The input of the box is connected to a pulse generator with a fixed-output amplitude and a fixed frequency of waveform period (frequency = 1/period). With the pulse generator adjusted to a fixed-output amplitude, you need only monitor the period of the fixed-frequency waveform produced by the generator, which is done with channel 1 of an oscilloscope connected in parallel with the pulse generator. Connect channel 2 to the output point of the circuit card circuit you want to monitor and record. Consistent with the procedure followed in the AC and DC analysis approach, a data sheet can be constructed with two vertical columns, (1) "Time" and (2) "Output voltage amplitude." First observe the channel 2 output on the oscilloscope and adjust the Time-base switch of the oscilloscope until the desired number of periods (length of time of the oscilloscope sweep) is to your liking. Record the amplitude and time at the first point of interest, next record the time and amplitude data at some fixed time interval from the first measurement, and continue this process. Then the data can be plotted and exactly portrays the waveform seen on the oscilloscope. The chart of the recorded data is shown in Table 1-3.

Table 1-3. Recorded Output Waveform Data

Time in us	Output Voltage
0	0
.1	1.766
.2	4.025
.3	4.750
.4	4.619
.5	4.406
.6	4.615
.7	4.555
.8	4.508
.9	4.567
1.0	4.548
2.0	4.545
2.2	4.545
2.4	0.522
2.5	-0.217
2.6	-0.067
2.7	0.136
2.8	-0.067
2.9	-0.016
3.0	0.039
4.0	0.000
4.1	1.757
4.2	4.025
4.3	4.750

This Time versus amplitude data is useful in determining the transient performance characteristics of a circuit. Information about the rise time, fall time, propagation delay, and amplitude of the output is obtainable. When requested to perform a transient analysis, SPICE produces an oscilloscopic display and the associated data. Such a plot of the recorded data is shown in Figure 1-3.

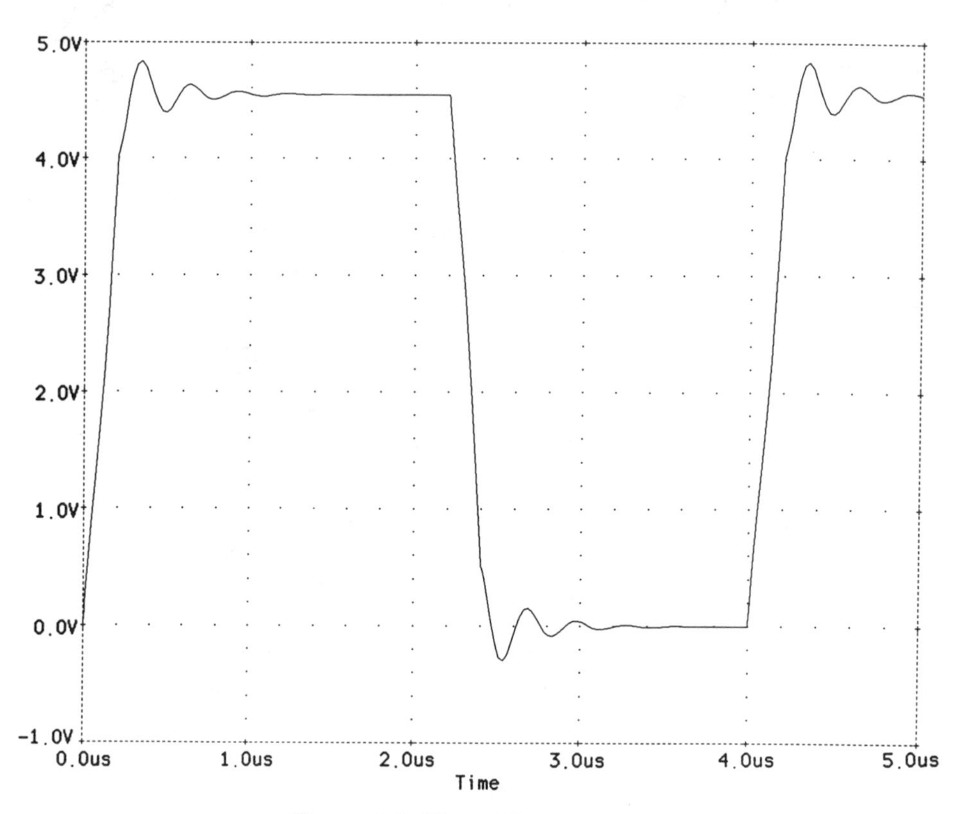

Figure 1-3. Plot of Table 3-1 Data

The main elements of a transient analysis are as follows:

1. It uses a time and amplitude defined input waveform source.
2. It simulates the waveform being applied to the input of the circuit and predicts the transient response of the chosen output.
3. It provides printed or plotted data that represents time versus amplitude.

It must be noted again that an AC analysis is a frequency-domain analysis and has no relationship to any time-dependent analysis. The transient analysis is a time-dependent analysis and has no relationship to any frequency variation of the input. The transient analysis is over a specified time interval.

1.3.2 ACCURACY OF AN ANALYSIS

The reader is provided, in this text, with all the component information and models necessary to allow a highly accurate analysis in SPICE to exist. However, the availability of a circuit analysis program does not guarantee an accurate analysis by itself. To properly make use of SPICE, the user must have some idea of the expected outcome of an analysis and realize that SPICE is primarily a verification tool and secondarily a design tool.

The accuracy of the analysis is based on the sophistication of the user more than the actual discrete models within SPICE. For example, assume it becomes necessary to model a simple RC circuit to determine the voltage across a capacitor as a function of time and the applied voltage. The user could model a step voltage source in series with a resistor (R1) in series with a capacitor (C1) to ground. The desired results could be realized by performing a transient analysis and observing the time-dependent waveform occurring across the capacitor. The analysis might predict a specific voltage across the capacitor at a specified time after the step voltage was applied. This voltage would be supported by traditional time constant equations associated with RC circuits. The user could take the circuit into the laboratory and attempt to validate the SPICE results. The laboratory results might show a lesser voltage across the capacitor, even though the capacitor and resistor are measured as having the values as used in the SPICE analysis.

So then, why is there a difference between the laboratory and the SPICE results? Figure 1-4 shows the difference between a pure capacitor and a realistic capacitor model; that is the problem is likely to be in the model of the capacitor. A real capacitor is a pure capacitance in series with an equivalent series resistance (ESR). In theory, if the frequency applied to a capacitor approaches infinity, then the reactance of a capacitor approaches zero.

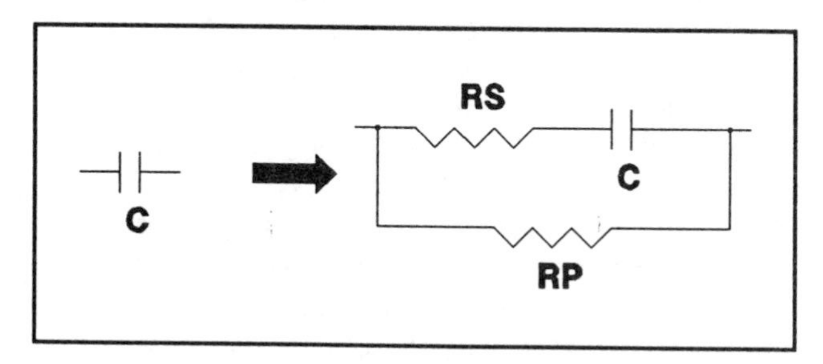

Figure 1-4. Real versus Theoretical Capacitor Model

In reality, the capacitor has a resistance of a few ohms because of the ESR. Additionally, no capacitor will hold a charge indefinitely. The charge is dissipated by a leakage resistance in parallel with the capacitor. This leakage resistance probably was the reason that the laboratory and SPICE results disagreed, especially if the resistor R1 in series with capacitor C1 is large. So does the problem lie with SPICE or the user? SPICE, in this case, provided a result consistent with the model of the capacitor presented, and, in fact, the problem of an inaccurate analysis lay exclusively with the user.

The important point here is that circuit analysis packages produce results only as good as the modeling. The user decides the sophistication of the modeling effort and trades off the time invested in modeling against the acceptable accuracy. Experience has shown that, in approximately 95 percent of the cases, the level of effort necessary to realize a 99 percent accuracy cannot be justified when considering the time that will have to be expended.

1.3.3 OVERCOMING MODELING LIMITATIONS

The information provided in this text is designed to be as complete and as encompassing as possible. Specifically, the solid state models provided are a result of hours of modeling refinement. The solid state models represent the proper characteristics of the devices, such that a high degree of accuracy in any area of analysis (DC, AC, transient) can be realized. Entire texts and theses have been written on modeling specific types of components. The accuracy of the model is therefore a function of how detailed the parameter selection has been. For example, many models are available to describe the proper parameters of a 2N2222 transistor and each model makes use of different parameters and different values for a given parameter. It is suggested that the application of the device, within a circuit, drives the parameter selection. Experience has shown that, for most applications, the modeling information provided, in this text, will meet virtually all applications the reader may encounter.

But what is to be done if the required device model is not provided? The approach recommended is to obtain the vendors data sheet describing the part. If the part is an integrated circuit, the vendor usually provides a schematic of the internal components and their interconnections. Suppose you are working with the MC1496 (balanced modulator/demodulator) made by Motorola. The schematic of the part can be obtained from a Motorola data book and can be treated as a discrete circuit. However, what if the vendor does not provide information on the types of transistors and diodes used? In modeling for some 15 years now, I have found that most diodes can be simulated by using a 1N914 or 1N3064 diode model and most transistors can be modeled using a 2N2222A (NPN) or 2N2904A (PNP) model. These choices will support most low-to-medium-speed analog circuit SPICE modeling simulations. Should higher speed be required, the 2N2369A (NPN) is an excellent choice. For a medium-speed complementary pair of transistors use the 2N3904 (NPN) and 2N3906 (PNP). The results obtained, by using the diodes and transistors listed, approach the published vendor parameters typically within 85 to 90 percent. In instances where success is not realized, the reader should consult library resources associated with modeling work for the required solution to the problem.

1.4 SUMMARY

The intent of this chapter has been to acquaint the reader with the capabilities and limitations of SPICE. The reader is encouraged to focus on the type of analyses available in SPICE. The analyses are summarized as follows.

1. Operating point analysis - An analysis that provides voltage and current flow information about the circuit related to a steady-state condition with no input stimulus applied.

2. DC Analysis - A DC-level-based analysis that provides data to predict the DC response of an output to a DC voltage at the input.

3. AC Analysis - A frequency based analysis that predicts the output amplitude or phase shift through a circuit as a function of a fixed-amplitude frequency applied to the input. This input frequency is over a specified range and thus frequency versus output response data is provided.

4. Transient Analysis - A time-based analysis that provides an oscilloscopic display of an output. The output display typically is a result of an input stimulus to the circuit and is in the form of time versus amplitude.

Having completed this section, the reader has, hopefully, started to develop a feel for the types of SPICE analyses that can be realized and the flexibility that the SPICE program provides in meeting the analysis and simulation capabilities associated with circuit design. As the text progresses, the use of SPICE, techniques to speed up and enhance the analysis, and the proper modeling of components will be our increasing focus.

CIRCUIT CONCEPTS

This chapter is designed to acquaint the reader with the concepts necessary to permit the creation, execution, and analysis of a basic SPICE circuit. To accomplish this it will be necessary to present (1) information for an understanding of correct modeling procedures, (2) an explanation of the basic components of SPICE concluding with the implementation of a sample circuit and (3) a review of the analysis data associated with the simulation of that circuit.

All SPICE packages have the same basic concept associated with the creation and analysis of a circuit. The steps necessary are as follows:

1. Create an input file for SPICE using an ASCII text editor
2. Submit the created file to SPICE for analysis
3. Receive an output file, from SPICE, which contains information about the requested analysis
4. Review and print the SPICE output file as necessary

Since SPICE packages vary, it is suggested, unless indicated in the SPICE documentation associated with the package, that the use of lower case characters should be avoided. Additionally, the editor should be carefully reviewed for the exclusion of certain key depressions not supported by the editor and for the inclusion of special characters, not recognized by SPICE, that the editor may make use of to realize its specific function.

2.1 DEFINING A CIRCUIT

A sample circuit (shown in Figure 2-1) will be used throughout this chapter. Although the circuit may appear complex, it's COMPLEXITY IS AN ILLUSION and one circuit may be analyzed and modeled as easily as the next. The circuit shown represents an example of a complex resistor array that would require an extensive analysis effort on the part of all but the most skilled designers. It will be shown that the circuit can be quickly simulated and analyzed using the basic SPICE capabilities.

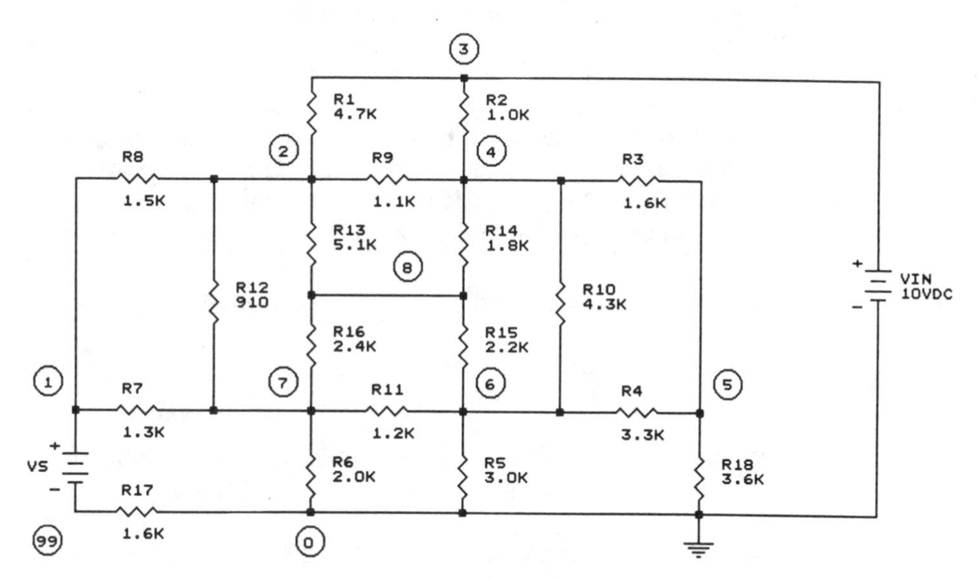

Figure 2-1. Resistor Array Sample Circuit

2.1.1 PROCEDURES

In reviewing the circuit in Figure 2-1, note that the circuit resembles a geometric structure where the cross members are a maze of resistors. Each resistor has a specified value and connected across the circuit is a battery (voltage source). The first step in the procedure is to define what you want from the analysis. This decision will drive the addition, to the schematic, of other sensor type components which make the use of SPICE more automated. For example, if we desired to measure the current through R17, then we might want to put an ammeter in series with R17. Adding an ammeter with a zero impedance does not adversely affect the results of the analysis and actually enhances the results of the analysis.

After determining what you want in an analysis, the next step is to assign names to each of the components. Each component must start with the appropriate designator for that class of components. In the case of resistors, use the letter "R." All component designators are well defined and will be discussed in the following sections. All component designators must begin with an *alpha character.* This character provides SPICE with the knowledge of the component type and the units of measure to assign to the components numeric value. In addition to the designator letter, assign a name or number(s) to facilitate your personal desires. Then assign nodes to the circuit. A *node* is defined as the connection of two or more leads. Each lead must originate from a component. Thus, if two resistors such as R1 and R2, are connected, the connection of these two components defines a node. All nodes must be integer numeric values. If a node is defined using other than integer numeric values, the SPICE program reports an error. The range of acceptable node numerical designators is 1 to 9999 (node number 0 is reserved as the ground node designation). Not only does this represent a virtually unlimited number of nodes, but allows a maximum degree of user flexibility in assigning nodes. Nodes need not be assigned or exist in a sequential fashion. It is completely acceptable to assign a node the number 1 and the next node the number 76. It is not necessary for nodes 2 through 75 to exist at all in the circuit.

In summary, the following initial steps are necessary for a SPICE analysis:

1. Obtain a copy of the schematic, which you desire to analyze, with component values identified

2. Determine what it is you want to analyze and incorporate mechanisms into the schematic to facilitate the SPICE analysis

3. Assign the appropriate designators to each applicable component shown on the schematic

4. Assign node numbers to the schematic

2.1.2 CONCEPT OF THE GROUND NODE

All circuits must have at least one ground node which is traditionally assigned to the ground reference point of the circuit and must always be assigned the node number "0" (zero). Failure to do this will result in the SPICE program aborting any attempts at analysis. Multiple points in a circuit may be at ground, and, if this is the case, all those points which are connected to ground must have been assigned the node number of zero. In some circuits, a ground may not exist. If this is the case, assign the reference node as node zero. A reference node is any node in the circuit that all operating point node voltages will be referenced against.

2.1.3 DC PATH RESTRICTIONS

There are a few restrictions that the SPICE program places on the user. One of those restrictions is that every node must have a DC path to ground (node 0). This restriction is imposed to assure that SPICE can identify the DC voltages at every node in the circuit. For example, assume that you connect two capacitors in series in a circuit. The node that is formed between the two capacitors has no DC path to ground and this will cause SPICE to generate an error and abort the analysis. To overcome this problem, it is recommended that a high value of resistance (100 Meg Ω) be placed from this node to ground (node 0) thus satisfying the SPICE requirements without introducing any significant circuit degradation. Other restrictions are that two or more voltage sources may not be connected in parallel and two or more current sources may not be connected in series. This problem can be overcome by placing a small resistor (0.001 Ω) between the voltage or current sources. This will introduce no adverse affect on the accuracy of the analysis.

2.1.4 CURRENT FLOW RESTRICTIONS

SPICE provides the user with maximum flexibility in the use of the input file format. Except for diodes, transistors, independent sources, and subcircuits, all of which will be discussed later, the user is free to enter components in any orientation desired. SPICE will determine the direction of current flow for any specified circuit. SPICE does, however, maintain an internal electron flow convention from the ground node (node 0) to the positive node. Thus, if a voltage source is providing current from its positive node, through a circuit, to ground then SPICE will represent this current as a negative current indicating electron flow direction. This internal convention is used by SPICE and the user need only be sensitive to the convention when determining the direction of current flow.

2.1.5 NUMERICAL CONVENTIONS

A number, used in SPICE to describe a component value, may be an integer field (12, -44), a floating point field (3.14159), either an integer or floating point number followed by an integer exponent (1E-14, 2.65E3), or either an integer or a floating point number followed by one of the scale factors listed in Table 2-1.

Letters immediately following a number that are not scale factors are ignored, and letters immediately following a scale factor are ignored. Hence, 10, 10V, 10VOLTS, and 10HZ all represent the same number, and M, MA, MSEC, and MMHOS all represent the same scale factor. Note that a thousand ohm resistor may be specified using any of the following designations:

1000OHMS, 1000.0, 1E3, 1E+3, 1E+03, 1.0E3, 1K, 1.000K, or 0.001MEG

SPICE recognizes the use of "E" in the context of 10^x. As a point of interest, SPICE expects an "E" format as an input, whereas output produced by SPICE is typically in a "D" format, such that 2.3E+3 = 2.3D+3. For example, assume we were to specify the value of 1.5 ma. In scientific notation it would be represented as 1.5×10^{-3}, and in SPICE the number is specified as 1.5E-3. If 1.5 ma were the output value, then SPICE would specify the output as 1.5D-03.

Table 2-1 Numerical conventions

Symbol	Notation	Example	SPICE Format
T	1E12		3.34T or 3.34E12
G	1E9	2.3 GHz	2.3G or 2.3E9
MEG	1E6	1.6 MHz	1.6MEG or 1.6E6
K	1E3	4.7 KΩ	4.7K or 4.7E3
M	1E-3	2.3 ms	2.3M or 2.3MS
U	1E-6	18 μF	18U or 18UF
N	1E-9	1.2 ns	1.2NS or 1.2N
P	1E-12	150 pF	150P or 150PF
F	1E-15		3.3F or 3.3E-15

2.2 BASIC COMPONENTS

This section is designed to introduce the reader to the basic components of an analysis. The components discussed in this section are resistors, capacitors, inductors, current-dependent inductors, voltage-dependent capacitors, independent voltage sources, and independent current sources. Each component will be discussed as necessary to permit the use of these components in a circuit.

Before looking at the individual components, we must revisit the concept of the node. As you recall, a node is defined as the connection of two or more components in a circuit and nodes must be specified using only numeric values. As previously mentioned, all components must start with a specified alphabetical character. This character tells the SPICE program (1) the units to apply to the component (ohms, volts, amps, Farads, Henrys, etc.) and (2) the format to expect for the description of the component. Each component starts with a specified letter followed by up to *seven* alpha-numeric characters. Thus we might choose to name a resistor R1 or we might choose to name it RLOAD. Both are valid descriptions in that the first character in the name was the character "R." The following sections, in discussing each of the components, show the general form of the input and several examples of how the component might be described. This, in conjunction with Figure 2-1, allows a programming concept to develop for the describing of a SPICE circuit. All *terms shown in brackets are optional* and are not required to satisfy the format expected by SPICE.

2.2.1 RESISTORS

The general form of the resistor is as follows:

RXXXXXXX N1 N2 VALUE [TC = TC1[,TC2]]

where N1 and N2 are element nodes, VALUE is the resistance in ohms (positive or negative but not 0), and TC1 and TC2 are temperature coefficients. Looking at Figure 2-1, assume we wanted to specify the 5.1 KΩ resistor, which was designated R13. Since the resistor has a node connection at node 2 and node 8, we could specify the resistor in the following ways:

```
R13 2 8 5.1K
R13 2 8 5100
R13 8 2 5.1K
R13 8 2 5100
R13 2 8 5.1E3
```

All of the above forms are valid descriptions of a 5.1 KΩ resistor. As discussed previously, the designator "K" means 1E+3 and therefore 5100 Ω is the same as 5.1 KΩ. Also the reader should note that the node orientation is not important. The nodes specifying a resistor, capacitor, or inductor are left to the discretion of the user.

Assume you have a resistor described as

```
RC1 12 17 9.1K
```

which is a resistor whose name is RC1 connected between nodes 12 and 17 with a resistance of 9.1 KΩ. Had the resistor been specified as 9.1E3 opposed to 9.1K, the designation would still convey the same value to SPICE.

Real resistors, however, have a temperature coefficient. That is to say the value of the resistor varies as the ambient temperature varies. Care should be exercised in the use of temperature coefficients (TC1 and TC2) due to the form of the equation shown.

$$Y = R*[[TC2(TEMP-Tnom)^2] + [TC1(TEMP-Tnom)] + 1]$$

For example, assume the resistor RC1 above varied at 0.1 percent/°C and you analyzed the circuit at 127°C. Many times the temperature coefficient of a resistor is specified in parts per million per degree Centigrade. To determine the value of TC1 using this specification, the value in parts per million per degree Centigrade is divided by 1E6. Based on the previous equation, we would assume the following conditions in computing the value of the resistor at 127°C.

Tnom = 27° C
TEMP = 127° C
TC1 = 0.1%/100 = 0.001

Thus by computing the value of resistance at 127° C, the resistance would be

$$(9.1\text{K}\Omega) * [1 + (.001)*(127-27)] = 10.01\ \text{K}\Omega$$

The resistor would be specified as

```
RC1 12 17 9.1K TC1=.001
```

It should be noted that the TC1 value is optional and if not included then SPICE assumes a default value of 0. If a negative coefficient is desired, then the value of TC1 and/or TC2 should be negative. An example of this might be in the modeling of a thermistor in a temperature sensitive circuit. Should the user desire to use the TC2 temperature coefficient, then the equation takes on the form of

$$Y = R*(AX^2+BX+C)$$

where X is the temperature being analyzed minus 27° C, A=TC2, B=TC1, and C=1. As can be seen, the effect of using TC2 is that the value is multiplied by the temperature difference squared, thus producing a rapidly changing value of resistance for a small change in temperature.

2.2.2 CAPACITORS

Capacitors, like resistors, start with a specific letter and carry no polarity sensitivity. Capacitors must be specified using the letter "C." SPICE assumes the capacitance units to be Farads. The general form is as follows:

CXXXXXXX N+ N- CVALUE [IC=INCOND]

where N+ is the positive element node, N- is the negative element node, CVALUE is the value of capacitor in Farads, and IC specifies the initial condition in volts.

For the capacitor, the value of IC is the initial (time-zero) value of capacitor voltage (in volts). Initial conditions is a parameter used exclusively with transient analyses which is discussed in Chapter 6. It may be desirable, from a theoretical standpoint, to set the value of a capacitor to a known value. A typical circuit in has a resistor in parallel with a capacitor. The exponential decay characteristics are of interest. By setting the value of IC to some voltage, the analysis starts with the voltage across the capacitor at that value and the associated decay may be observed when performing a transient (time-dependent) analysis. The use of this parameter will be discussed further in Chapter 6 in conjunction with a "use initial conditions" (UIC) option which is part of the transient analysis.

Possible examples of the designation of a capacitor are shown below. Assume a capacitor is (1) assigned the name CBYP, (2) connected between nodes 13 and ground, and (3) given a value of 1μF. The capacitor could be described in a SPICE circuit as one of the following:

```
CBYP 13 0 1UF
CBYP 0 13 1U
CBYP 13 0 1E-6
```

The reader is directed toward the three previous examples. All three examples describe the same capacitor. The designation of "F" for Farads is not necessary. SPICE assumes that all capacitors carry the units of Farads. Also, 1U is the same as 1E-6 as discussed before. The choice of which form to use is left to the user. Continuing to another example, assume a capacitor is (1) assigned the name of CTC, (2) connected between nodes 17 and 23, and (3) given a value of 150 pF. The capacitor, at the start of the analysis (initial condition), has 3 V of DC across it. The capacitor might be described as follows:

```
CTC 17 23 150P IC=3V
```

2.2.3 INDUCTORS

Inductors start with the letter "L" and carry no polarity sensitivity. SPICE assumes the inductance units to be in Henrys. The general form is follows:

LXXXXXXX N+ N- LVALUE [IC=INCOND]

where N+ is positive element node, N- is the negative element node, LVALUE is the value of inductor in Henrys, and IC is the value of the initial condition in amperes. For the inductor, the value of IC is the initial (time-zero) value of inductor current (in Amperes) that flows from N+ through the inductor to N-. Note that the initial conditions (if any) apply only if the UIC option is specified in a transient analysis as discussed in Chapter 6.

Possible examples of an inductor are shown below. Assume an inductor is (1) appointed the name LBYP, (2) connected between nodes 13 and ground, and (3) given a value of 1 μH. The inductor could be described in a SPICE circuit as follows:

```
LBYP 13 0 1UH
LBYP 0 13 1U
LBYP 13 0 1E-6
```

The reader is directed toward the three examples above. All three examples describe the same inductor. If we are to use the IC capability associated with an inductor, assume an inductor is (1) designated name of LUCY, (2) connected between nodes 44 and 69, and (3) given a value of 1.23 mH. The inductor, at the time of the analysis (initial condition), has 10 mA flowing through it. The inductor could be described as follows:

```
LUCY 44 69 1.23MH IC=10MA
```

2.2.4 NONLINEAR CAPACITORS AND INDUCTORS

Real capacitors, like inductors, vary in value as a function of the voltage applied across a capacitor or the current flowing through an inductor. Typically, the value of an inductor decreases as the current through it increases, and the value of a capacitor increases as the voltage across it increases. SPICE allows the user to make use of this phenomenon by providing an alternate form for describing an inductor or capacitor. The form is that of a polynomial expression which permits the description of a Nth order polynomial. The general form is shown as follows:

CXXXXXXX N+ N- POLY C0 C1 C2 ... Cn [IC = INCOND]

LYYYYYYY N+ N- POLY L0 L1 L2 ... Ln [IC = INCOND]

where C0 is the initial value of the capacitor, C1 is the linear constant, C2 is the quadratic constant, and Cn is the highest order constant; and correspondingly, L0 is the initial value of the inductor, L1 is the linear constant, L2 is the quadratic constant, and Ln is the highest order constant.

$$\mathrm{CVALUE} = C0 + C1*V + C2*V^2 + \dots Cn*V^n$$

$$\mathrm{LVALUE} = L0 + L1*I + L2*I^2 + \dots Ln*I^n$$

As can be seen by the two equations, the number of polynomial terms is unlimited; however, it was asserted that inductance decreases as the current increases, which is not supported by these equations. To overcome this problem, the user need only specify negative numbers for L1, L2, and so on to produce a decreasing inductance with increasing current. As an illustration of the power of a polynomial expression, Figure 2-2 represents a plot of Voltage versus Capacitance for the values shown. The reader should note that the value of C4, though small is taken to the 4th power of voltage and therefore does have a substantial effect on the capacitance at increasing voltage. The equation for the plot shown is as follows:

$$\mathrm{CVALUE} = 1\mathrm{UF} + 1*V + 0.01*V^2 + 0.01*V^3 - 0.001*V^4$$

2.2.5 COUPLED (MUTUAL) INDUCTORS

A mutually coupled inductor, also known as a transformer, can be modeled successfully by SPICE. To understand how this is accomplished, assume that you have a transformer with a single primary winding and a single secondary winding. The primary winding has an inductance of 1 mH and the secondary winding has an inductance of 100 mH. Since a transformer is nothing more than two inductors whose energy is coupled from one to the other, the transformer can first be modeled as two separate inductors. Assume the inductors are labeled LP and LS.

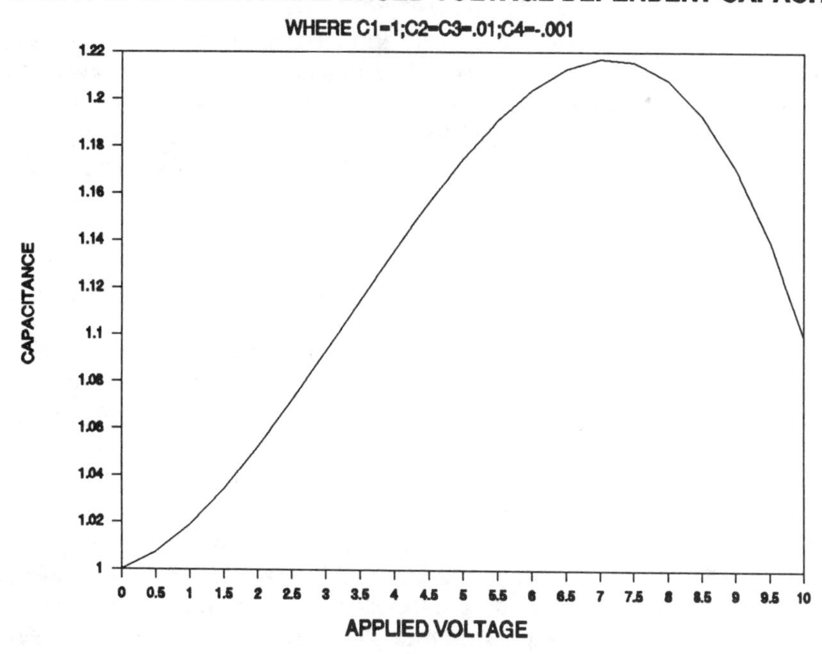

Figure 2-2. Plot of capacitance based on a polynomial expression

Based on the discussions of Section 2.2.3, we could describe the inductors as follows:

```
LP 1 0 1MH
LS 2 3 100MH
```

In any transformer, there is a phase relationship that exists between the primary and secondary. This phase relationship is maintained in SPICE by using a standard "dot" convention. Assume one dot is associated with node 1 and the other associated with node 2. If a positive signal is applied to node 1 then an in-phase (positive) signal will appear at node 2. In our example, LP has node 1 and LS has node 2 and these nodes are the first of the two nodes for each inductor.

Thus the "dot" convention always follows the first node specified for each of the inductors making up a transformer. For proper transformer operation, energy must be coupled between inductors. A coupling coefficient must therefore exist. In the simplest of terms, this coefficient represents the coupling efficiency between the inductors in a lossless transformer design. In a lossless design, a coupling coefficient of 1.0 (100 percent) might be used. A more typical value is .99 (99 percent). As the reader may recall from AC theory fundamentals, the physical spacing of two inductors determines the amount of energy coupled between the windings. Specifically, the coupling efficiency determines the mutual inductance of the transformer.

Real transformers have more than just inductance in their primaries and secondaries. Inductors are made of wire, and wire has a finite resistance per foot; therefore inductors are better modeled as an inductor in series with a resistor. Inductors are normally wound on a coil form and the windings have a capacitance between each turn known as interwinding capacitance. A typical inductor is shown in Figure 2-3.

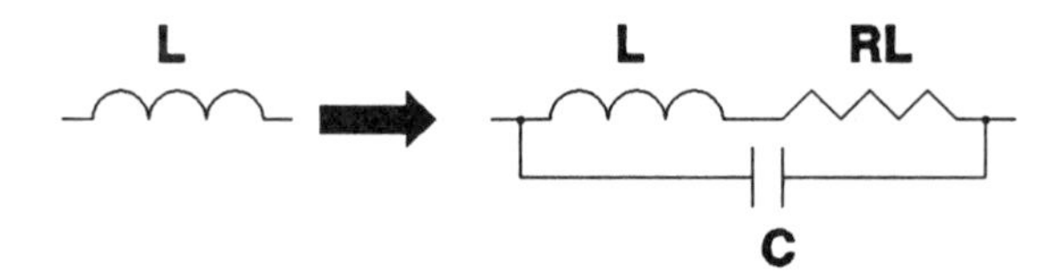

Figure 2-3. Practical inductor model

Proper modeling of a transformer would suggest the use of not only a resistor in series with the inductor but also a capacitor in parallel with the series RL combination. The use of this expanded model will provide far better results when high-frequency signals are applied to the windings of the transformer.

From the previous discussion, a transformer theoretically consists of a minimum of two or more inductors and a coupling coefficient. Many times the inductance of one of the windings may not be explicitly known. Normally the turns ratio of a transformer is known, and thus the following relationship is supported by SPICE in transformer performance evaluation.

$$LA * NB^2 = LB * NA^2$$

where LA is the winding A inductance, LB is the winding B inductance, and NA and NB are the number of turns of wire in windings A and B, respectively.

A transformer can be defined using a "K" designator much like the R designator used for resistors. Assume the transformer is designated as KX, the transformer could be described to SPICE as follows:

```
KX LP LS 0.99
```

Notice that no node numbers are used in the description of the transformer. The nodes have already been defined in the descriptions of inductors LP and LS. The KX designation simply couples the two inductors together to form a transformer. The general form is shown as follows:

KXXXXXXX LYYYYYYY LZZZZZZZ VALUE

LYYYYYYY and LZZZZZZZ are the names of the two coupled inductors. The VALUE is the coefficient of coupling, which must be greater than 0 and less than or equal to 1. Using the "dot" convention, place a "dot" on the first node of each inductor.

If we expand the concept further, we might envision a transformer with a single primary and a dual center tapped secondary such as might be found in a power transformer. Assume the primary is designated LP and has an inductance of 1 mH. The secondaries are designated LS1 and LS2 and each has an inductance of 100 mH. Assume a coupling coefficient of 0.95 exists between all windings. The transformer could be specified as follows:

```
LP 1 0 1MH
LS1 2 3 100MH
LS2 3 4 100MH
K1 LP LS1 0.95
K2 LP LS2 0.95
K3 LS1 LS2 0.95
```

Several things should be apparent in looking at this example and Figure 2-4.

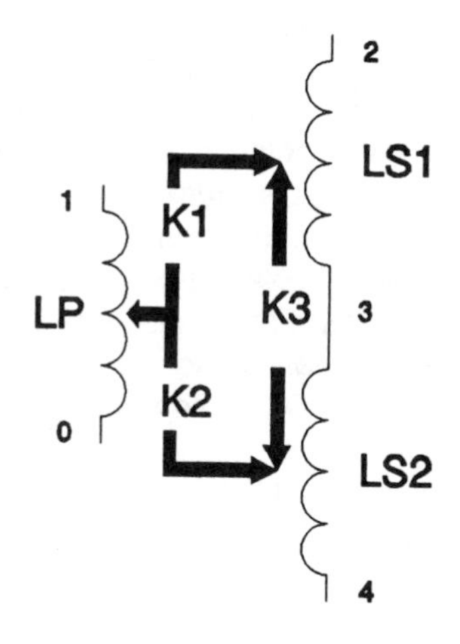

Figure 2-4. Transformer model

1. LS1 and LS2 form the center tap of the secondary at node 3. Using this convention eliminates the need for a shorting resistor between the two windings to form the combined secondary.

2. Three definitions of K terms are required. The number of K designators required will be [N*(N−1)]/2 where N is the number of windings (separate inductors) that make up the transformer.

3. In defining a multiple (more than 2) winding transformer, the efficiency of each winding could vary and hence the coupling

coefficient need not be the same value for each pair of windings. A failure to define all winding pairs with the K designators will result in an erroneous result produced by the SPICE analysis.

Additional information on transformer modeling is available but is not in the scope of this book. The reader is again reminded that transformers require that critical modeling techniques be employed. This is essential when considering saturable transformer designs with specific performance characteristics. The use of improperly modeled transformers in circuit analysis environments represents one of the largest error risks to the modeler.

2.2.6 INDEPENDENT VOLTAGE SOURCES

An independent voltage source can initially be viewed at as a zero impedance power supply or battery. In reality, the independent voltage source is much more than a battery. Such a source may act as an AC sinusoidal frequency source or as a pulse source (see Chapters 3, 4, 5, and 6). To continue the notion of a power supply or battery, assume such a component has two leads extending from it. One lead is labeled + and the other –. This is also true when dealing with an independent voltage source. The general form is shown below:

VXXXXXXX N+ N– [DC] VALUE

where N+ is the positive node, N– is the negative node, and VALUE is the output voltage. In the general form, an independent voltage source starts with a "V" designation followed by up to seven alpha-numeric characters. The N+ carries with it the polarity of the VALUE. The N– node in-turn represents the negative node. Positive current is assumed to flow from the positive node, through the source, to the negative node. Neither node of the voltage source need be grounded. Assume you were going to specify the voltage source shown in Figure 2-1. The source is labeled VIN and is connected with its positive node to node 3 and its negative node to node 0 (ground). The magnitude of the voltage is shown to be 10 V DC. The voltage source could be specified as follows:

```
VIN 3 0 10
VIN 3 0 10V
VIN 0 3 -10
VIN 0 3 -10V
VIN 3 0 DC 10
```

All five examples accurately describe the same supply. In the first two examples, the difference lies in the use of the V after the voltage. SPICE ignores the V and any use of verbiage after the 10 except as shown in Section 2.1.6. This use of verbiage is for the convenience of the user. Note that no space should exist between the number (10) and the letter (V) as SPICE uses spaces to detect the separation of parameters. As shown above, the first referenced node carries the polarity of the voltage. In the last two examples node 0 is negative with respect to node 3 or restated node 3 is positive with respect to node 0. The two statements are equal.

The last example is the preferred method to designate a voltage source. It indicates that VIN is truly a DC voltage source. This will become significant later when we make use of a voltage source as both a DC and AC source. A voltage magnitude will be associated with each of the DC and AC designators. It is, therefore, recommended that the DC form as in Example 5 be used if the voltage source will act as more than just a DC power supply during the analysis.

2.2.7 INDEPENDENT CURRENT SOURCES

Independent current sources have the same input format as independent voltage sources but they differ in two ways:

1. The first letter of the name is I rather than V.

2. SPICE will assume that the units are Amperes rather than Volts.

The independent current source can better be viewed as a constant current source. It is useful in many operational amplifier models that use a constant current source in the emitter circuit of the differential amplifier section. The general format is as follows:

IXXXXXXX N+ N- [DC] VALUE

where N+ is the positive node, N- is the negative node, and VALUE is the output current. In the general form, an independent current source starts with an I designation followed by up to seven alpha-numeric characters. Again the N+ is the positive node of the source and carries with it the polarity of the VALUE. The N- node in-turn represents the negative terminal of the source. Current is assumed to flow from the negative node to the positive node. Neither node of the current source need be grounded. Assume you were going to specify a constant current source named ISOURCE for use in a differential amplifier. The source might be connected from the common junction of the emitters and connected to the negative supply point. The common emitters are node 2 and the negative supply is node 75. The current source is designed to be a 2 mA source. The current source as shown in Figure 2-5 could be specified as follows:

```
ISOURCE 75 2 2M
ISOURCE 75 2 2MA
ISOURCE 2 75 -2M
ISOURCE 75 2 2E-3
ISOURCE 75 2 DC 2M
```

All five examples accurately describe the same current source.

2.2.8 AMMETER APPLICATIONS

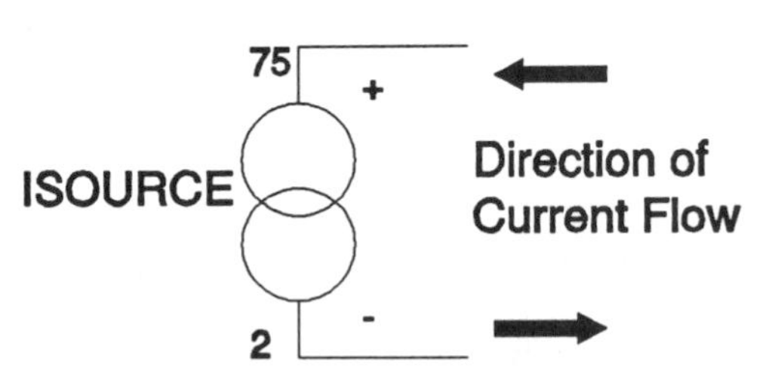

Figure 2-5. Current Source Model

Note that current flow through a component can only be directly calculated by SPICE using an ammeter in series with that component. In the case of SPICE, an ammeter is represented by an independent voltage source with a voltage of zero. When SPICE performs an operating-point analysis (as discussed in detail in Section 2.3.1) the currents flowing through each voltage source are computed and displayed. This provides a simple means of determining the current through any component in a circuit. Looking at resistor R17 of Figure 2-1, it should be noted that a supply VS is in series with the resistor. Node 99 has been defined for the intersection of VS and R17. The supply VS does not affect the analysis results because the voltage sources have zero resistance/impedance, which, in conjunction with a zero voltage potential, allows VS to act as a perfect ammeter.

An alternative to using ammeters is to compute the current in a conventional fashion. This could be accomplished, in the case of R17, by subtracting the voltage at node 1 from that at node 0 and dividing the difference by the resistance of R17. Although this lacks the sophistication of an ammeter this approach may be more timely. Other applications of ammeters lie in the area of AC analysis. Where an AC voltage source provides an input signal source to a circuit, the data resulting from an ammeter placed between the signal source and the input to the circuit is that of applied voltage and the current flowing from the signal source as a function of frequency. By dividing the current into the voltage, it is possible to plot input impedance against frequency for any AC circuit or input resistance against applied voltage for a DC circuit analysis. This application will be further elaborated upon in Chapter 3.

2.3 SAMPLE CIRCUIT

This section shows the commands necessary to describe the schematic in Figure 2-1. Before discussing the commands, we need to know (1) what information is to be extracted and (2) how the desired extraction of the information is to occur. To accomplish this, the concepts of Operating point information and Transfer function are discussed at length.

2.3.1 OPERATING POINT ANALYSIS

An Operating point analysis, as discussed in Section 1.3.1, represented the a power supply connected to the circuit where no stimulus was applied to the input. The operating point analysis provides voltage information on every node of the circuit with respect to ground. SPICE performs an operating point analysis before implementing a DC, AC, or Transient analysis. Thus, requesting an operating point analysis may be redundant if one of the three analysis types is being requested. The operating point analysis should be requested if the user wants information only on the circuit in a static environment.

The operating point analysis is realized by putting a .OP command in the SPICE input file. Including this statement in an input file will force SPICE to determine the DC operating point of the circuit with inductors shorted and capacitors opened. An example of an operating point analysis is shown in Figure 2-6.

```
RESISTOR ARRAY

****     SMALL SIGNAL BIAS SOLUTION       TEMPERATURE =   27.000 DEG C
******************************************************************************
NODE   VOLTAGE     NODE   VOLTAGE     NODE   VOLTAGE     NODE   VOLTAGE

(    1)    2.0339  (    2)    3.5559  (    3)   10.0000  (    4)    4.6718

(    5)     .0128  (    6)    2.0999  (    7)    2.3675  (    8)    3.2250

(   99)    2.0339


    VOLTAGE SOURCE CURRENTS
    NAME          CURRENT

    VIN          -6.699E-03
    VS            1.271E-03

    TOTAL POWER DISSIPATION   6.70E-02  WATTS
```

Figure 2-6. Example of an operating point analysis

2.3.2 TRANSFER FUNCTION ANALYSIS

A Transfer function analysis may be requested on any circuit. It requires that you specify an input voltage/current source and a node of interest. The analysis provides three pieces of information:

1. The resistance seen by the input voltage/current source (delta input voltage)/(delta input current)

2. The resistance seen by the node of interest looking back into the circuit

3. The ratio realized between the node of interest and the input voltage/current source (delta input source change/delta output node change)

The general form of the transfer function ".TF" is shown below. The reader is reminded that the values computed for the transfer function are based solely on the operating point information and are not applicable to DC, AC, or transient analysis solutions.

.TF OUTPUT-NODE/SOURCE INPUT-SOURCE

Assume, as in the case of Figure 2-1, that you want to know the resistance seen by VIN, the resistance seen by node 8 and the ratiomatic relationship between the voltage at node 8 and VIN. The transfer function command could be specified as follows:

.TF V(8) VIN

Assume, instead, that you want to determine the resistance seen by VS and the relationship of the current flowing through VS and the voltage at VIN. This could be accomplished as follows:

.TF I(VS) VIN

Where the output is a specified voltage node V(8) and the input is a specified voltage source VIN, the ratio of V(8)/VIN is actually the differential gain of the circuit, because SPICE performs a small incremental change to VIN and looks to see the resulting change in V(8). This occurs regardless of the type of input source or output specification. The results of a **.TF** statement are shown in Figure 2-7.

```
RESISTOR ARRAY
 ****     OPERATING POINT INFORMATION      TEMPERATURE =    27.000 DEG C
******************************************************************************

****     SMALL-SIGNAL CHARACTERISTICS

     V(8)/VIN =  3.225E-01

     INPUT RESISTANCE AT VIN =  1.493E+03

     OUTPUT RESISTANCE AT V(8) =  9.204E+02
```

Figure 2-7. Results of using a .TF statement

2.3.3 SOLUTION TO SAMPLE CIRCUIT

At this point in the process, it makes sense to program and analyze an actual circuit. The circuit shown in Figure 2-1 will be used for this task. The circuit represents a complex resistor array whose node voltages, currents, power dissipation and input/output resistance are desired pieces of information. Before starting the process, two absolute rules associated with all circuit analyses must be established.

1. All circuits must start with a title statement as the first statement. If you forget to include a title statement, SPICE will assume that your first entry in the file is the title statement and the circuit may incorrectly run. The title statement may be anything. The title statement must, however, be less than 80 characters in length.

2. All circuits must end with an ".END" statement. If the user fails to include an .END statement as the last entry in the input file then SPICE will produce an error indicating that a .END statement was not found or that a component in the circuit was redefined. In either case an error statement of some type will occur.

In addition to the two restrictions above, the user has the opportunity to insert a comment statement(s) anywhere between the Title and .END statements. The comment is invoked using an asterisk as the first entry. For example, the following comment might be entered into the file.

* This is where I placed the load resistance.

The user must take care to keep the comments on any line to 78 characters or less. Comment statements are also very useful in turning an analysis command into a comment so that SPICE will not act on the command. At this point we are ready to enter the first circuit. The input file for Figure 2-1 is shown as follows.

```
RESISTOR ARRAY
* THIS REPRESENTS A MAJOR RESISTOR ARRAY IN WHICH THE NODE
* VOLTAGES, CURRENTS, INPUT RESISTANCE SEEN BY VIN, OUTPUT
* RESISTANCE SEEN AT NODE 8 AND THE GAIN BETWEEN VIN AND
* NODE 8 WILL BE EVALUATED.
VIN 3 0 DC 10
R1 2 3 4.7K
R2 3 4 1K
R3 4 5 1.6K
R4 5 6 3.3K
R5 6 0 3K
R6 7 0 2K
R7 1 7 1.3K
R8 1 2 1.5K
R9 2 4 1.1K
R10 4 6 4.3K
R11 7 6 1.2K
R12 7 2 910
R13 2 8 5.1K
R14 8 4 1.8K
R15 8 6 2.2K
R16 7 8 2.4K
R17 99 0 1.6K
VS 1 99 0
* NOTE THE AMMETER VS IN SERIES WITH R17
R18 5 0 3.6
.OP
.TF V(8) VIN
.END
```

Some observations can be made about this circuit description.

1. The title of the circuit is "RESISTOR ARRAY."

2. The comment associated with the circuit is "* THIS REPRESENTS A MAJOR RESISTOR ARRAY IN WHICH THE NODE VOLTAGES, CURRENTS, INPUT RESISTANCE SEEN BY VIN, OUTPUT RESISTANCE SEEN AT NODE 8 AND THE GAIN BETWEEN VIN AND NODE 8 WILL BE EVALUATED." Notice that multiple comments can occur as long as the line containing the comment begins with an asterisk.

3. The resistor and voltage source definitions may be made in any sequence desired. The order of items between the Title statement and the ".END" statement is free form. It does not matter, for example, if the .OP statement is placed as the first statement after the title. The user must, however, be sensitive to the use of spaces for parameter separation.

4. The current through R17 is desired and thus an ammeter (VS) is placed in series with the resistor. Because the resistor, without the ammeter, is connected between node 1 and ground (node 0), the inserted ammeter (VS) is from node 1 to node 99 and the resistor is from node 99 to ground. This has the same electrical affect as inserting the resistor between nodes 1 and 0.

5. All node voltages are desired, and thus a .OP statement is included. The resistance seen by VIN, the resistance seen by node 8, and the ratio of node 8 to VIN is desired and, hence, a .TF V(8) VIN statement is placed in the file.

The circuit is submitted to SPICE with the following results occurring.

```
 RESISTOR ARRAY

 ****     CIRCUIT DESCRIPTION

******************************************************************************
* THIS REPRESENTS A MAJOR RESISTOR ARRAY IN WHICH THE NODE
* VOLTAGES, CURRENTS, INPUT RESISTANCE SEEN BY VIN, OUTPUT
* RESISTANCE SEEN AT NODE 8 AND THE GAIN BETWEEN VIN AND
* NODE 8 WILL BE EVALUATED.
VIN 3 0 DC 10
R1 2 3 4.7K
R2 3 4 1K
R3 4 5 1.6K
R4 5 6 3.3K
R5 6 0 3K
R6 7 0 2K
R7 1 7 1.3K
R8 1 2 1.5K
```

```
R9 2 4 1.1K
R10 4 6 4.3K
R11 7 6 1.2K
R12 7 2 910
R13 2 8 5.1K
R14 8 4 1.8K
R15 8 6 2.2K
R16 7 8 2.4K
R17 99 0 1.6K
VS 1 99 0
* NOTE THE AMMETER VS IN SERIES WITH R17
R18 5 0 3.6
.OP
.TF V(8) VIN
.END
```

```
 RESISTOR ARRAY

 ****     SMALL SIGNAL BIAS SOLUTION       TEMPERATURE =   27.000 DEG C

 ******************************************************************************
 NODE   VOLTAGE      NODE   VOLTAGE     NODE   VOLTAGE      NODE   VOLTAGE

(    1)    2.0339  (    2)    3.5559  (    3)   10.0000  (    4)    4.6718

(    5)     .0128  (    6)    2.0999  (    7)    2.3675  (    8)    3.2250

(   99)    2.0339

    VOLTAGE SOURCE CURRENTS
    NAME          CURRENT

    VIN          -6.699E-03
    VS            1.271E-03

    TOTAL POWER DISSIPATION   6.70E-02  WATTS
```

```
RESISTOR ARRAY
  ****     OPERATING POINT INFORMATION      TEMPERATURE =   27.000 DEG C
******************************************************************************

 ****     SMALL-SIGNAL CHARACTERISTICS

      V(8)/VIN =  3.225E-01

      INPUT RESISTANCE AT VIN =  1.493E+03

      OUTPUT RESISTANCE AT V(8) =  9.204E+02
```

2.3.4 ANALYZING THE SAMPLE CIRCUIT DATA

The data produced by SPICE consists of three basic pieces:

1. Input file listing

2. Small signal bias solution (from the .OP command)

3. Small signal characteristics (from the .TF command)

Looking at the small signal bias solution, note that the voltages for each node, with respect to node 0, are shown. The voltage at node 8 is 3.2250 V. The voltage at node 3 is 10.0000 V which is the voltage defined for the voltage source VIN.

The current flowing through VIN is -6.99E-3, which represents the current flowing from node 0 of the voltage source, through the circuit, to node 3. The current is therefore 6.99 mA. The current is shown as a negative value, even though it clearly flows from the positive node of VIN to the negative node (ground) of VIN. The reader is reminded that this is due to the electron flow convention used by SPICE. The current flowing through R17 is shown by the current flowing through VS which is 1.271 mA. We could have arrived at this answer by dividing the voltage at node 1 by 1.6 KΩ. For example, (3.5559 V/1.6 KΩ) = 1.27 mA.

The small signal characteristics data shows that the ratio of the voltages between node 8 and VIN is 0.3225. Realizing that VIN is 10 V, 10V * .3225 = 3.225 V which is the voltage at node 8. The input resistance seen by VIN is 1.493 KΩ ohms and the resistance seen by node 8 is 920.4 Ω.

The information in this simple analysis is extensive and quickly obtained using SPICE. The effort to derive this amount of information using traditional calculation methods should convince the reader that the low effort expended by the user using the SPICE program is truly worth the time investment to learn SPICE.

2.4 DIAGNOSTIC OPTIONS

SPICE provides a means of controlling the resulting analysis format by use of the ".OPTIONS" command. The .OPTIONS command has many choices but is extremely useful in the area of circuit diagnostics. When constructing and modeling large complex circuits, the probability of failing to connect a component to a node or typing the wrong value is high. The .OPTIONS command reduces the pain of finding the error. The general form of the .OPTIONS command is shown below:

.OPTIONS OPT1 OPT2 ... (or OPT=OPTVAL ...)

The list of options is shown in Table 2-2 for purposes of completeness. The .OPTIONS statement allows the user to reset program control and user options for specific simulation purposes. Any combination of the options may be included, and they may be in any order.

Table 2-2 Options Available with SPICE

Option	Effect
ACCT	Prints accounting information
LIST	Prints summary list of input data
NOMOD	Suppresses printout of model parameters
NOPAGE	Suppresses page ejects
NODE	Prints the node list
OPTS	Prints present values of options
GMIN=x •	Resets the value of GMIN, the minimum conductance allowed by the
RELTOL=x •	Resets the relative error tolerance of the program. The default value is
ABSTOL=x •	Resets the absolute current error tolerance of the program. The default
VNTOL=x •	Resets the absolute voltage error tolerance of the program. The default
TRTOL=x •	Resets the transient error tolerance. The default value is 7.0. This parame-
CHGTOL=x •	Resets the charge tolerance of the program. The default value is 1.0E-14.
PIVTOL=x •	Resets the absolute minimum value for a matrix entry to be accepted as a
PIVREL=x •	Resets the relative ratio between the largest column entry and an accept-
NUMDGT=x •	Sets number of significant digits for printout. Default is 4.
TNOM=x •	Resets the nominal temperature. The default value is 27° C (300° K).
ITL1=x •	Resets the DC iteration limit. The default is 100.
ITL2=x •	Resets the DC transfer curve iteration limit. The default is 50.
ITL3=x •	Sets the lower transient analysis iteration limit. The default is 4.
ITL4=x •	Sets the transient analysis time point iteration limit. The default is 10.
ITL5=x •	Resets the transient analysis total iteration limit. The default is 5000. Set
LIMPTS=x •	Sets the total number of points that can be printed or plotted. The default
DEFL=x •	Resets the value for MOS channel length; the default is 100.0 μm.
DEFW=x •	Resets the value for MOS channel width; the default is 100.0 μm.
DEFAD=x •	Resets the value for MOS drain diffusion area; the default is 0.0.
DEFAS=x •	Resets the value for MOS source diffusion area; the default is 0.0.

• x is some positive number.

A typical sequence to place in every analysis as follows:

```
.OPTIONS NODE LIST ITL5=0 LIMPTS=5001
```

This sequence produces the following:

1. A node list, that is, a list of all circuit nodes with the elements connected to each of the nodes. This listing is extremely useful when performing a troubleshooting activity on a complex circuit.

2. A list of all circuit elements used, their node orientation and their values.

3. ITL5=0 sets the iteration limit for a transient analysis to infinity, thus avoiding a premature analysis termination on complex analyses.

4. LIMPTS=5001 sets the number of plottable or printable points to 5001 and overrides the normal SPICE default limit of 201 points.

The actual results from such a command will be shown in the AC analysis presented in Chapter 3.

2.5 SUMMARY

In summary, this section has attempted to present the concept of the basic SPICE elements and the concept of the operating point and transfer function analysis. The summary of those commands used in this section is shown below.

1. Resistors

   RXXXXXXX N1 N2 VALUE [TC = TC1[,TC2]]

2. Capacitors

   CXXXXXXX N+ N- CVALUE [IC=INCOND]

3. Inductors

   LXXXXXXX N+ N- LVALUE [IC=INCOND]

4. Voltage-dependent capacitors

   CXXXXXXX N+ N- POLY C0 C1 C2 ... [IC = INCOND]

5. Current-dependent inductors

   LYYYYYYY N+ N- POLY L0 L1 L2 ... [IC = INCOND]

6. Transformers

KXXXXXXX LYYYYYYY LZZZZZZZ VALUE

7. Independent voltage sources

VXXXXXXX N+ N- [DC] VALUE

8. Independent current sources

IXXXXXXX N+ N- [DC] VALUE

9. Operating point

.OP

10. Transfer function

.TF OUTPUT-NODE INPUT-SOURCE

11. Options

.OPTIONS OPT1 OPT2 ... (or OPT=OPTVAL ...)

AC ANALYSIS

This chapter focuses on the commands unique to an AC analysis and the proper use of those commands. The reader should review Section 1.3.1 for an understanding of what an AC analysis represents and its significance in a laboratory environment.

3.1 SAMPLE AC CIRCUIT

Figure 3-1 shows the sample AC circuit used as an example throughout this Chapter. A review of the circuit reveals that it is made up of three inductors, three capacitors, and two resistors. This circuit consists of two series LC circuits and a parallel LC circuit. The interaction of the elements provides three peak resonant points, which show up across the load resistor RL. The analysis of this circuit is not difficult at a given frequency using complex variables (Rx ± Iy); however, determining the frequencies of resonance might present a difficult chore using conventional manual analysis techniques.

An AC analysis requires three pieces of information:

1. The AC stimulation source and its amplitude

2. The points (nodes), within the circuit, to be analyzed and how they are they to be analyzed

3. The frequency sweep range of the AC stimulation source will be viewed at the output.

In this chapter, we show the interrelationship of these three elements and the appropriate use of these elements in obtaining the required analysis accuracy.

3.2 INDEPENDENT AC VOLTAGE SOURCE

As discussed in Sections 2.2.5 and 2.2.6, the independent voltage and current sources can provide either DC voltage/current or other more advanced capability. The general form of the independent voltage and current sources is shown below.

VXXXXXXX N+ N- DC DCVALUE AC ACVALUE ACPHASE

IXXXXXXX N+ N- DC DCVALUE AC ACVALUE ACPHASE

where N+ and N- are the positive and negative nodes, DCVALUE is the DC voltage or current, ACVALUE is the AC voltage or current, and ACPHASE is the AC phase shift.

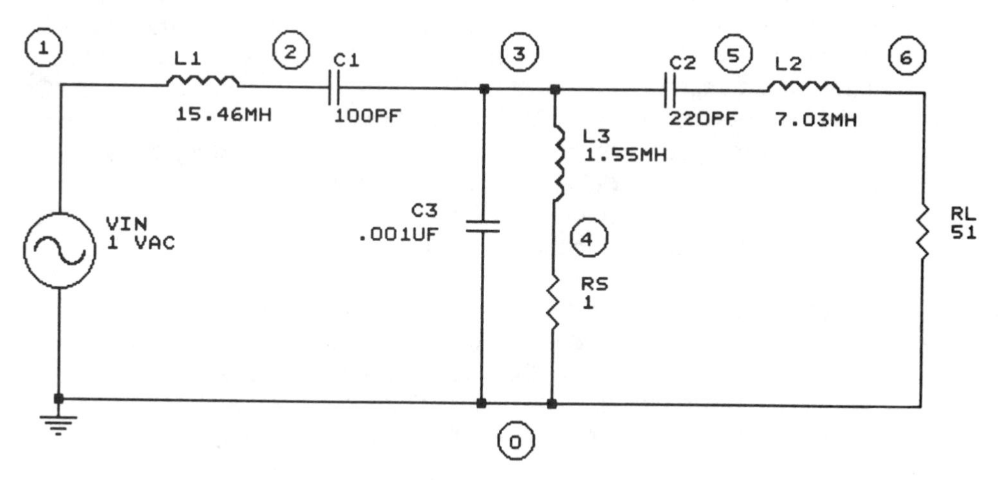

Figure 3-1. Sample AC Circuit

The use of both DC and AC values for voltage or current might at first seem strange. Imagine an amplifier circuit whose input is directly coupled (no DC blocking capacitor) and whose gain is a function of the internal circuit biasing due to the DC voltage or current applied. If the DC voltage or current is changed then the internal biasing of circuit semiconductor components might change. Such a change might affect the overall gain of the circuit at any specified node. Looking at this in a slightly different manner, assume the input to the amplifier is a DC offset voltage with a sine wave riding on the DC offset. If the DC offset were 1 Volt DC and the AC amplitude were 1 Volt AC then the voltage source might be modeled as:

```
VIN 1 0 DC 1 AC 1
```

The physical realization of this would be the placing of a 1 V DC battery in series with a 1 V AC source with a 0 V DC offset.

The SPICE AC analysis begins by computing a DC operating point, as discussed in Section 2.3.1, which determines all the DC node voltages in the circuit. After determining this operating point, SPICE will then run an AC analysis as specified by the user. The AC analysis will not show any DC level on an output but simply uses the DC voltage to establish the circuits operating point. Looking again at Figure 3-1, because this circuit is purely passive and the capacitors and inductors are not voltage or current dependent, you will see no use of the DC is required. The input source may be defined as:

```
VIN 1 0 DC 0 AC 1    or        VIN 1 0 AC 1
```

Either form is applicable and both provide the same results. If SPICE does not detect a DC value, then it assumes 0 V for the DC component. The reader is directed to the value of the AC signal. In an AC analysis, clipping and distortion of waveforms are not taken into account by SPICE. In this light, the use of 1 V AC makes sense by providing the user with additional information on the performance of the circuit. Assume you were to inject 1 Volt AC into the circuit input and read 3.24 Volts AC at the output node. This immediately tells the user that the gain at this point is 3.24 (gain = output voltage/input voltage). Thus, the use of the normalized 1 Volt AC input level makes sense in simplifying the analysis.

The use of a single voltage or current source is not mandatory. Assume that the analysis requires the use of a balanced modulator that has both a carrier and modulating input. SPICE will allow the use of multiple voltage and/or current sources as stimuli to the circuit. If a 1 Volt AC source named VX must be 43 degrees out of phase with a source named VIN, then the following definition for VX might be used.

```
VX 23 5 DC 0 AC 1 43
```

SPICE recognizes the first numerical value after the AC statement as the magnitude and the second as the phase. If no value is specified for the phase, then the phase is assumed to have an angle of zero degrees. In a similar sense, if no AC component is listed at all then the AC magnitude and phase are considered to be 0.

3.3 PRINT COMMAND

SPICE requires the use of a print or plot command for every analysis performed except the operating point and transfer function. All print commands have certain characteristics, which will become more clear as the reader progresses through the text. Before discussing the specific format of the print command, we will discuss the five output formats: Magnitude, Phase, decibels, Real values, and Imaginary values.

3.3.1 MAGNITUDE

Magnitude is a voltage measurement at a node or a current flow measurement through a voltage source (ammeter). Assume that a 1 Volt AC signal is injected into a circuit and a 5.2 Volt AC signal is measured at some node. Whether this is 5.2 V Peak to Peak, 5.2V root mean square, or 5.2 V peak does not matter. If you assume the input is in Volts rms, then the output is in Volts rms, and any measurement made in the circuit is in Volts rms. Resolution of the issue lies with the user deciding what the input really represents.

As discussed in Section 3.2, the use of a 1 Volt AC input source permits the direct viewing of magnitude as circuit gain rather than voltage or current, which may allow the user to assess circuit performance better. If, for example, an amplifier were to be modeled in SPICE and was powered by a 10 Volt DC power supply, assuming no voltage amplification due to capacitor or inductor action, one would expect the maximum peak to peak output to be 10 Volts. If the amplifier exhibits a gain of 100 at some frequency, the maximum input amplitude would be $\leq$ (10 V/100) or 0.1 V peak to peak. Because SPICE does not recognize the presence of an actual time-dependent wave form in an AC analysis, any input amplitude is multiplied by the gain, with the output being the result of this multiplication. If 1 Volt AC was applied to our example, then SPICE would report a magnitude of 100 Volts AC. Since this is unrealistic in an actual implementation of the design, the focus should be on looking at the magnitude as gain rather than a voltage.

The ability to measure the magnitude of an input current to a circuit has significance with a unity voltage-input source. If the input magnitude of the current is viewed versus input frequency, the result may be viewed as transconductance or the reciprocal of input impedance. The result is an analysis of input transconductance versus frequency which may be useful in assessing impedance-matching problems between circuitry.

The reader is reminded that, if the magnitude of an output voltage or current is plotted against frequency, the result is in the form of a Bode plot, which may find significant use in the determination of specific breakpoints and the associated slopes of an output as a function of the circuit components and overall design.

3.3.2 PHASE

Phase is very different from magnitude. All electronic circuits using capacitive, inductive, or solid state components may experience some phase shift between the input AC source and the output node or source. Assume that the circuit of interest is an inverting amplifier. Because it is inverting, one would assume that the output would be 180 degrees out of phase with the input at some frequency. The phase shift would be 180 degrees at the output with respect to the input. SPICE will analyze this phase shift when phase information is requested. The use of phase shift in conjunction with gain may provide valuable information about the overall performance and stability of the circuit. In the case of a stereo amplifier, if a phase shift variation is present in the frequency range for audible intelligence, a distortion of the intelligence may exist that is unacceptable to the listener and acceptable reproduction of the audio signal is not achieved.

3.3.3 DECIBELS

Alternating current circuits may have a gain or loss that can be expressed in terms of decibels. Decibel voltage is expressed in Volts DB = 20 log (Vout). Assuming that the input amplitude is set to unity (1.0) then the output represents true circuit gain or loss in decibels. The following statements are valid only for a normalized input amplitude.

1. If the decibel value is negative then the circuit is said to have a loss (output voltage is less than input voltage).

2. If the decibel value is 0 then the circuit has unity gain (output voltage is equal to input voltage).

3. If the decibel value is positive then the circuit has gain (output voltage is greater than input voltage).

SPICE can provide information about an output node or current with respect to an input source in decibels. The information can be either in the form of voltage or current. Decibel values may serve as a useful expression of an output by readily providing information related to slope (e.g. dB/decade), actual gain or loss, or the determination of half-power points (3-dB points).

3.3.4 REAL

As the reader may recall, from AC fundamentals, any complex circuit (L-R-C) has a real and potentially an imaginary component, associated with a specific node voltage or current. SPICE provides information about an output voltage node or voltage source relative to the real and imaginary components of that output. *Real* components are represented as positive components of a circuit and represent the resistive portion (real) of the circuit.

3.3.5 IMAGINARY

The imaginary component of a node can be either positive or negative and is a reflection of whether the circuit is capacitive or inductive at a specific frequency. This type of information, while not normally requested in a standard analysis is available as a SPICE output parameter.

3.3.6 MULTI-NODE DESIGNATIONS

SPICE permits the analysis of multiple nodes or currents in a circuit. No practical limit exists on the number of nodes that can be analyzed to the manner in which they are analyzed. For example, assume that it is of interest to look at the magnitude of the AC voltage at node 4 with respect to ground, designated VM(4) or VM(4,0) both of which represent the same command. If the voltage between nodes 2 and 5 were of interest then the designation would be VM(2,5). Current measurement is different from voltage measurement in that *current flow is measured through a voltage source.* Assume that the current flow through a capacitor were of interest, then an ammeter (in the form of a zero volt voltage source) would be placed in series with the capacitor. Assume that the name of the ammeter is VCAP. The current flowing through VCAP, as a function of frequency, could be specified as IM(VCAP) which is the magnitude of the current flowing through VCAP. Traditionally then, voltage is measured at a node and current is measured through a voltage source.

3.3.7 PRINT COMMAND FORMAT

The print command format for the AC analysis is as follows:

.PRINT AC OV1 [OV2 ... OV8]

This statement defines the contents of a tabular listing of one of eight output variables. The specified outputs require an AC analysis. The form for voltage or current output variables (OV1 ...) is as follows:

VX(N1[,N2])

Where VX can be any of the five forms of output as discussed in Sections 3.3.1 through 3.3.6. The options for VX are as follows:

VR. Real part of the voltage
VI. Imaginary part of the voltage
VM. Magnitude of the voltage
VP. Phase of the voltage
VDB. 20 * $\log_{10}$ (V(X))

Current can also be specified in the general form:

IY(VXXXXXX)

where IY can be any of the five forms of output in a current sense as discussed previously.

The distinction between voltage and current is that voltage is related to a node and current is related to a voltage source.

IR.	Real part of the current
II.	Imaginary part of the current
IM.	Magnitude of the current
IP.	Phase of the current
IDB.	20 * $\log_{10}$ (I(VXX))

For example, assume that it is desired to find the magnitude, phase, and decibel gain or loss at nodes 2, 3, 4, and between 5 and 6. The total requested output nodes exceed the eight maximum allowed nodes. It is, therefore, necessary to break the request into two .PRINT statements as follows:

```
.PRINT AC VM(2) VM(3) VM(4) VM(5,6) VP(2) VP(3) VP(4) VP(5,6)
.PRINT AC VDB(2) VDB(3) VDB(4) VDB(5,6)
```

The existence of more than one .PRINT AC statements is acceptable to SPICE, which simply treats each statement as a separate request.

The .PRINT AC statement will cause SPICE to compute lists of values presented in tabular form. The first .PRINT AC statement, above, would produce the following list of column titles and associated data.

```
FREQ   VM(2)  VM(3)  VM(4)  VM(5,6)  VP(2)  VP(3)  VP(4)  VP(5,6)
```

Should it be desirable to show the magnitude, phase, and decibels of the current through a voltage source VCAP then the following .PRINT AC statement would be given:

```
.PRINT AC IM(VCAP) IP(VCAP) IDB(VCAP) VM(5,6) VP(5,6) VDB(5,6)
```

In the .PRINT AC command, the output was in the form of frequency on the far left column followed by the printed data of up to 8 voltage nodes/current sources. The reader should note that the current and voltage parameters may be intermixed and need not be segmented for proper SPICE operation.

3.4 PLOT COMMANDS

The plot command is syntactical identical to the print command. Rather then specifying .PRINT AC VM(6), the plot command is evoked by using .PLOT AC VM(6). The difference lies in the form of the output. Using the .PRINT AC command causes the output to take the form of frequency on the far left column followed by the printed data of up to 8 voltage nodes/current sources. The .PLOT AC command uses the output format of the frequency on the far left of the page and the first requested output node/current source to the immediate right followed by a dot matrix plot of up to eight sources as specified in the .PLOT AC command. The plot routine uses multiple scales to display the plotted data with a separate character on the plot to represent the requested node/source. In the case where two or more plotted data points coincide, SPICE places an uppercase **X** to show this intersection.

3.5 AC FREQUENCY SWEEP PARAMETER DEFINITION

Up to now we have looked at the format of the voltage or current source definition of the AC signal source, its amplitude, and the ways in which we might get printed or plotted material from the analysis. The remaining element is the frequency range of the analysis. The manner in which the frequency is swept can be specified in three different formats.

1. Linear sweep

2. Decade sweep

3. Octave sweep

All three formats have a specific usefulness depending on the analysis being performed. In the next three sections, each of the three types will be reviewed in some detail. It is important to note that any given AC analysis will only calculate data at a finite set of specified frequencies and this can become significant if a given resonance point has a peak between two of the frequencies being computed. Should this occur, then the true amplitude of the peak or the existence of any peak may be missed by SPICE altogether. This will be further elaborated upon in the sample AC circuit problem of Figure 3-1.

3.5.1 LINEAR SWEEP

The linear sweep, as the name implies, is a linear frequency sweep (equally spaced frequency points) over a specified frequency range. Suppose for example it were desirable to perform an AC sweep over a range of 100 Hz to 200 Hz. If 100 points were desired over this range (100 Hz, 101 Hz, ...), then it would appear reasonable to request a linear sweep of 100 points to realize each frequency. This assumption is a pitfall for many users. The range of 100 to 200 Hz, assuming 1 Hz increments actually includes (200-100) + 1 = 101 points, because the number

200 must also be included in the calculations. SPICE computes the frequency interval using the following equation:

(F end - F start)/(# points - 1) = Frequency increment

The linear sweep is most useful over less than a decade of frequency range. As a rule of thumb, if the ending frequency is less than ten times the starting frequency then the linear sweep option should be used for the best overall accuracy.

3.5.2 DECADE SWEEP

As the name implies, a decade sweep is designed to provide a sweep of a number of points over a specific decade. For example, assume it is desired to sweep the frequency over a 100 Hz to 100 KHz range. This range encompasses three decades of variation (100 KHz/100 Hz) = 1000 = 10^X where X=3. SPICE does not place the frequency points of a decade sweep in a linear fashion, rather it uses the following equation to compute the point spacing and uses the computed number as a multiplier.

$$X^{ND} = 10$$

The value of ND is the number of requested points and X is the multiplier for the frequency spacing. For example, the first frequency would be at 100 Hz, the second at (100 Hz)*(X), the third at (100 Hz)*(X)*(X) and so on until a decade of frequency variation or the number of requested points plus 1 have been computed. The process then starts over for the next decade using the same value of ND which is based solely on the requested number of points. An example of a decade sweep at 100 points/decade would yield a value of X = 1.0233. Thus, the first analyzed frequency is 100 Hz, the second is 102.33 Hz, and the third is 104.71 Hz.

3.5.3 OCTAVE SWEEP

The octave sweep performs identically to the decade sweep, except the sweep is on an octave rather than decade basis and the multiplier is computed as follows:

$$X^{NO} = 2$$

The value of NO is the number of requested points and the value of X is the multiplier. Thus a frequency change of 100 Hz to 200 Hz represents one octave of frequency change. The frequencies to be analyzed are computed in the same fashion as in the decade sweep.

3.5.4 ACCURACY OF THE AC ANALYSIS

The accuracy of an AC analysis is based on the number of requested data points. Normally SPICE allows only 201 total points to be computed for a given analysis. The use of the .OPTIONS LIMPTS=5001 statement in the input file allows this limit to be overridden and up to 5001 points can be computed. The user should always try to plot the full 5001 points to get the maximum accuracy that SPICE can deliver. Assume the user wants to sweep the frequency over a 100 Hz to 10 MHz range at 1000 points per decade. Since there are five decades, the number of points that will be computed and displayed are (5 decades at 1000 points/decade) + 1 = 5001. This causes two problems: (1) the time involved to needed compute 5001 points (if the circuit being analyzed is complex, then the time required may not be acceptable to the user); (2) the size of the output data file and the resulting number of pages of printed or plotted material that might be generated. At 50 lines per page, the hardcopy output would be at least 100 pages of data per .PLOT or .PRINT command.

If the conservation of time and paper is an issue, then the alternative might be to perform a limited point per decade sweep. For example, first look at the data to determine the frequency range of interest; once the desired frequency range is identified, then a second, more limited, frequency range sweep should be performed. Such a procedure can provide the same accuracy but at a significant savings in time and paper.

3.5.5 AC FREQUENCY SWEEP COMMAND

The statement that will evoke the AC frequency sweep, discussed previously, is shown below.

.AC DEC NP FS FE

.AC OCT NP FS FE

.AC LIN NP FS FE

where DEC initiates a decade sweep, OCT initiates an octave sweep, LIN initiates a linear sweep, NP are the number of points per sweep type, and FS and FE are the starting and ending frequencies, respectively.

If this statement is included in the file, SPICE will perform an AC analysis of the circuit over the specified frequency range in the format specified (LIN, DEC, OCT). Note that, for this analysis to be meaningful, at least one independent source must have been specified with an AC value. A failure to specify a source in SPICE will result in an AC source value of 0 Volts AC.

For example, it may be desired to perform an AC sweep using a decade variation over a frequency range of 1 Hz to 10 KHz at a computed rate of 10 points per decade. The following statement would realize this requirement.

```
.AC DEC 10 1 10K
```

Other examples are shown below.

```
.AC OCT 10 1K 100MEG
```

The above example indicates an analysis which will sweep the input source from 1 KHz to 100 MHz in an octave variation with 10 points computed per octave. The final example, shows an analysis, with the frequency of the input source ranging from 100 Hz to 400 Hz in 101 steps. This will produce an increment of 3 Hz per step for a total of 101 computed points.

```
.AC LIN 101 100 400HZ
```

3.5.6 AC COMMAND GROUPINGS

In reviewing the material presented so far in the Chapter, it should now be clear that three statements are necessary to realize an AC analysis:

```
VIN 1 2 AC 1                a source definition
.AC DEC 100 100 10MEG       a sweep range and type
.PLOT AC VM(6)              a plot output definition
                                 and/or
.PRINT AC VM(6)             a print output definition
```

The print and plot statements realize the same result with the presentation format being the variant. Notice that the following statements must exist (1) a voltage/current source, (2) an .AC statement, and (3) a .PLOT AC and/or a .PRINT AC, to accomplish an AC analysis. If any of the three are missing the AC analysis will be unsuccessful.

3.6 SAMPLE AC ANALYSIS

The circuit shown in Figure 3-1 can be analyzed for the existence of three resonant peaks across RL (node 6) over a 80 KHz to 200 KHz range. It is desired to determine the frequencies of these three peaks and the magnitude, phase shift, and decibel gain or loss at the three frequencies. Additionally, the .OPTIONS statement, presented in Section 2.4, should be used to assure diagnostic information is provided by SPICE and that sufficient data point calculations are permitted. The resulting circuit description is shown as follows:

```
AC ANALYSIS OF A THREE PEAK FILTER
* THIS CIRCUIT HAS PEAKS BETWEEN 80KHZ AND 200KHZ
* SET THE AC VOLTAGE SOURCE TO 1 VOLT AC & 1 VOLT DC
VIN 1 0 DC 1 AC 1
.OPTIONS NODE LIST LIMPTS=5001
L1 1 2 15.46MH
C1 2 3 100PF
C2 3 5 220PF
L2 5 6 7.03MH
L3 3 4 1.55MH
RS 4 0 1
C3 3 0 .001UF
RL 6 0 51
* SET THE AC ANALYSIS SWEEP FOR A SWEEP FROM 80KHZ TO
* 200KHZ
.AC LIN 200 80K 200K
* PRINT & PLOT THE VOLTAGE MAGNITUDE, PHASE, & DB AT
* NODE 6
.PRINT AC VM(6) VP(6) VDB(6)
.PLOT AC VM(6) VP(6) VDB(6)
.END
```

Notice that a linear frequency sweep has been chosen that will generate 201 points of computed data. Also a plot of the data as well as a printed listing of the data has been requested.

3.7 AC ANALYSIS SOLUTIONS

After submitting the file, in Section 3.6, to SPICE for analysis, the following results are obtained. This section presents the analysis results and discusses the significance of the results considering the goal of the analysis.

```
AC ANALYSIS OF A THREE PEAK FILTER

****     CIRCUIT DESCRIPTION

******************************************************************************
* THIS CIRCUIT HAS PEAKS BETWEEN 80KHZ AND 200KHZ
* SET THE AC VOLTAGE SOURCE TO 1 VOLT AC & 1 VOLT DC
VIN 1 0 DC 1 AC 1
.OPTIONS NODE LIST LIMPTS=5001
L1 1 2 15.46MH
C1 2 3 100PF
C2 3 5 220PF
L2 5 6 7.03MH
L3 3 4 1.55MH
RS 4 0 1
C3 3 0 .001UF
RL 6 0 51
* SET THE AC ANALYSIS SWEEP FOR A SWEEP FROM 80KHZ TO
* 200KHZ
.AC LIN 200 80K 200K
* PRINT & PLOT THE VOLTAGE MAGNITUDE, PHASE, & DB AT
* NODE 6
.PRINT AC VM(6) VP(6) VDB(6)
.PLOT AC VM(6) VP(6) VDB(6)
.END
```

```
AC ANALYSIS OF A THREE PEAK FILTER

****     ELEMENT NODE TABLE

******************************************************************************
      0    RS       RL       C3       VIN

      1    L1       VIN

      2    C1       L1

      3    C1       C2       C3       L3

      4    RS       L3

      5    C2       L2

      6    RL       L2
```

```
AC ANALYSIS OF A THREE PEAK FILTER

****     CIRCUIT ELEMENT SUMMARY

******************************************************************************
**** RESISTORS

     NAME         NODES     MODEL      VALUE       TC1        TC2        TCE

     RS           4    0             1.00D+00   .00D+00    .00D+00    .00D+00
     RL           6    0             5.10D+01   .00D+00    .00D+00    .00D+00

**** CAPACITORS AND INDUCTORS

     NAME         NODES     MODEL    IN COND     VALUE

     C1           2    3              .00D+00  1.00D-10
     C2           3    5              .00D+00  2.20D-10
     C3           3    0              .00D+00  1.00D-09
     L1           1    2              .00D+00  1.55D-02
     L2           5    6              .00D+00  7.03D-03
     L3           3    4              .00D+00  1.55D-03

**** INDEPENDENT SOURCES

     NAME         NODES   DC VALUE   AC VALUE   AC PHASE   TRANSIENT

     VIN          1    0  1.00D+00   1.00D+00    .00D+00
```

```
AC ANALYSIS OF A THREE PEAK FILTER

****     SMALL SIGNAL BIAS SOLUTION       TEMPERATURE =   27.000 DEG C

******************************************************************************
 NODE   VOLTAGE      NODE   VOLTAGE      NODE   VOLTAGE      NODE   VOLTAGE

(  1)    1.0000     (  2)    1.0000     (  3)     .0000     (  4)     .0000

(  5)     .0000     (  6)     .0000
```

AC ANALYSIS OF A THREE PEAK FILTER

**** AC ANALYSIS TEMPERATURE = 27.000 DEG C

****

FREQ	VM(6)	VP(6)	VDB(6)
8.000E+04	1.477E-03	-9.090E+01	-5.661E+01
8.060E+04	1.584E-03	-9.093E+01	-5.600E+01
8.121E+04	1.701E-03	-9.095E+01	-5.538E+01
8.181E+04	1.830E-03	-9.098E+01	-5.475E+01
8.241E+04	1.972E-03	-9.102E+01	-5.410E+01
8.302E+04	2.129E-03	-9.105E+01	-5.344E+01
8.362E+04	2.303E-03	-9.109E+01	-5.275E+01
8.422E+04	2.497E-03	-9.113E+01	-5.205E+01
8.482E+04	2.714E-03	-9.118E+01	-5.133E+01
8.543E+04	2.957E-03	-9.123E+01	-5.058E+01
8.603E+04	3.232E-03	-9.128E+01	-4.981E+01
8.663E+04	3.544E-03	-9.134E+01	-4.901E+01
8.724E+04	3.901E-03	-9.141E+01	-4.818E+01
8.784E+04	4.312E-03	-9.149E+01	-4.731E+01
8.844E+04	4.789E-03	-9.158E+01	-4.640E+01
8.905E+04	5.347E-03	-9.168E+01	-4.544E+01
8.965E+04	6.009E-03	-9.180E+01	-4.442E+01
9.025E+04	6.803E-03	-9.194E+01	-4.335E+01
9.085E+04	7.771E-03	-9.210E+01	-4.219E+01
9.146E+04	8.972E-03	-9.231E+01	-4.094E+01
9.206E+04	1.050E-02	-9.256E+01	-3.958E+01
9.266E+04	1.249E-02	-9.289E+01	-3.807E+01
9.327E+04	1.520E-02	-9.332E+01	-3.636E+01
9.387E+04	1.908E-02	-9.394E+01	-3.439E+01
9.447E+04	2.504E-02	-9.489E+01	-3.203E+01
9.508E+04	3.534E-02	-9.651E+01	-2.904E+01
9.568E+04	5.715E-02	-9.993E+01	-2.486E+01
9.628E+04	1.306E-01	-1.117E+02	-1.768E+01
9.688E+04	2.858E-01	1.394E+02	-1.088E+01
9.749E+04	8.890E-02	1.028E+02	-2.102E+01
9.809E+04	5.225E-02	9.696E+01	-2.564E+01
9.869E+04	3.777E-02	9.468E+01	-2.846E+01
9.930E+04	3.010E-02	9.346E+01	-3.043E+01
9.990E+04	2.538E-02	9.270E+01	-3.191E+01
1.005E+05	2.221E-02	9.218E+01	-3.307E+01
1.011E+05	1.995E-02	9.180E+01	-3.400E+01
1.017E+05	1.828E-02	9.151E+01	-3.476E+01
1.023E+05	1.701E-02	9.128E+01	-3.539E+01
1.029E+05	1.602E-02	9.109E+01	-3.591E+01
1.035E+05	1.524E-02	9.093E+01	-3.634E+01
1.041E+05	1.462E-02	9.079E+01	-3.670E+01
1.047E+05	1.413E-02	9.068E+01	-3.700E+01
1.053E+05	1.375E-02	9.057E+01	-3.724E+01
1.059E+05	1.345E-02	9.048E+01	-3.743E+01
1.065E+05	1.322E-02	9.040E+01	-3.757E+01
1.071E+05	1.306E-02	9.032E+01	-3.768E+01
1.077E+05	1.295E-02	9.025E+01	-3.775E+01
1.083E+05	1.290E-02	9.019E+01	-3.779E+01
1.089E+05	1.289E-02	9.012E+01	-3.779E+01
1.095E+05	1.293E-02	9.006E+01	-3.777E+01
1.102E+05	1.301E-02	9.001E+01	-3.771E+01
1.108E+05	1.314E-02	8.995E+01	-3.763E+01
1.114E+05	1.330E-02	8.990E+01	-3.752E+01
1.120E+05	1.351E-02	8.984E+01	-3.739E+01

****

FREQ	VM(6)	VP(6)	VDB(6)
1.126E+05	1.377E-02	8.979E+01	-3.722E+01
1.132E+05	1.407E-02	8.974E+01	-3.703E+01
1.138E+05	1.442E-02	8.968E+01	-3.682E+01
1.144E+05	1.483E-02	8.963E+01	-3.658E+01
1.150E+05	1.530E-02	8.957E+01	-3.631E+01
1.156E+05	1.583E-02	8.951E+01	-3.601E+01
1.162E+05	1.644E-02	8.945E+01	-3.568E+01
1.168E+05	1.714E-02	8.938E+01	-3.532E+01
1.174E+05	1.793E-02	8.931E+01	-3.493E+01
1.180E+05	1.884E-02	8.923E+01	-3.450E+01
1.186E+05	1.988E-02	8.915E+01	-3.403E+01
1.192E+05	2.109E-02	8.906E+01	-3.352E+01
1.198E+05	2.250E-02	8.896E+01	-3.296E+01
1.204E+05	2.414E-02	8.884E+01	-3.234E+01
1.210E+05	2.609E-02	8.871E+01	-3.167E+01
1.216E+05	2.843E-02	8.855E+01	-3.092E+01
1.222E+05	3.128E-02	8.837E+01	-3.009E+01
1.228E+05	3.482E-02	8.815E+01	-2.916E+01
1.234E+05	3.931E-02	8.787E+01	-2.811E+01
1.240E+05	4.519E-02	8.752E+01	-2.690E+01
1.246E+05	5.320E-02	8.704E+01	-2.548E+01
1.252E+05	6.472E-02	8.636E+01	-2.378E+01
1.258E+05	8.265E-02	8.532E+01	-2.165E+01
1.264E+05	1.143E-01	8.348E+01	-1.884E+01
1.270E+05	1.844E-01	7.939E+01	-1.468E+01
1.276E+05	4.520E-01	6.313E+01	-6.897E+00
1.282E+05	5.898E-01	-5.387E+01	-4.586E+00
1.288E+05	2.084E-01	-7.800E+01	-1.362E+01
1.294E+05	1.241E-01	-8.291E+01	-1.812E+01
1.301E+05	8.842E-02	-8.499E+01	-2.107E+01
1.307E+05	6.878E-02	-8.613E+01	-2.325E+01
1.313E+05	5.639E-02	-8.686E+01	-2.498E+01
1.319E+05	4.788E-02	-8.736E+01	-2.640E+01
1.325E+05	4.168E-02	-8.774E+01	-2.760E+01
1.331E+05	3.697E-02	-8.802E+01	-2.864E+01
1.337E+05	3.328E-02	-8.825E+01	-2.956E+01
1.343E+05	3.031E-02	-8.844E+01	-3.037E+01
1.349E+05	2.788E-02	-8.860E+01	-3.109E+01
1.355E+05	2.586E-02	-8.873E+01	-3.175E+01
1.361E+05	2.415E-02	-8.884E+01	-3.234E+01
1.367E+05	2.269E-02	-8.895E+01	-3.288E+01
1.373E+05	2.144E-02	-8.904E+01	-3.338E+01
1.379E+05	2.035E-02	-8.912E+01	-3.383E+01
1.385E+05	1.940E-02	-8.919E+01	-3.425E+01
1.391E+05	1.856E-02	-8.926E+01	-3.463E+01
1.397E+05	1.782E-02	-8.932E+01	-3.498E+01
1.403E+05	1.717E-02	-8.938E+01	-3.530E+01
1.409E+05	1.659E-02	-8.943E+01	-3.560E+01
1.415E+05	1.608E-02	-8.948E+01	-3.588E+01
1.421E+05	1.562E-02	-8.953E+01	-3.613E+01
1.427E+05	1.521E-02	-8.958E+01	-3.636E+01
1.433E+05	1.485E-02	-8.962E+01	-3.657E+01
1.439E+05	1.453E-02	-8.966E+01	-3.676E+01
1.445E+05	1.424E-02	-8.971E+01	-3.693E+01
1.451E+05	1.399E-02	-8.975E+01	-3.708E+01
1.457E+05	1.378E-02	-8.979E+01	-3.722E+01
1.463E+05	1.359E-02	-8.983E+01	-3.734E+01

****

FREQ	VM(6)	VP(6)	VDB(6)
1.469E+05	1.343E-02	-8.986E+01	-3.744E+01
1.475E+05	1.329E-02	-8.990E+01	-3.753E+01
1.481E+05	1.319E-02	-8.994E+01	-3.760E+01
1.487E+05	1.310E-02	-8.998E+01	-3.765E+01
1.493E+05	1.304E-02	-9.002E+01	-3.769E+01
1.499E+05	1.301E-02	-9.006E+01	-3.772E+01
1.506E+05	1.300E-02	-9.010E+01	-3.772E+01
1.512E+05	1.301E-02	-9.014E+01	-3.772E+01
1.518E+05	1.305E-02	-9.019E+01	-3.769E+01
1.524E+05	1.311E-02	-9.023E+01	-3.765E+01
1.530E+05	1.320E-02	-9.028E+01	-3.759E+01
1.536E+05	1.331E-02	-9.033E+01	-3.751E+01
1.542E+05	1.346E-02	-9.038E+01	-3.742E+01
1.548E+05	1.364E-02	-9.044E+01	-3.730E+01
1.554E+05	1.386E-02	-9.049E+01	-3.717E+01
1.560E+05	1.411E-02	-9.056E+01	-3.701E+01
1.566E+05	1.440E-02	-9.062E+01	-3.683E+01
1.572E+05	1.474E-02	-9.069E+01	-3.663E+01
1.578E+05	1.514E-02	-9.077E+01	-3.640E+01
1.584E+05	1.560E-02	-9.086E+01	-3.614E+01
1.590E+05	1.613E-02	-9.095E+01	-3.585E+01
1.596E+05	1.675E-02	-9.106E+01	-3.552E+01
1.602E+05	1.746E-02	-9.117E+01	-3.516E+01
1.608E+05	1.830E-02	-9.131E+01	-3.475E+01
1.614E+05	1.929E-02	-9.146E+01	-3.429E+01
1.620E+05	2.046E-02	-9.164E+01	-3.378E+01
1.626E+05	2.186E-02	-9.185E+01	-3.321E+01
1.632E+05	2.358E-02	-9.210E+01	-3.255E+01
1.638E+05	2.570E-02	-9.240E+01	-3.180E+01
1.644E+05	2.838E-02	-9.278E+01	-3.094E+01
1.650E+05	3.187E-02	-9.327E+01	-2.993E+01
1.656E+05	3.658E-02	-9.392E+01	-2.873E+01
1.662E+05	4.326E-02	-9.484E+01	-2.728E+01
1.668E+05	5.342E-02	-9.624E+01	-2.545E+01
1.674E+05	7.064E-02	-9.861E+01	-2.302E+01
1.680E+05	1.058E-01	-1.035E+02	-1.951E+01
1.686E+05	2.089E-01	-1.186E+02	-1.360E+01
1.692E+05	3.891E-01	1.576E+02	-8.199E+00
1.698E+05	1.467E-01	1.112E+02	-1.667E+01
1.705E+05	8.199E-02	1.021E+02	-2.172E+01
1.711E+05	5.590E-02	9.849E+01	-2.505E+01
1.717E+05	4.200E-02	9.659E+01	-2.754E+01
1.723E+05	3.339E-02	9.541E+01	-2.953E+01
1.729E+05	2.756E-02	9.461E+01	-3.120E+01
1.735E+05	2.334E-02	9.402E+01	-3.264E+01
1.741E+05	2.017E-02	9.358E+01	-3.391E+01
1.747E+05	1.768E-02	9.324E+01	-3.505E+01
1.753E+05	1.570E-02	9.296E+01	-3.608E+01
1.759E+05	1.407E-02	9.273E+01	-3.703E+01
1.765E+05	1.272E-02	9.254E+01	-3.791E+01
1.771E+05	1.157E-02	9.237E+01	-3.873E+01
1.777E+05	1.059E-02	9.223E+01	-3.950E+01
1.783E+05	9.746E-03	9.211E+01	-4.022E+01
1.789E+05	9.008E-03	9.200E+01	-4.091E+01
1.795E+05	8.359E-03	9.191E+01	-4.156E+01
1.801E+05	7.784E-03	9.182E+01	-4.218E+01
1.807E+05	7.272E-03	9.174E+01	-4.277E+01

****

FREQ	VM(6)	VP(6)	VDB(6)
1.813E+05	6.814E-03	9.167E+01	-4.333E+01
1.819E+05	6.401E-03	9.161E+01	-4.388E+01
1.825E+05	6.027E-03	9.155E+01	-4.440E+01
1.831E+05	5.687E-03	9.150E+01	-4.490E+01
1.837E+05	5.378E-03	9.145E+01	-4.539E+01
1.843E+05	5.094E-03	9.140E+01	-4.586E+01
1.849E+05	4.834E-03	9.136E+01	-4.631E+01
1.855E+05	4.595E-03	9.132E+01	-4.675E+01
1.861E+05	4.374E-03	9.129E+01	-4.718E+01
1.867E+05	4.169E-03	9.125E+01	-4.760E+01
1.873E+05	3.979E-03	9.122E+01	-4.800E+01
1.879E+05	3.802E-03	9.119E+01	-4.840E+01
1.885E+05	3.638E-03	9.116E+01	-4.878E+01
1.891E+05	3.484E-03	9.114E+01	-4.916E+01
1.897E+05	3.340E-03	9.111E+01	-4.952E+01
1.904E+05	3.206E-03	9.109E+01	-4.988E+01
1.910E+05	3.079E-03	9.106E+01	-5.023E+01
1.916E+05	2.960E-03	9.104E+01	-5.057E+01
1.922E+05	2.848E-03	9.102E+01	-5.091E+01
1.928E+05	2.743E-03	9.100E+01	-5.124E+01
1.934E+05	2.643E-03	9.098E+01	-5.156E+01
1.940E+05	2.549E-03	9.097E+01	-5.187E+01
1.946E+05	2.460E-03	9.095E+01	-5.218E+01
1.952E+05	2.376E-03	9.093E+01	-5.248E+01
1.958E+05	2.296E-03	9.092E+01	-5.278E+01
1.964E+05	2.220E-03	9.090E+01	-5.307E+01
1.970E+05	2.148E-03	9.089E+01	-5.336E+01
1.976E+05	2.079E-03	9.087E+01	-5.364E+01
1.982E+05	2.014E-03	9.086E+01	-5.392E+01
1.988E+05	1.951E-03	9.085E+01	-5.419E+01
1.994E+05	1.892E-03	9.084E+01	-5.446E+01
2.000E+05	1.835E-03	9.082E+01	-5.473E+01

```
AC ANALYSIS OF A THREE PEAK FILTER

****     AC ANALYSIS                      TEMPERATURE =   27.000 DEG C

******************************************************************************
LEGEND:

*: VM(6)
+: VP(6)
=: VDB(6)

    FREQ       VM(6)

(*)------------ 1.000D-03    1.000D-02    1.000D-01    1.000D+00    1.000D+01
                      - - - - - - - - - - - - - - - - - - - - - - - - - -

(+)------------ -2.000D+02   -1.000D+02    .000D+00    1.000D+02    2.000D+02
                      - - - - - - - - - - - - - - - - - - - - - - - - - -

(=)------------ -6.000D+01   -4.000D+01   -2.000D+01    .000D+00    2.000D+01
                      - - - - - - - - - - - - - - - - - - - - - - - - - -
 8.000D+04  1.477D-03 . X          .+          .            .           .
 8.060D+04  1.584D-03 .  X         .+          .            .           .
 8.121D+04  1.701D-03 .  X         .+          .            .           .
 8.181D+04  1.830D-03 .  X         .+          .            .           .
 8.241D+04  1.972D-03 .   X        .+          .            .           .
 8.302D+04  2.129D-03 .   X        .+          .            .           .
 8.362D+04  2.303D-03 .    X       .+          .            .           .
 8.422D+04  2.497D-03 .    X       .+          .            .           .
 8.482D+04  2.714D-03 .     X      .+          .            .           .
 8.543D+04  2.957D-03 .     X      .+          .            .           .
 8.603D+04  3.232D-03 .      X     .+          .            .           .
 8.663D+04  3.544D-03 .      X     .+          .            .           .
 8.724D+04  3.901D-03 .       X    .+          .            .           .
 8.784D+04  4.312D-03 .       X    .+          .            .           .
 8.844D+04  4.789D-03 .        X   .+          .            .           .
 8.905D+04  5.347D-03 .        X   .+          .            .           .
 8.965D+04  6.009D-03 .         X  .+          .            .           .
 9.025D+04  6.803D-03 .          X .+          .            .           .
 9.085D+04  7.771D-03 .           X.+          .            .           .
 9.146D+04  8.972D-03 .           X.+          .            .           .
 9.206D+04  1.050D-02 .            X+          .            .           .
 9.266D+04  1.249D-02 .            .X          .            .           .
 9.327D+04  1.520D-02 .            .+X         .            .           .
 9.387D+04  1.908D-02 .            .+  X       .            .           .
 9.447D+04  2.504D-02 .            .+   X      .            .           .
 9.508D+04  3.534D-02 .            +      X    .            .           .
 9.568D+04  5.715D-02 .            +         X .            .           .
 9.628D+04  1.306D-01 .          + .           . X          .           .
 9.688D+04  2.858D-01 .            .           .     X      .    +      .
 9.749D+04  8.890D-02 .            .          X.            +           .
 9.809D+04  5.225D-02 .            .        X  .            +           .
 9.869D+04  3.777D-02 .            .       X   .           +.           .
 9.930D+04  3.010D-02 .            .     X     .           +.           .
 9.990D+04  2.538D-02 .            .    X      .           +.           .
 1.005D+05  2.221D-02 .            .    X      .           +.           .
 1.011D+05  1.995D-02 .            .   X       .           +.           .
 1.017D+05  1.828D-02 .            .  X        .           +.           .
 1.023D+05  1.701D-02 .            .  X        .           +.           .
 1.029D+05  1.602D-02 .            .  X        .           +.           .
```

```
    FREQ        VM(6)

(*)----------- 1.000D-03     1.000D-02     1.000D-01     1.000D+00     1.000D+01
                     - - - - - - - - - - - - - - - - - - - - - - - - - -

(+)----------- -2.000D+02    -1.000D+02      .000D+00     1.000D+02     2.000D+02
                     - - - - - - - - - - - - - - - - - - - - - - - - - -

(=)----------- -6.000D+01    -4.000D+01    -2.000D+01      .000D+00     2.000D+01
                     - - - - - - - - - - - - - - - - - - - - - - - - - -
 1.035D+05  1.524D-02 .           . X          .          +.            .
 1.041D+05  1.462D-02 .           . X          .          +.            .
 1.047D+05  1.413D-02 .           . X          .          +.            .
 1.053D+05  1.375D-02 .           . X          .          +.            .
 1.059D+05  1.345D-02 .           . X          .          +.            .
 1.065D+05  1.322D-02 .           . X          .          +.            .
 1.071D+05  1.306D-02 .           . X          .          +.            .
 1.077D+05  1.295D-02 .           .X           .          +.            .
 1.083D+05  1.290D-02 .           .X           .          +.            .
 1.089D+05  1.289D-02 .           .X           .          +.            .
 1.095D+05  1.293D-02 .           .X           .          +.            .
 1.102D+05  1.301D-02 .           .X           .          +.            .
 1.108D+05  1.314D-02 .           . X          .          +.            .
 1.114D+05  1.330D-02 .           . X          .          +.            .
 1.120D+05  1.351D-02 .           . X          .          +.            .
 1.126D+05  1.377D-02 .           . X          .          +.            .
 1.132D+05  1.407D-02 .           . X          .          +.            .
 1.138D+05  1.442D-02 .           . X          .          +.            .
 1.144D+05  1.483D-02 .           . X          .          +.            .
 1.150D+05  1.530D-02 .           . X          .          +.            .
 1.156D+05  1.583D-02 .           .  X         .          +.            .
 1.162D+05  1.644D-02 .           .  X         .          +.            .
 1.168D+05  1.714D-02 .           .  X         .          +.            .
 1.174D+05  1.793D-02 .           .  X         .          +.            .
 1.180D+05  1.884D-02 .           .   X        .          +.            .
 1.186D+05  1.988D-02 .           .   X        .          +.            .
 1.192D+05  2.109D-02 .           .   X        .          +.            .
 1.198D+05  2.250D-02 .           .    X       .          +.            .
 1.204D+05  2.414D-02 .           .    X       .          +.            .
 1.210D+05  2.609D-02 .           .    X       .          +.            .
 1.216D+05  2.843D-02 .           .     X      .          +.            .
 1.222D+05  3.128D-02 .           .     X      .         + .            .
 1.228D+05  3.482D-02 .           .      X     .         + .            .
 1.234D+05  3.931D-02 .           .       X    .         + .            .
 1.240D+05  4.519D-02 .           .        X   .         + .            .
 1.246D+05  5.320D-02 .           .        X   .         + .            .
 1.252D+05  6.472D-02 .           .          X .         + .            .
 1.258D+05  8.265D-02 .           .           X.         + .            .
 1.264D+05  1.143D-01 .           .            .X        + .            .
 1.270D+05  1.844D-01 .           .            . X      +  .            .
 1.276D+05  4.520D-01 .           .            .      +X   .            .
 1.282D+05  5.898D-01 .           .     +      .        X  .            .
 1.288D+05  2.084D-01 .           .  +         .  X        .            .
 1.294D+05  1.241D-01 .           . +          .X          .            .
 1.301D+05  8.842D-02 .           . +         X.           .            .
 1.307D+05  6.878D-02 .           . +        X .           .            .
 1.313D+05  5.639D-02 .           . +       X  .           .            .
 1.319D+05  4.788D-02 .           . +      X   .           .            .
 1.325D+05  4.168D-02 .           . +     X    .           .            .
```

```
      FREQ        VM(6)

(*)----------- 1.000D-03     1.000D-02     1.000D-01     1.000D+00     1.000D+01
                   - - - - - - - - - - - - - - - - - - - - - - - - - - - - - - -

(+)----------- -2.000D+02   -1.000D+02      .000D+00     1.000D+02     2.000D+02
                   - - - - - - - - - - - - - - - - - - - - - - - - - - - - - - -

(=)----------- -6.000D+01   -4.000D+01    -2.000D+01      .000D+00     2.000D+01
                   - - - - - - - - - - - - - - - - - - - - - - - - - - - - - - -
  1.331D+05  3.697D-02 .            . +    X      .             .             .
  1.337D+05  3.328D-02 .            . +    X      .             .             .
  1.343D+05  3.031D-02 .            . +   X       .             .             .
  1.349D+05  2.788D-02 .            .+    X       .             .             .
  1.355D+05  2.586D-02 .            .+   X        .             .             .
  1.361D+05  2.415D-02 .            .+   X        .             .             .
  1.367D+05  2.269D-02 .            .+   X        .             .             .
  1.373D+05  2.144D-02 .            .+  X         .             .             .
  1.379D+05  2.035D-02 .            .+  X         .             .             .
  1.385D+05  1.940D-02 .            .+  X         .             .             .
  1.391D+05  1.856D-02 .            .+ X          .             .             .
  1.397D+05  1.782D-02 .            .+ X          .             .             .
  1.403D+05  1.717D-02 .            .+ X          .             .             .
  1.409D+05  1.659D-02 .            .+ X          .             .             .
  1.415D+05  1.608D-02 .            .+ X          .             .             .
  1.421D+05  1.562D-02 .            .+ X          .             .             .
  1.427D+05  1.521D-02 .            .+X           .             .             .
  1.433D+05  1.485D-02 .            .+X           .             .             .
  1.439D+05  1.453D-02 .            .+X           .             .             .
  1.445D+05  1.424D-02 .            .+X           .             .             .
  1.451D+05  1.399D-02 .            .+X           .             .             .
  1.457D+05  1.378D-02 .            .+X           .             .             .
  1.463D+05  1.359D-02 .            .+X           .             .             .
  1.469D+05  1.343D-02 .            .+X           .             .             .
  1.475D+05  1.329D-02 .            .+X           .             .             .
  1.481D+05  1.319D-02 .            .+X           .             .             .
  1.487D+05  1.310D-02 .            .+X           .             .             .
  1.493D+05  1.304D-02 .            .X            .             .             .
  1.499D+05  1.301D-02 .            .X            .             .             .
  1.506D+05  1.300D-02 .            .X            .             .             .
  1.512D+05  1.301D-02 .            .X            .             .             .
  1.518D+05  1.305D-02 .            .+X           .             .             .
  1.524D+05  1.311D-02 .            .+X           .             .             .
  1.530D+05  1.320D-02 .            .+X           .             .             .
  1.536D+05  1.331D-02 .            .+X           .             .             .
  1.542D+05  1.346D-02 .            .+X           .             .             .
  1.548D+05  1.364D-02 .            .+X           .             .             .
  1.554D+05  1.386D-02 .            .+X           .             .             .
  1.560D+05  1.411D-02 .            .+X           .             .             .
  1.566D+05  1.440D-02 .            .+X           .             .             .
  1.572D+05  1.474D-02 .            .+X           .             .             .
  1.578D+05  1.514D-02 .            .+X           .             .             .
  1.584D+05  1.560D-02 .            .+ X          .             .             .
  1.590D+05  1.613D-02 .            .+ X          .             .             .
  1.596D+05  1.675D-02 .            .+ X          .             .             .
  1.602D+05  1.746D-02 .            .+ X          .             .             .
  1.608D+05  1.830D-02 .            .+ X          .             .             .
  1.614D+05  1.929D-02 .            .+  X         .             .             .
  1.620D+05  2.046D-02 .            .+  X         .             .             .
```

```
     FREQ        VM(6)

(*)------------   1.000D-03   1.000D-02   1.000D-01   1.000D+00   1.000D+01
                       - - - - - - - - - - - - - - - - - - - - - - - - - -

(+)------------  -2.000D+02  -1.000D+02    .000D+00   1.000D+02   2.000D+02
                       - - - - - - - - - - - - - - - - - - - - - - - - - -

(=)------------  -6.000D+01  -4.000D+01  -2.000D+01    .000D+00   2.000D+01
                       - - - - - - - - - - - - - - - - - - - - - - - - - -
  1.626D+05  2.186D-02 .           .+  X        .           .            .
  1.632D+05  2.358D-02 .           .+   X       .           .            .
  1.638D+05  2.570D-02 .           .+   X       .           .            .
  1.644D+05  2.838D-02 .           .+    X      .           .            .
  1.650D+05  3.187D-02 .           .+     X     .           .            .
  1.656D+05  3.658D-02 .           .+     X     .           .            .
  1.662D+05  4.326D-02 .           .+      X    .           .            .
  1.668D+05  5.342D-02 .           +        X   .           .            .
  1.674D+05  7.064D-02 .           +          X .           .            .
  1.680D+05  1.058D-01 .           +            X           .            .
  1.686D+05  2.089D-01 .         + .            .   X       .            .
  1.692D+05  3.891D-01 .           .            .       X   .      +     .
  1.698D+05  1.467D-01 .           .            . X         .+           .
  1.705D+05  8.199D-02 .           .           X.           +            .
  1.711D+05  5.590D-02 .           .         X  .           +            .
  1.717D+05  4.200D-02 .           .       X    .           +            .
  1.723D+05  3.339D-02 .           .      X     .          +.            .
  1.729D+05  2.756D-02 .           .     X      .          +.            .
  1.735D+05  2.334D-02 .           .    X       .          +.            .
  1.741D+05  2.017D-02 .           .   X        .          +.            .
  1.747D+05  1.768D-02 .           .  X         .          +.            .
  1.753D+05  1.570D-02 .           .  X         .          +.            .
  1.759D+05  1.407D-02 .           . X          .          +.            .
  1.765D+05  1.272D-02 .           .X           .          +.            .
  1.771D+05  1.157D-02 .           .X           .          +.            .
  1.777D+05  1.059D-02 .           X            .          +.            .
  1.783D+05  9.746D-03 .           X            .          +.            .
  1.789D+05  9.008D-03 .          X.            .          +.            .
  1.795D+05  8.359D-03 .          X.            .          +.            .
  1.801D+05  7.784D-03 .          X.            .          +.            .
  1.807D+05  7.272D-03 .         X .            .          +.            .
  1.813D+05  6.814D-03 .         X .            .          +.            .
  1.819D+05  6.401D-03 .        X  .            .          +.            .
  1.825D+05  6.027D-03 .        X  .            .          +.            .
  1.831D+05  5.687D-03 .        X  .            .          +.            .
  1.837D+05  5.378D-03 .       X   .            .          +.            .
  1.843D+05  5.094D-03 .       X   .            .          +.            .
  1.849D+05  4.834D-03 .       X   .            .          +.            .
  1.855D+05  4.595D-03 .       X   .            .          +.            .
  1.861D+05  4.374D-03 .      X    .            .          +.            .
  1.867D+05  4.169D-03 .      X    .            .          +.            .
  1.873D+05  3.979D-03 .      X    .            .          +.            .
  1.879D+05  3.802D-03 .      X    .            .          +.            .
  1.885D+05  3.638D-03 .     X     .            .          +.            .
  1.891D+05  3.484D-03 .     X     .            .          +.            .
  1.897D+05  3.340D-03 .     X     .            .          +.            .
  1.904D+05  3.206D-03 .     X     .            .          +.            .
  1.910D+05  3.079D-03 .    X      .            .          +.            .
  1.916D+05  2.960D-03 .    X      .            .          +.            .
```

```
      FREQ        VM(6)

(*)----------- 1.000D-03    1.000D-02    1.000D-01    1.000D+00    1.000D+01
               - - - - - - - - - - - - - - - - - - - - - - - - - - - - - - -

(+)----------- -2.000D+02   -1.000D+02    .000D+00    1.000D+02    2.000D+02
               - - - - - - - - - - - - - - - - - - - - - - - - - - - - - - -

(=)----------- -6.000D+01   -4.000D+01   -2.000D+01    .000D+00    2.000D+01
               - - - - - - - - - - - - - - - - - - - - - - - - - - - - - - -
 1.922D+05  2.848D-03 .     X      .          .          +.          .
 1.928D+05  2.743D-03 .     X      .          .          +.          .
 1.934D+05  2.643D-03 .    X       .          .          +.          .
 1.940D+05  2.549D-03 .    X       .          .          +.          .
 1.946D+05  2.460D-03 .    X       .          .          +.          .
 1.952D+05  2.376D-03 .    X       .          .          +.          .
 1.958D+05  2.296D-03 .    X       .          .          +.          .
 1.964D+05  2.220D-03 .    X       .          .          +.          .
 1.970D+05  2.148D-03 .   X        .          .          +.          .
 1.976D+05  2.079D-03 .   X        .          .          +.          .
 1.982D+05  2.014D-03 .   X        .          .          +.          .
 1.988D+05  1.951D-03 .   X        .          .          +.          .
 1.994D+05  1.892D-03 .   X        .          .          +.          .
 2.000D+05  1.835D-03 .  X         .          .          +.          .
               - - - - - - - - - - - - - - - - - - - - - - - - - - - - - - - -
```

3.8 INTERPRETATION OF THE AC ANALYSIS

The data shown represents more than might be requested with a normal AC analysis. For example, the .OPTIONS statement, placed in the file, produced two outputs which are the element node table and the circuit element summary. The element node table provides information on each node of the circuit model. Specifically, the node number appears in the far left column with the connected component(s) designators to the right. Node 2, for example, has the components C1 and L1 connected to it. This information may serve as an invaluable diagnostic tool in determining how components of a circuit are interconnected. If a given node has fewer components connected to it than expected, then an extra node may exist in the circuit and a connection between the two nodes is necessary for proper operation. As a general practice, if this condition arises, the user has two choices.

1. Correct the node numbers in the circuit.

2. Place a .001 Ω resistor between the two nodes to create a short.

Either approach is proper. Rarely, if ever, will a 0.001 Ω resistor effect circuit analysis results. In the first case, if file descriptive accuracy is important to the user, then corrections should be made. If time is of consideration and the user is inclined toward efficiency rather than circuit description correctness, then the second approach is most appropriate.

The circuit element summary part of the analysis is exactly as the name portrays. Each circuit element is listed by its designated name, the nodes it is connected to are defined, and its value is given. Any special attributes, such as temperature coefficients or parameter definitions, are also listed. For example, in the analysis, the voltage source VIN is listed as being connected from node 1 to 0 with a DC value of 1 V, an AC value of 1 V, and no phase shift. This information, while useful, can be as readily obtained from a review of the input circuit description. Notice that the DC voltages, like the AC voltages, are listed as 1.00D+00. The term **D** is the same as **E** and simply means times 10 to the XX power, where XX is the 00 value to the right of the D.

An operating point analysis is next presented with all DC node voltages computed. A review of the circuit input file will reveal no .OP command exists to yield this nodal analysis. The analysis came about due to the need, of the SPICE program, to compute the DC operating point prior to every analysis. Note: the operating point analysis is different from the one produced with a .OP command in that it does not list the voltage sources and their currents or the total power dissipation of the circuit! The reader should notice that the voltage at nodes 3 through 6 is 0 V. If 1 V of DC voltage is applied to the input, then why did a positive voltage not show up on nodes 3 through 6? The answer lies in a review of the circuit. Notice that between nodes 2 and 3 is capacitor C1. Capacitors block DC voltage, thus no DC voltage could occur at those nodes.

The last of the analysis is the printed and plotted data. The data portion consists of four columns. The first column contains the frequency points at which the output node information was calculated followed by the requested data in the print statement [VM(6) VP(6) VDB(6)]. The plot presents all the printed material in a graphic form. The plot is lengthwise down the page and has for a vertical axis, the three requested parameters plotted based on separate scaling. The magnitude, phase, and decibels are all plotted against frequency. Figure 3-2, 3-3, and 3-4 are high resolution graphical representation of the plotted data. The high resolution plots can usually be obtained using most conventional SPICE packages with a post processor piece of software. The reader is encouraged to examine the formats of the .PRINT AC and .PLOT AC outputs and to compare the .PLOT AC results against the high resolution plots.

The three resonant frequencies, looking at the printed or plotted data, exist at 96.88 KHz, 128.2 KHz, and 169.2 KHz. The magnitude, phase, and decibel data, at each of the three frequencies, is found using the .PRINT AC generated data. The data is shaded in the analysis and summarized in Table 3-1.

Table 3-1. AC Analysis Summary Data

Frequency	96.88 KHz	128.2 KHz	169.2 KHz
Magnitude	285.8 mV	589.8 mV	389.1 mV
Phase	139.4 deg	-53.87 deg	157.6 deg
DB	-1.088 dB	-4.586 dB	-8.199 dB

The three resonant peaks shown may not be the true peaks because of the nature of the pre-computed frequencies of an AC analysis. Higher accuracy might be obtained by three analysis runs that performed a linear sweep over a frequency range defined by the upper and lower frequencies surrounding the perceived center frequency. The first resonant peak is seen to be 96.88 KHz. A review of the .PRINT AC VM(6) output data provides a frequency on either side of the 96.88 KHz which are 96.28 KHz and 97.49 KHz. If the analysis was rerun using the following sweep definition, then significantly greater precision might be obtained in determining the true resonant peak frequency and associated output data. The .AC statement would provide a frequency step size of about 12 Hz which would provide a high resolution of accuracy when compared to the frequency of the peak.

```
.AC LIN 100 96.28K 97.49K
```

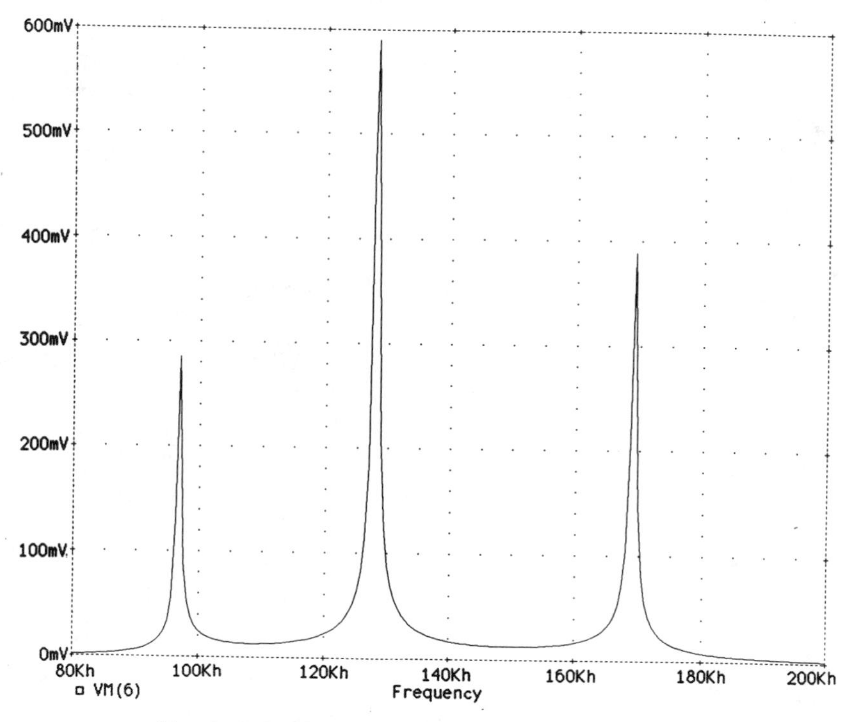

Figure 3-2 Magnitude versus Frequency Plot.

3.9 OTHER ANALYSIS TRICKS

The AC analysis capability can be further enhanced by adding a zero voltage source (VS 3 4 0) in series with an input. The source can detect the current flowing into the input. The .PRINT AC command should include an output definition such as IM(VS). If the SPICE analysis will permit, a plot of VS/IM(VS) can result.

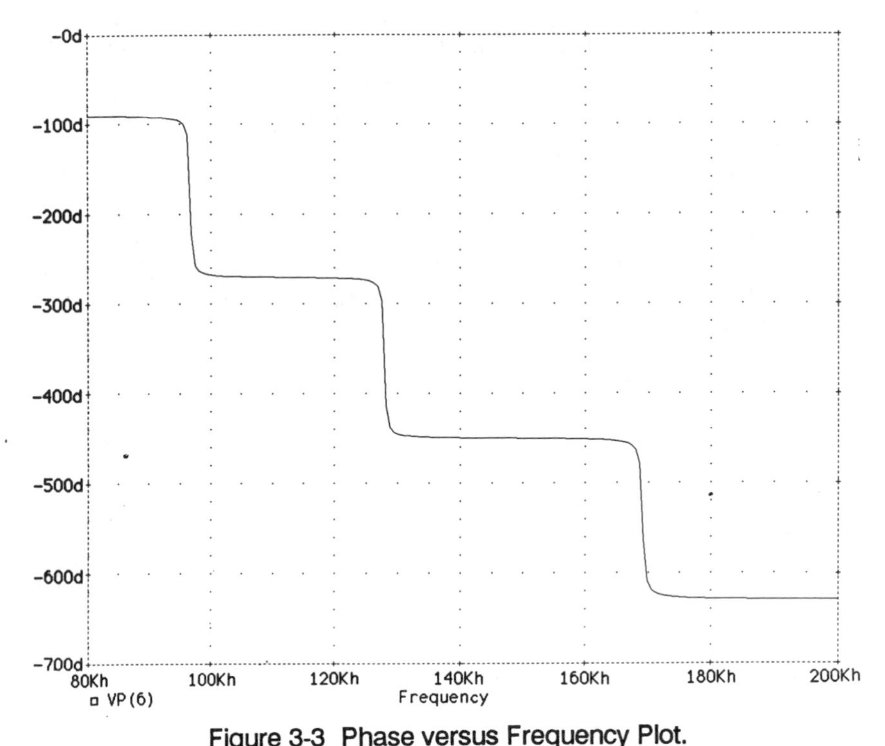

Figure 3-3 Phase versus Frequency Plot.

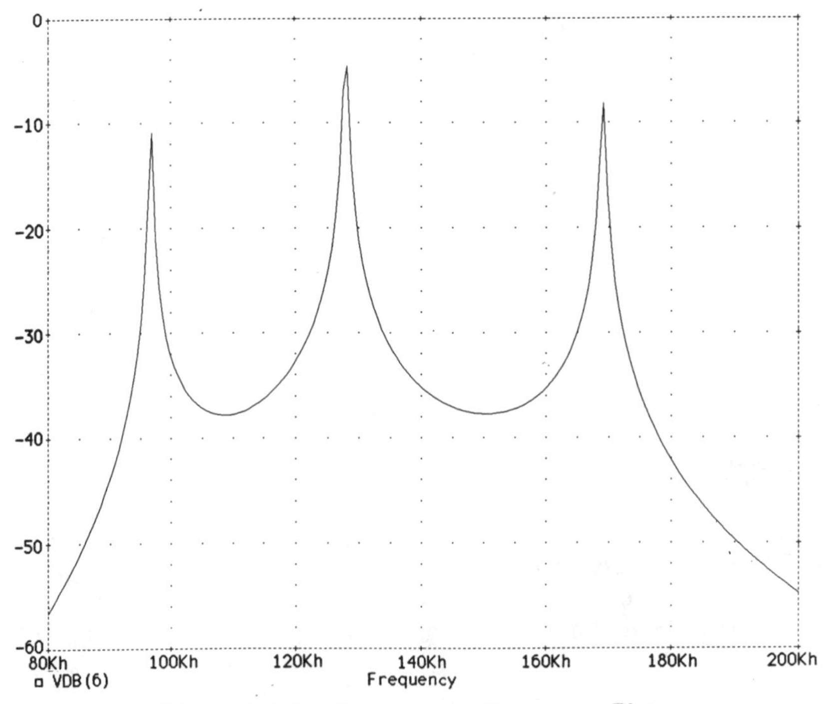

Figure 3-4. Decibels versus Frequency Plot.

This ratio represents the AC input impedance and can be plotted as a function of frequency. With a little imagination, the user can realize the plot of any number of parameters versus frequency. If such a capability as plotting V/I does not exist, then by normalizing the AC input voltage to 1 V, plotting the current will yield a plot of transconductance verses frequency. This approach is valid since V/I=R and if V=1 then I/1 = 1/R = transconductance.

As another example, assume that it is desired to plot output power as a function of frequency. If the load resistance is known then a plot of a V^2/R function or a I^2*R function can be performed. Notice that both calculations are a function of R due to the possible phase shift that may exist between voltage and current.

If this phase shift exists, then the computed power may not represent the true power. Since a resistor has no phase shift capability, the problem is eliminated and the results provided represent real power. Both of these functions represent power and the power delivery capabilities of a circuit might be analyzed. SPICE places, in the hands of the user, an unlimited ability to analyze a multitude of circuits in any number of ways. The ability is limited only by the users creativity and ingenuity.

3.10 SUMMARY

In this Chapter, information on the use of the following commands has been presented. The reader is encouraged to revisit the commands and attempt an exploration of their capabilities and applications outside the scope of this section

1. Independent voltage source

   VXXXXXXX N+ N- DC DCVALUE AC ACVALUE ACPHASE

2. Independent current source

   IXXXXXXX N+ N- DC DCVALUE AC ACVALUE ACPHASE

3. Print statement for ac analysis

   . PRINT AC OV1 [OV2 ... OV8]

4. Plot statement for ac analysis

   . PLOT AC OV1 [OV2 ... OV8]

5. AC sweep definition for AC analysis

   .AC DEC NP FS FE
   .AC OCT NP FS FE
   .AC LIN NP FS FE

DC ANALYSIS

Thus far, we have discussed the concepts of operating point, transfer function, and AC analysis. The next logical discussion is in the area of DC analysis. DC analysis was initially described in Section 1.3.1. and represents an analysis for which effects are observed on an output node/source of a circuit using a DC stimulation applied to some input point on the circuit. To fully appreciate the power of a DC analysis, an understanding of the solid state is necessary, including the following concepts:

1. Sensitivity analysis
2. DC analysis
3. The use and definition of diodes
4. The use and definition of transistors
5. The use and definition of FETS and MOSFETS

Although this represents a large amount of subject matter if studied in depth, the focus of this chapter is to familiarize the reader with the basic concepts and their applications.

Any DC analysis requires three basic pieces of information.

1. The DC source

2. The nodes/sources to be analyzed and how are they to be analyzed.

3. The voltage or current sweep range that must be applied to the input and analyzed at the output.

In this chapter, the interrelationship of these three pieces is demonstrated.

4.1 SAMPLE DC CIRCUIT

A sample DC circuit is shown in Figure 4-1. The circuit represents a differential amplifier with a single-end input, and a differential output. This circuit serves as the basis to illustrate the use of solid-state devices, sensitivity analysis, and DC analysis parameters.

DC analysis is best used for implementing with active components. It could be performed for the effects of various input voltages on a **passive circuit**, but this could be accomplished by an operating point analysis. For example, if a purely passive circuit is modeled and an operating point analysis is performed, the result is a relationship between a voltage or current source and each node in the circuit. It is reasonable to assume that a gain/loss constant exists between a supply and a node. Once the relationship is established, the ability to predict an output node voltage/current for any input source value becomes trivial.

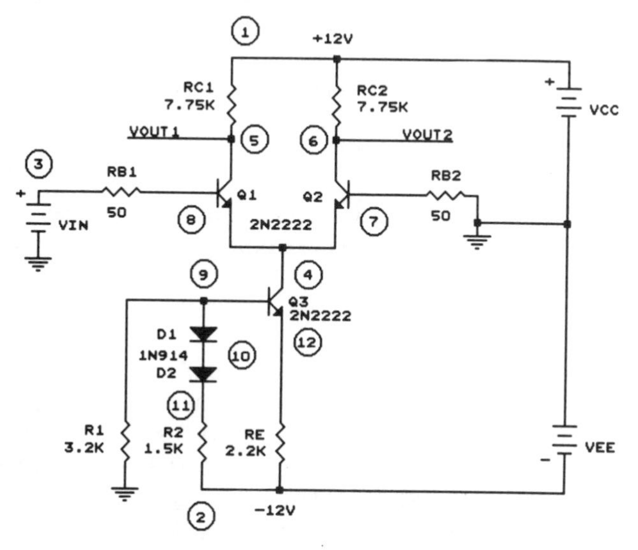

Figure 4-1. Sample DC Circuit.

DC analysis is most often used for nonlinear circuit evaluation, such as the differential amplifier. It may become desirable to predict the area of linearity of an amplifier circuit to enable the user to (1) restrict the range of the input to the circuit or (2) adjust the component values of the circuit to realize a maximally linear output swing (the maximum peak-to-peak voltage swing with the least amount of nonlinearity).

The circuit in Figure 4-1 will be analyzed to determine the following:

1. The most sensitive components of the circuit relative to their effect on the output of the amplifier?

2. The optimal output voltage swing (peak-to-peak) that can be realized without introducing a major nonlinear element?

3. The node voltages at each node of the circuit.

4.2 SENSITIVITY ANALYSIS

A ***Sensitivity analysis*** is an analysis performed on one or more outputs that provides information on the affects of every circuit component and that component's parameters on each specified output. If the circuit contains many solid-state devices (say 20) and these devices each have an average of 15 parameters, then, as a minimum, the analysis for a single node will contain at least 300 to 500 lines of data or approximately 6 to 10 pages of data per output. If many outputs have been identified, then a large amount of data and resulting hardcopy may be generated. The user is strongly encouraged to review the number of outputs of interest and restrict the scope of the analysis to essential outputs.

The general form for the implementation of a sensitivity analysis is as follows:

.SENS OV1 [OV2 ...]

Assume, for example, that you want to realize a sensitivity analysis on nodes 5 and 6 of the circuit in Figure 4-1. A statement could be placed in the input file as follows:

```
.SENS V(5) V(6)
```

This will cause a sensitivity analysis to be performed on every component and parameter of the circuit relative to it's influence on the output voltage at node 5 and node 6. It should be of no surprise to the reader, at this point, that a sensitivity of the current through an element, using an ammeter, can also be realized. The form of the sensitivity analysis output is shown in Section 4.10.

A partial sensitivity analysis, as shown in Table 4-1, involves the listing of the output node(s), each element name, the element value, element sensitivity, and normalized sensitivity. Look at the last row of Table 4-1; the last component shown, R1, has a value of 3.2 KΩ. The element sensitivity is 1.287 mV/unit. Because resistor units are in ohms, the change is 1.287 mV/Ω, which indicates that node 5 will increase 1.287 mV for every ohm increase of R1 and will decrease the same amount for every ohm decrease of R1. Assume, for a moment that the DC bias voltage at node 5 is 5 V at the bias point of the sensitivity analysis. Assume further that the DC supply voltage supplying the circuit is +12 V. If we want to increase the value of R1 by 7500 Ω, multiply 1.287 mV/Ω by 7500, which yields an increase of 9.65 V. Since the output was at 5 V, adding the 9.65 V produces an output voltage of 14.65 V with a resistance R1 of 10.7 KΩ. Clearly, this is not possible because the supply voltage is only 12 V. Therefore, sensitivity of any node/source, to a specified component, is meaningful only over a narrow range of values due to changes in circuit biasing.

The normalized sensitivity is simply another way of looking at the data. In the discussed example R1, above, the normalized sensitivity is 41.19 mV/percent change in R1. If R1 were to increase in value 1%, then 3200 Ω * 0.01 = 32 Ω. The increase of 32 Ω * 1.287 mV = 41.19 mV which is the value of the normalized sensitivity. A negative element sensitivity or normalized sensitivity indicates that an increase in resistance, as in the case of RC1, will cause the voltage at the specified node to drop by that amount.

In summary, the sensitivity analysis is based solely on the operating point of the circuit and has no validity when viewed with a DC or AC analysis. Restated, the sensitivity analysis purely attempts to describe the influence, of each circuit component, on a node based on the DC operating point of the circuit.

Table 4-1. Partial Sensitivity Analysis Example

```
DC SENSITIVITIES OF OUTPUT V(5)


      ELEMENT          ELEMENT        ELEMENT        NORMALIZED
       NAME             VALUE       SENSITIVITY     SENSITIVITY
                                   (VOLTS/UNIT)  (VOLTS/PERCENT)

      RC1            7.750D+03      -8.723D-04      -6.761D-02
      RC2            7.750D+03      -3.399D-05      -2.634D-03
      RB1            5.000D+01       7.390D-04       3.695D-04
      RB2            5.000D+01      -7.390D-04      -3.695D-04
      R1             3.200D+03       1.287D-03       4.119D-02
```

4.3 ACTIVE COMPONENTS

There are four classes of active components available to the SPICE user:

1. Diode
2. Transistors (bipolar junction)
3. FET (field effect)
4. MOSFET

The circuit components described to this point typically require only a few parameters. However, the models for semiconductor devices, included in the SPICE program require many parameters. Often, many devices in a circuit are defined by the same set of device model parameters. For example, a circuit may contain six transistors of the 2N2222A variety. For this reason, a set of device model parameters is defined by a separate .MODEL statement and assigned a unique model name allowing the parameters associated with that device to be specified only once. The device element statements in SPICE refer only to the model name. This implementation alleviates the need to specify all of the model parameters in each device element statement.

Each semiconductor device element statement contains the device name, the nodes to which the device is connected, and the device model name. In addition, other optional parameters may be specified for some devices (e.g., geometric factors and an initial condition).

4.3.1 DIODES

Diodes are specified as a two-node device. The first specified node is the anode and the second is the cathode. All diodes, whether signal, power, or zener diodes, are specified from anode to cathode. All diodes must start with a "D" and may have up to seven alphanumeric characters following the D. The general form of the diode is shown below:

DXXXXXXX NA NC MNAME [AREA] [OFF] [IC=VD]

To further elaborate on the general form, NA and NC are the positive and negative nodes, respectively. NA defines the anode of the diode and NC defines the cathode of the diode. MNAME is the model name which may be any name desired providing it begins with an alphabetical character. AREA is the optional area factor which indicates how many such devices are connected in parallel. The area parameter is useful in increasing the current handling capability of a diode without having to redefine a complete set of diode parameters. If the AREA factor is omitted, a value of 1 is assumed. The OFF parameter indicates an optional starting condition on the device for a DC analysis. This OFF parameter is useful in realizing the successful convergence of a circuit and the resulting operating point definition. Specifically, the OFF parameter turns the diode off (places a switch in series with the diode) while the operating point is being computed.

The optional initial condition specification using IC = VD is intended for use with the UIC option on the .TRAN statement, discussed later in Chapter 6, when a transient analysis is desired starting from other than the quiescent operating point. Assume that we were to describe diode D1 of the circuit in Figure 4-1. The diode is a 1N914 and will be assigned a name D1N914. The diode is a single diode (Area = 1) and is assumed to be active (not off). The diode device element statement could then be specified as follows:

D1 9 10 D1N914

Describing the connection of a diode is not sufficient to describe the parameters of the diode. Diodes come in all shapes and sizes and carry with them specific characteristics related to power-handling capacity, switching speed, forward-voltage drop, reverse breakdown voltage, and other dynamic parameters. Because of this, model parameters on each diode type used in SPICE must be defined. If a model definition is attempted, without the use of parameters, then SPICE uses a set of default values. The resulting diode is a nearly perfect diode and possesses characteristics that may produce unrealistic results.

The general form for describing the parameters of the diode is shown below. If a specific diode type, for example a 1N914, is used more than once in a circuit, the model parameters need only be described and defined once.

.MODEL MNAME D [PNAME1=PVAL1 PNAME2=PVAL2 ...]

The .MODEL statement specifies a set of model parameters that will be used by one or more diode device element statements. MNAME is the model name used with the DXXXXXXX diode description discussed above. The letter D, before the terms in brackets, specifies that this is a model of a diode.

The DC characteristics of a diode are determined by the parameters IS and N. An ohmic resistance is specified as RS. Charge storage effects are modeled by a parameter TT (transit time). The nonlinear depletion layer capacitance is determined by the parameters CJO, VJ, and M. The temperature dependence of the saturation current is defined by the parameters EG (the energy gap) and XTI (the saturation current temperature exponent). Reverse breakdown is modeled by an exponential increase in the reverse diode current and is determined by the parameters BV and IBV (both of which are positive numbers). Table 4-2 reflects a listing of the available diode parameters that define the characteristics of a diode. The default column specifies the value that SPICE will use if no value is specified.

Table 4-2. List of Diode Model Parameters

	Name	Parameter	Unit	Default	Example	Area
1	IS	Saturation current	A	1.0E-14	1.0E-14	*
2	RS	Ohmic resistance	Ω	0	10	*
3	N	Emission coefficient	-	1	1.0	
4	TT	Transit time	S	0	0.1 ns	
5	CJO	Zero-bias junction capacitance	F	0	2 pF	*
6	VJ	Junction potential	V	1	0.6	
7	M	Grading coefficient	-	0.5	0.5	
8	EG	Activation energy	eV	1.11	1.11 Si 0.69 Sbd 0.67 Ge	
9	XTI	Saturation current temperature exponent	-	3.0	3.0 jn	
10	KF	Flicker noise coefficient	-	0		
11	AF	Flicker noise exponent	-	1		
12	FC	Coefficient for forward-bias depletion capacitance formula	-	0.5		
13	BV	Reverse breakdown voltage	V	∞	40.0	
14	IBV	Current at breakdown voltage	A	1.0E-10		
15	ISR	Recombination current parameter	A	0		
16	NR	Emission coefficient of NR	-	2		
17	IKF	High-injection 'knee' current	A	∞		
18	TIKF	IKF temperature coefficient (linear)	1/°C	0		
19	TRS1	RS temperature coefficient (linear)	1/°C	0		
20	TRS2	RS temperature coefficient (quadratic)	$1/°C^2$	0		

A set of parameters that may not be supported by all versions of SPICE are shown as shaded items in the Table. The reader is encouraged to research thoroughly which parameters are supported by the SPICE package in use.

For example, normally the value of EG defaults to 1.11, which is the energy gap of a silicon diode. If this value is 0.69, the energy gap is associated with a Schottky barrier diode. If the value is 0.67, a germanium diode is specified.

Assume that you want to specify a 1N914 diode, as in the circuit of Figure 4-1. The model information could be specified as follows:

```
.MODEL D1N914 D (IS=1E-13 RS=16 CJO=2.2PF TT=12NS BV=100
+ IBV=1E-13)
```

Where the 1N914 is a silicon diode with a reverse breakdown voltage of 100 V and a series resistance of 16 Ω. The model could have been extracted from vendor data sheets at the expense of much time and effort. A description of the extraction process would require many chapters and is outside the scope of this text; specific material on solid state modeling may be found in most libraries. To expedite the user's ability to make maximum use of SPICE, a limited model library has been provided in Appendix A, including information on a variety of diodes and transistors. These parameters listed in Appendix A describe the specific device in sufficient detail to permit maximum model accuracy when performing a DC, AC, or transient analysis.

The reader is directed to the use of the + sign on the second line of the diode description (.MODEL D1N914 above). When the description of any component exceeds approximately 70 columns, a carriage return should be implemented after the last full parameter description and the new line should start with a "+" followed by a space. This is mandatory in most systems when diodes and transistors are being specified and indicates, to SPICE, that the description of the data is being extended to two or more lines.

In table 4-2 an asterisk in the "Area" column indicates that the specified parameter will be affected by the area statement of the diode DXXXXXXX statement. For example, the parameters affected by the area are the IS, RS, and CJO parameters. If two or more diodes are placed in parallel using the area option, the value of IS will increase, the value of RS will decrease, and the value of CJO will increase. This should be obvious to the reader as resistors in parallel decrease in value, capacitors in parallel add in value, and leakage currents in parallel add in value.

4.3.2 ZENER DIODES

Zener diodes are a special case of a regular diode. The specification of the nodes (anode and cathode) of the zener are the same as for a regular diode. Reverse breakdown voltage is mostly what defines the zener action and resulting characteristics of the zener and differentiates it from a signal diode. The parameters of a 1N751 that is a 5.1 V general-purpose zener diode are shown below. Note that the BV (reverse breakdown) parameter is specified as 5.13 V at a reverse current of 77.21 mA. If the zener is a 5.1 V, then why specify the BV at 5.13 V? The answer is complex and is associated with the RS, IS, and IBV parameters to assure that the zener voltage at the specified test current (20 mA in this case), yields a zener that optimally fits all it's manufacturers specifications.

```
.MODEL D1N751 D(IS=880.5E-18 N=1 RS=0.25 CJO=175P M=0.5516
+ VJ=0.75 BV=5.13 IBV=77.21M TT=1.443M)
```

Additional zener modeling information can be found in Appendix A.

4.3.3 TRANSISTORS

Transistors are specified as a three- or four-lead device. The fourth lead represents the substrate of the transistor and is rarely available for access. Traditionally, the substrate of a transistor is tied to ground (node 0) and is not of concern to the user. The three major nodes of the transistor are the collector, base, and emitter. The three nodes must be specified in exactly the order of CBE when describing the connections of a transistor. When a transistor is described in a circuit, the first letter of the name must be Q. The letter "Q", like the diode and all other components discussed so far, may be followed by up to seven alphanumeric characters. The general form for specifying a transistor is as follows:

QXXXXXXX NC NB NE [NS] MNAME [AREA] [OFF] [IC=VBE,VCE]

NC, NB, and NE are the collector, base, and emitter nodes respectively. NS is the optional substrate node. MNAME is the model name and, as in the case of the diode, may be anything starting with an alphabetical character. The AREA is the area factor, which indicates the number of devices connected in parallel. AREA acts in the same fashion described for the diode, but will affect a greater number of parameters. If the area factor is omitted, a value of 1 is assumed.

The optional OFF parameter, as in the case of the diode, indicates a removal of the transistor from the circuit, during the computation of an initial operating point condition. The optional initial condition specification, using [IC = VBE, VCE], is intended for use with the UIC option on the .TRAN statement, to be discussed later, when a transient analysis is desired starting from other than the quiescent operating point.

Possible examples of a transistor designation are shown below:

```
Q23 10 24 13 QMOD IC=0.6,5.0.
```

This description is of a transistor Q23 with the collector connected to node 10, the base connected to node 24, and the emitter connected to node 13. The transistor is described in a .MODEL QMOD statement with initial conditions of VBE = 0.6 V and VCE = 5 v.

```
Q50A 11 26 4 20 MOD1 OFF.
```

This description is of a transistor Q50A with the collector connected to node 11, the base connected to node 26, the emitter connected to node 4, and the substrate connected to node 20. The transistor is described in the .MODEL MOD1 statement and is excluded from the determination of the operating point calculation.

A transistor, simply connected into a circuit with the CBE nodes defined, has obtained no identity relative to its parameters or gender (NPN or PNP). This information is provided, as in the case of the diode, by the use of a .MODEL statement. The .MODEL statement for a NPN transistor is as follows:

.MODEL MNAME NPN [PNAME1=PVAL1 PNAME2=PVAL2 ...]

The .MODEL statement specifies the type of transistor (NPN or PNP) and a set of model parameters that may be used by one or more device element statements.

In a similar fashion a PNP is defined as follows:

.MODEL MNAME PNP [PNAME1=PVAL1 PNAME2=PVAL2 ...]

Parameter values are defined by appending the parameter name, as shown in Table 4-3, followed by an equal sign and the parameter value. Model parameters that are not given a value are assigned the default values shown in Table 4-3 for each model type. The following describes a 2N2222A transistor with all necessary parameters:

```
.MODEL Q2N2222A NPN(IS=14.34F VAF=74.03 BF=255.9
+ NE=1.307 ISE=14.34F IKF=0.2847 XTB=1.5 BR=6.092 IKR=0
+ RC=1 CJC=7.306P MJC=0.3416 FC=0.5 CJE=22.01P MJE=0.377
+ TR=46.91N TF=411.1P ITF=0.6 VTF=1.7 XTF=3 RB=10)
```

A PNP transistor, such as the 2N2907A, may be described as follows:

```
.MODEL Q2N2907A PNP(IS=650.6E-18 VAF=115.7 BF=231.7
+ NE=1.829 ISE=54.81F IKF=1.079 XTB=1.5 BR=3.563 IKR=0
+ RC=0.715 CJC=14.76P MJC=0.5383 FC=0.5 CJE=19.82P
+ MJE=0.3357TR=111.3NTF=603.7PITF=0.65VTF=5XTF=1.7RB=10)
```

If a matched pair of transistors (NPN and matched complement PNP) is desired then the user may select a matched pair from Appendix A. If this is not fruitful, the alternative is to pick a PNP or NPN device that closely approximates one of the desired devices and create a complement, for example,

```
.MODEL QN NPN(IS=14.34F VAF=74.03 BF=255.9
+ NE=1.307 ISE=14.34F IKF=0.2847 XTB=1.5 BR=6.092 IKR=0
+ RC=1 CJC=7.306P MJC=0.3416 FC=0.5 CJE=22.01P MJE=0.377
+ TR=46.91N TF=411.1P ITF=0.6 VTF=1.7 XTF=3 RB=10)

.MODEL QP PNP(IS=14.34F VAF=74.03 BF=255.9
+ NE=1.307 ISE=14.34F IKF=0.2847 XTB=1.5 BR=6.092 IKR=0
+ RC=1 CJC=7.306P MJC=0.3416 FC=0.5 CJE=22.01P MJE=0.377
+ TR=46.91N TF=411.1P ITF=0.6 VTF=1.7 XTF=3 RB=10)
```

In this instance the parameters of the 2N2222A transistor were chosen as the desired characteristics. Two separate devices were created and named QN and QP for the NPN and PNP versions of the device. Notice that the only difference between the two devices is the NPN and PNP designations in front of the parenthesis.

Table 4-3. Modified Gummel-Poon Transistor Model Parameters

	Name	Parameter	Unit	Default	Example	Area
1	IS	Transport saturation current	A	1.0E-16	1.0E-15	*
2	BF	Ideal maximum forward beta	-	100	100	
3	NF	Forward current emission coefficient	-	1.0	1	
4	VAF	Forward Early voltage	V	∞	200	
5	IKF	Corner for forward beta high current roll-off	A	∞	0.01	*
6	ISE	B-E leakage saturation current	A	0	1.0E-13	*
7	NE	B-E leakage emission coefficient	-	1.5	2	
8	BR	Ideal maximum reverse beta	-	1	0.1	
9	NR	Reverse current emission coefficient	-	1	1	
10	VAR	Reverse Early voltage	V	∞	200	
11	IKR	Corner for reverse beta high current roll-off	A	∞	0.01	*
12	ISC	B-C leakage saturation current	A	0	1.0E-13	*

	Name	Parameter	Unit	Default	Example	Area
13	NC	B-C leakage emission coefficient	-	2	1.5	
14	RB	Zero bias base resistance	Ω	0	100	*
15	IRB	Current where base resistance falls halfway to minimum value	A	∞	0.1	*
16	RBM	Minimum base resistance at high currents	Ω	base resistance	10	*
17	RE	Emitter resistance	Ω	0	1	*
18	RC	Collector resistance	Ω	0	10	*
19	CJE	B-E zero-bias depletion capacitance	F	0	2PF	*
20	VJE	B-E built-in potential	V	0.75	0.6	
21	MJE	B-E junction exponential factor	-	0.33	0.33	
22	TF	Ideal forward transit time	Sds	0	0.1ns	
23	XTF	Coefficient for bias dependence of TF	-	0		
24	VTF	Voltage describing VBC dependence of TF	V	∞		
25	ITF	High-current parameter for affect on TF	A	0		*
26	PTF	Excess phase at freq = 1.0/(TF*2π)Hz	°	0		
27	CJC	B-C zero-bias depletion capacitance	F	0	2pF	*
28	VJC	B-C built-in potential	V	0.75	0.5	
29	MJC	B-C junction exponential factor	-	0.33	0.5	
30	XCJC	Fraction of B-C depletion capacitance connected to internal base node	-	1		
31	TR	Ideal reverse transit time	S	0	10Ns	
32	CJS	Zero-bias collector-substrate capacitance	Fs	0	2pF	*
33	VJS	Substrate junction built-in potential	V	0.75		

	Name	Parameter	Unit	Default	Example	Area
34	MJS	Substrate junction exponential factor	-	0	0.5	
35	XTB	Forward and reverse beta temperature exponent	-	0		
36	EG	Energy gap for temperature affect on IS	eV	1.11		
37	XTI	Temperature exponent for affect on IS	-	3		
38	KF	Flicker-noise coefficient	-	0		
39	AF	Flicker-noise exponent	-	1		
40	FC	Coefficient for forward-bias depletion capacitance formula	-	0.5		
41	ISS	substrate p-n saturation current	A	0		
42	NS	substrate p-n emission coefficient	-	1		
43	TRE1	RE temperature coefficient (linear)	1/°C	0		
44	TRE2	RE temperature coefficient (quadratic)	$1/°C^2$	0		
45	TRB1	RB temperature coefficient (linear)	1/°C	0		
46	TRB2	RB temperature coefficient (quadratic)	$1/°C^2$	0		
47	TRM1	RBM temperature coefficient (linear)	1/°C	0		
48	TRM2	RBM temperature coefficient (quadratic)	$1/°C^2$	0		
49	TRC1	RC temperature coefficient (linear)	1/°C	0		
50	TRC2	RC temperature coefficient (quadratic)	$1/°C^2$	0		

The bipolar junction transistor (BJT) parameters used in the modified Gummel Poon model are listed in Table 4-3. The parameter names used in versions of SPICE discussed earlier. Parameters 41 to 50 may not be supported by all versions of SPICE. The reader is encouraged to research thoroughly which parameters are supported by the SPICE package in use.

The BJT model in SPICE is an adaptation of the integral charge control model of Gummel and Poon. This modified Gummel-Poon model extends the original model to include several affects at high bias levels. The model will automatically simplify to the simpler Ebers-Molls model when certain parameters are not specified. The parameter names used in the modified Gummel-Poon model have been chosen to be more easily understood by the user and to reflect better both physical and circuit design thinking.

The DC model is defined by the parameters IS, BF, NF, ISE, IKF, and NE which determine the forward current gain characteristics, IS, BR, NR, ISC, IKR, and NC which determine the reverse current gain characteristics, and VAF and VAR which determine the output conductance for forward and reverse regions. Three ohmic resistances RB, RC, and RE are included, where RB is dependent on the high current. Base charge storage is modeled by forward and reverse transit times, TF and TR (the forward transit time TF being bias dependent if desired), and by non linear depletion layer capacitances which are determined by CJE, VJE, and MJE for the BE junction, CJC, VJC, and MJC for the BC junction and CJS, VJS, and MJS for the CS (collector substrate) junction. The temperature dependence of the saturation current, IS, is determined by the energy gap, EG, and the saturation current temperature exponent, XTI. Additionally base current temperature dependence is modeled by the beta temperature exponent XTB in the new model.

A major error made by first-time users of SPICE, when analyzing any solid state device, especially transistors, is that they do not specify or specify incorrectly all essential parameters, in which case, the analysis will run normally but the results will be inconsistent with actual circuit performance. The problem is accentuated by the significant interactions of parameters. When one parameter is negated, the default value for that parameter is assumed. A summary of key transistor parameters and their significance to specification parameters and each other is shown in Table 4-4.

4.3.4 JUNCTION FIELD-EFFECT TRANSISTORS

The junction-field-effect (JFET) transistor is treated much in the same fashion as the bi polar transistor. There is no substrate node but there exists a drain, gate, and source. The nodes must be specified in the order DGS to realize proper circuit description and operation.

The JFET is designated by a "J" followed by the usual seven characters. The general form is

JXXXXXXX ND NG NS MNAME [AREA] [OFF] [IC=VDS,VGS]

Table 4-4. Transistor Parameter Interdependence

Parameters	Description and implementation
NE, NC	Normally set to 4
BF, ISE, IKF	These are adjusted to give the nominal beta versus collector current curve. BF controls the mid-range beta, ISE/IS controls the low-current roll-off, and IKF controls the high-current rolloff.
ISC	Set to ISE.
IS, RB, RE, RC	These are adjusted to give the nominal VBE versus IC and VCE versus IC curves in saturation. IS controls the low-current value of VBE, RB plus RE controls the rise of VBE with IC, and RE plus RC controls the rise of VCE with IC. RC is typically set to 0.
VAF	Using the voltages specified on the data sheet, VAF is set to give the nominal output impedance (RO from the .OP output) on the data sheet.
CJC, CJE	Using the voltages specified on the data sheet, CJC and CJE are set to give the nominal input and output capacitances (CPI and CMU from the .OP output and Cibo and Cobo on the data sheet).
TF	Using the voltages and currents specified on the data sheet for FT, TF is adjusted to produce the nominal value of FT from the .OP output.
TR	Using the rise and fall time circuits on the data sheet, TR and/or TF are adjusted to give a transient analysis which shows the nominal values of the turn-on delay, rise time, storage time, and fall time.
KF, AF	These parameters are only set if the data sheet has a specification for noise. Then, AF is set to 1 and KF is set to produce a total noise at the collector which is greater than the generator noise at the collector by the rated number of decibels.

Where ND, NG, and NS are the drain, gate, and source nodes respectively. MNAME is the model name, AREA is the area factor, which indicates the number of devices that are connected in parallel, and OFF indicates an initial condition of the device for quiescent operating point determination. If the area factor is omitted, a value of 1.0 is assumed. The optional initial condition specification, using {IC=VDS, VGS}, is intended for use with the UIC option on the .TRAN statement, to be discussed later. The UIC is deployed when a transient analysis is desired to start from other than the quiescent operating point.

For example, a sample JFET, J1, can be written:

```
J1 7 2 3 JM1 OFF
```

With the drain connected to node 7, the gate connected to node 2, and the source connected to node 3. The parameters, of the JFET, are described in a model JM1. The initial conditions show the JFET in the OFF state.

No designation, to whether the JFET is a P or N channel device, has been defined as yet. This designation is performed in the .MODEL statement. The designation of an N channel JFET is as follows:

```
.MODEL MNAME NJF [PNAME1=PVAL1 PNAME2=PVAL2 ... ]
```

The designation of the P channel JFET is similar and as follows:

```
.MODEL MNAME PJF [PNAME1=PVAL1 PNAME2=PVAL2 ... ]
```

The JFET model is derived from the FET model of Shichman and Hodges. The DC characteristics are defined by the parameters (1) VTO and BETA, which determine the variation of drain current with gate voltage, (2) LAMBDA, which determines the output conductance, and (3) IS, which is the saturation current of the two gate junctions. Two ohmic resistances, RD and RS, are also included. Charge storage is modeled by non linear depletion layer capacitances for both gate junctions. This stored charge may vary as the -0.5 power of junction voltage and is defined by the parameters CGS, CGD, and PB.

```
.MODEL J2N3819 NJF(BETA=1.3M RD=1 RS=1 LAMBDA=2.25M
+ VTO=-3 IS=33.6F CGD=1.6P FC=0.5 CGS=2.4P KF=9.9E-18)
```

The parameters associated with the JFET are shown in Table 4-5.

Table 4-5. JFET Parameters

	Name	Parameter	Units	Default	Example	Area
1	VTO	Threshold voltage	V	-2.0	-2.0	
2	BETA	Transconductance parameter	A/V^2	1.0E-4	1.0E-3	*
3	Lambda	Channel length modulation parameter	1/V	0	1.0E-4	
4	RD	Drain ohmic resistance	Ω	0	100	*
5	RS	Source ohmic resistance	Ω	0	100	*
6	CGS	Zero-bias G-S junction capacitance	F	0	5pF	*
7	CGD	Zero-bias G-D junction capacitance	F	0	1pF	*
8	PB	Gate junction potential	V	1	0.6	

	Name	Parameter	Units	Default	Example	Area
9	IS	Gate junction saturation current	A	1.0E-14	1.0E-14	*
10	KF	Flicker noise coefficient	-	0		
11	AF	Flicker noise exponent	-	1		
12	FC	Coefficient for forward-bias depletion capacitance formula	-	0.5		
13	N	Gate p-n emission coefficient	-	1		
14	ISR	Gate p-n recombination current parameter	A	0		
15	NR	Emission coefficient for ISR	-	2		
16	ALPHA	Ionization coefficient	1/V	0		
17	VK	ionization 'knee' voltage	V	0		
18	M	Gate p-n grading coefficient	-	.5		
19	VTOTC	VTO temperature coefficient	V/°C	0		
20	BETATCE	Beta exponential temperature coefficient	%/°C	0		
21	XTI	ISA temperature coefficient	-	3		

The parameters shown as shaded may not be supported by all versions of SPICE. The reader is encouraged to research thoroughly which parameters are supported by the SPICE package in use.

4.3.5 MOSFET

The field effect (MOSFET) transistor is treated much in the same fashion as the JFET. There exists a drain, gate, and source plus a substrate node designated NB. The nodes must be specified in the order DGSB to realize proper circuit description and operation. The JFET is designated by an "M" followed by the usual seven characters as desired. The general form is shown below:

MXXXXXXX ND NG NS NB MNAME [L=VAL] [W=VAL]
+ [AD=VAL] [AS=VAL] [PD=VAL] [PS=VAL]
+ [NRD=VAL] [NRS=VAL] [OFF] [IC=VDS,VGS,VBS]

where		
	L	= channel length
	W	= channel width
	AD	= drain diffusion in square meters
	AS	= source diffusion in square meters
	PD	= perimeter of the drain junction in meters
	PS	= perimeter of the source junction in meters
	NRD	= equivalent number of squares of drain diffusion
	NRS	= equivalent number of squares of source diffusion
	VDS	= initial condition of drain-source voltage
	VGS	= initial condition of gate-source voltage
	VBS	= initial condition of substrate-source voltage

ND, NG, NS, and NB are the drain, gate, source, and bulk (substrate) nodes, respectively. MNAME is the model name. L and W are the channel length and width, in meters. AD and AS are the areas of the drain and source diffusions, in square meters.

Note that the suffix μ specifies microns (1E-6 m) and P specifies square-microns (1E-12 square-m). If any of L, W, AD, or AS are not specified, default values are used. The use of defaults simplifies input file preparation, as well as the editing required if device geometries are to be changed. PD and PS are the perimeters of the drain and source junctions, in meters. NRD and NRS designate the equivalent number of squares of the drain and source diffusions. These values multiply the sheet resistance RSH specified in the .MODEL statement for an accurate representation of the parasitic series drain and source resistance of each transistor.

PD and PS default to 0.0, whereas NRD and NRS default to 1.0. OFF indicates an initial condition on the device for DC analysis. The optional initial condition specification using {IC = VDS, VGS, VBS} is intended for use with the UIC option on the .TRAN statement, to be discussed later. Examples of a MOSFET are shown below:

```
M1 2 9 3 0 MOD1 L=10U W=5U AD=100P AS=100P PD=40U PS=40U
```

This example defines a MOSFET, designated M1, with the drain connected to node 2, the gate connected to node 9, the source connected to node 3, and the substrate connected to node 0 (ground). The MOSFET is described by the parameters associated with model MOD1. The MOSFET has a channel width (W) of 5 microns, a channel length (L) of 10 microns, a drain and source diffusion area (AD and AS) of 0.1E-9 m^2, and a drain and source junction perimeter (PD and PS) of 40 μ.
For example,

```
M31 2 17 6 10 IRF150
```

defines a MOSFET designated M31 with the drain connected to node 2, the gate connected to node 17, the source connected to node 6, and the substrate connected to node 10. Furthermore, the MOSFET is described by the parameters associated with model IRF150.

The LEVEL (1, 2, 3, or 4) of the MOSFET, discussed later, is defined in the .MODEL statement. The two designations, for a MOSFET, are NMOS for an N channel MOSFET and PMOS for a P-channel MOSFET. The general form of each is shown below:

```
.MODEL MNAME NMOS [PNAME1=PVAL1 PNAME2=PVAL2 ... ]

.MODEL MNAME PMOS [PNAME1=PVAL1 PNAME2=PVAL2 ... ]
```

An example of the model description for MOSFET device M31, being a IRF150, is shown below. Notice that the parameters associated with the MOSFET M1 description example are absent in the M31 device description. This is because those parameters (W, L, ...) are included in the MOSFET .MODEL definition.

```
.MODEL IRF150 NMOS(LEVEL=3 GAMMA=0 DELTA=0 ETA=0 THETA=0
+ KAPPA=0 VMAX=0 XJ=0 TOX=100N UO=600 PHI=0.6 RS=1.624M
+ KP=20.53U W=0.3 L=2U VTO=2.831 RD=1.031M RDS=444.4K
+ CBD=3.229N PB=0.8 MJ=0.5 FC=0.5 CGSO=9.027N CGDO=1.679N
+ RG=13.89 IS=194E-18 N=1 TT=288N)
```

SPICE provides four MOSFET device models, which differ in the formulation of the I-V characteristic. The variable LEVEL specifies the model to be used:

Level 1 → Shichman-Hodges model
Level 2 → Geometry-based analytic model
Level 3 → Semiempirical short-channel model
Level 4 → BSIM model

The DC characteristics of the Level 1 through Level 3 MOSFETs (for parameters see Table 4-6) are defined by the device parameters VTO, KP, LAMBDA, PHI and GAMMA. These parameters are computed by SPICE if process parameters (NSUB, TOX, ...) are supplied by the user. User specified values always override. VTO is positive for enhancement mode and negative for depletion mode N-channel devices, but it is negative for enhancement mode and positive for depletion mode P-channel devices. Charge storage is modeled by three constant capacitors: CGSO, CGDO, and CGBO. These capacitances represent overlap capacitances produced by the nonlinear thin-oxide capacitance. This thin-oxide capacitance is created (1) by the gate, source, drain, and bulk regions, and (2) by the non linear depletion-layer capacitances for both substrate junctions divided into bottom and periphery. The nonlinear depletion-layer capacitances vary as the MJ and MJSW power of junction voltage. The capacitances are determined by the parameters

CBD, CBS, CJ, CJSW, MJ, MJSW and PB. Charge storage effects are modeled a piece-wise linear voltage-dependent capacitance model. The thin-oxide charge storage effects are treated slightly different for the Level 1 model. These voltage-dependent capacitances are included only if TOX is specified in the parameter description.

There is some overlap among the parameters describing the junctions. For example, the reverse current can be defined as IS (in amperes) or as JS (in amperes/square meter). IS is an absolute value and JS is multiplied by AD and AS to give the reverse current of the drain and source junctions respectively. This methodology was chosen because there is little justification in always relating junction characteristics with AD and AS entered on the device statement, the reason being that the areas can be defaulted. The same idea applies to the zero-bias junction capacitances CBD and CBS (in Farads) and CJ (in Farads/square meter). The parasitic drain and source series resistance can be expressed as either RD and RS (in ohms) or RSH (in Ω/square). RSH is multiplied by the number of squares of NRD and NRS as input on the device statement.

The Level 1 through Level 3 MOSFET parameters are shown in Table 4-6.

Table 4-6. Level 1 through Level 3 MOSFET Parameters

	Name	Parameter	Units	Default	Example
1	LEVEL	Model index	-	1.0	
2	VTO	Zero-bias threshold voltage	V	0.0	1.0
3	KP	Transconductance parameter	A/V^2	2.0E-5	3.1E-5
4	Gamma	Bulk threshold parameter	$V^{.5}$	0.0	0.37
5	PHI	Surface potential	V	0.6	0.65
6	Lambda	Channel-length modulation (Level 1 or 2 only)	1/V	0.0	0.02
7	RD	Drain ohmic resistance	Ω	0.0	1.0
8	RS	Source ohmic resistance	Ω	0.0	1.0
9	CBD	Zero-bias bulk-drain junction capacitance	F	0.0	20 FF
10	CBS	Zero-bias bulk-source junction capacitance	F	0.0	20 FF
11	IS	Bulk junction saturation current	A	1.0E-14	1.0E-15
12	PB	Bulk junction potential	V	0.8	0.87

	Name	Parameter	Units	Default	Example
13	CGSO	Gate-source overlap capacitance per meter channel width	F/m	0.0	4.0E-11
14	CGDO	Gate-drain overlap capacitance per meter channel width	F/m	0.0	4.0E-11
15	CGBO	Gate-bulk overlap capacitance per meter channel length	F/m	0.0	2.0E-10
16	RSH	Drain and source diffusion sheet resistance	Ω/sq.	0.0	10.0
17	CJ	Zero-bias bulk junction bottom capacitance per square-meter of junction area	F/m^2	0.0	2.0E-4
18	MJ	Bulk junction bottom grading coefficient	-	0.5	0.5
19	CJSW	Zero-bias bulk junction sidewall capacitance per meter of junction perimeter	F/m	0.0	1.0E-9
20	MJSW	Bulk junction sidewall grading coefficient 0.50 = level 1 0.33 = level 2 0.33 = level 3			
21	JS	Bulk junction saturation current per sq-meter of junction area	A/m^2	0.0	
22	TOX	Oxide thickness	m	1.0E-7 for Level 1 and 2	1.0E-7
23	NSUB	Substrate doping	$1/cm^3$	0.0	4.0E15
24	NSS	Surface state density	$1/cm^2$	0.0	1.0E10
25	NFS	Fast surface state density	$1/cm^2$	0.0	1.0E10
26	TPG	Type of gate material: +1 = opposite of substrate -1 = same as substrate 0 = aluminum	-	1.0	
27	XJ	Metallurgical junction depth	M	0.0	1μ
28	LD	Lateral diffusion	M	0.0	0.8μ

	Name	Parameter	Units	Default	Example
29	UO	Surface mobility	cm^2 /V·s	600	700
30	UCRIT	Critical field for mobility degradation (Level 2 only)	V/cm	1.0E4	1.0E4
31	UEXP	Critical field exponent in mobility degradation (Level 2 only)	-	0.0	0.1
32	UTRA	Transverse field coefficient of mobility degradation (not normally used)	-	0.0	0.3
33	VMAX	Maximum drift velocity of carriers	M/s	0.0	5.0E4
34	NEFF	Total channel charge (fixed and mobile) coefficient (Level 2 only)	-	1.0	5.0
35	KF	Flicker noise coefficient	-	0.0	1.0E-26
36	AF	Flicker noise exponent	-	1.0	1.2
37	FC	Coefficient for forward-bias depletion capacitance formula	-	0.5	
38	DELTA	Width affect on threshold voltage	-	0.0	1.0
39	THETA	Mobility modulation (Level 3 only)	1/V	0.0	0.1
40	ETA	Static feedback (Level 3 only)	-	0.0	1.0
41	KAPPA	Saturation field factor (Level 3 only)	-	0.2	0.5
42	WD	Lateral diffusion	M	0.0	
43	XQC	Fraction of channel charge attributed to drain	-	1.0	
44	L	Channel length	M	DEFL	
45	W	Channel width	M	DEFW	
46	RG	Gate ohmic resistance	Ω	0.0	
47	RB	Bulk ohmic resistance	Ω	0.0	
48	RDS	Drain-source shunt resistance	Ω	∞	

	Name	Parameter	Units	Default	Example
49	JSSW	Bulk junction saturation sidewall current/length	A/M	0.0	
50	N	Bulk junction emission coefficient	-	1.0	
51	PBSW	Bulk junction sidewall potential	V	PB	
52	TT	Bulk junction transit time	S	0.0	

A set of parameters which may not be supported by all versions of SPICE. The reader is encouraged to research thoroughly which parameters are supported by the SPICE package in use.

A summary of key MOSFET parameters and their significance to specification parameters and each other is shown in Table 4-7.

4.4 INDEPENDENT DC VOLTAGE SOURCE

As discussed in Sections 2.2.5 and 2.2.6, the independent voltage and current sources can provide either DC voltage/current or more advanced capabilities. The general form of the independent voltage and current sources is shown below.

VXXXXXXX N+ N- DC DCVALUE AC ACVALUE ACPHASE

IXXXXXXX N+ N- DC DCVALUE AC ACVALUE ACPHASE

where N+ is the positive node, N- is the negative node, DCVALUE is the DC voltage or current, ACVALUE is the AC voltage or current, and ACPHASE is the AC phase shift at the source.

The voltage source, in a DC analysis, serves only as a source of voltage variation for the analysis. Any DC component associated with the source only has significance when SPICE computes the quiescent operating point information. The voltage/current source, when utilized by the DC analysis, is controlled exclusively by the .DC command. This command will be discussed later. Again, it is important to note that the voltage/current source amplitude, used to stimulate the input of the circuit in a DC analysis, has no bearing on the DC voltage/current amplitude specified within the .DC command.

Table 4-7. MOSFET Parameter Interdependence

Parameter	Source/method of determination
LEVEL	Normally set to 3 (short-channel device).
TOX	Determined from gate ratings.
L, LD, W, WD	Assume that L=2μ. Calculate from input capacitance.
XJ, NSUB	Assume usual technology.
IS, RD, RB	Determined from "source-drain diode forward voltage" specification or curve (Idr versus Vsd).
RS	Determined from Rds(on) specification.
RDS	Calculated from Idss specification or curves.
VTO, UO, THETA	Determined from "output characteristics" curve family (Ids versus Vds, stepped Vgs).
ETA, VMAX, CBS	Set for null effect.
CBD, PB, MJ	Determined from "capacitance versus Vds" curves.
RG	Calculated from rise/fall time specification or curves.
CGSO, CGDO	Determined from gate-charge, turn-on/off delay and rise time specifications.

4.5 PRINT COMMAND

SPICE, as stated previously, supports a .PRINT command for every analysis performed except the operating point and transfer function. The .PRINT commands all have certain characteristics, which will become more evident as the reader progresses through this text. The .PRINT command format for the DC analysis is as follows:

.PRINT DC OV1 [OV2 ... OV8]

This statement defines the contents of a tabular listing of one to eight output variables. DC is the type of the analysis for which the specified outputs are desired. The form for voltage or current output variables (OV1 ...) is as follows:

V(N1 [N2])

Current can also be specified in the general form of

I(VXXXXXX)

For example, assume that it is desired to find the voltage at nodes 2, 3, 4, 5, and between 5 and 6, 2 and 4, and 5 and 3. If the total number of nodes requested for the output exceeds the eight maximum, as mentioned previously, it will become necessary to break the request into several .PRINT statements. This might be done as follows:

```
.PRINT DC V(2) V(3) V(4) V(5) V(6) V(5,6) V(5,3)
.PRINT DC V(2,4)
```

The existence of more than one .PRINT DC statement is accepted by SPICE in any type of analysis. SPICE simply treats each statement as a separate request.

The .PRINT DC statement will cause SPICE to compute values, which are listed in tabular form. The following output data column format would be produced if the first .PRINT DC statement above were used.

```
VIN  V(2)  V(3)  V(4)  V(5)  V(6)  V(5,6)  V(5,3)
```

If it is desired to show the current through a voltage source VDC then the following .PRINT DC command would be given:

```
.PRINT DC I(VDC)
```

The reader is reminded that the current and voltage parameters may be intermixed and need not be segmented for proper SPICE operation.

4.6 PLOT COMMAND

As in all analyses, the plot command is identical to the print command. Rather then specifying .PRINT DC V(6) the plot command is invoked by using .PLOT DC V(6). The fundamental difference lies in the form of the output as discussed at length in Section 3.4.

4.7 DC SWEEP PARAMETER DEFINITION

Up to now we have looked at the format of the voltage or current source for the definition of the DC signal source. Additionally, the format of how the plot and print commands might be invoked has been discussed. The remaining element is the definition of the sweep range of the DC input for the DC analysis. Any DC sweep voltage or current source must be told three pieces of information:

1. Starting DC voltage

2. Ending DC voltage

3. Increment the voltage is to be stepped

The general form of the command that performs this is shown below:

.DC SRC1 VALSTRT1 VALSTOP1 VALINCR1 [SRC2 START2 STOP2 INCR2]

This statement defines the DC input source and sweep limits. SRC1 is the name of an independent voltage or current source. VALSTRT1, VALSTOP1, and VALINCR1 are the starting, final, and incrementing values respectively.

For example, assume an input voltage source labeled VIN is to be swept from menus -100 mV to + 100 mV in 1 mV steps. This is to say, a linear step function will be applied to the input. The function goes from menus -0.1 V to +0.1 V. The command could be specified as

```
.DC VIN -0.1 0.1 1E-3
```

SPICE, seeing this step function, first performs an operating point analysis assuming a DC input voltage of -0.01 V. The voltages/current sources that represent the designated outputs are recorded. SPICE then increments the input to -0.009 V and repeats this process until the entire range of input steps have been evaluated. Restated, SPICE performs discrete operating point calculations based on the number of input steps. The number of steps computed is [(VALSTOP1-VALSTART1)/VALINCR1]+1. This value must be considered when deciding whether to use an .OPTIONS LIMPTS=5001 statement in the input file.

A second source (SRC2) may optionally be specified with associated sweep parameters. In this case, the first source will be swept over its range for each value of the second source. This option can be useful for obtaining semiconductor device output characteristics. An example of this optional form is as follows:

```
.DC VDS 0 10 0.5 VGS 0 5 1
```

This example specifies a DC ramp, designated previously as VDS, that sweeps from 0 V to 10 V in 0.5-V steps. This sweep is nested inside a secondary sweep of a second voltage source VGS which sweeps from 0 V to 5 V in 1 V steps. The last example

```
.DC VCE 0 10 .25 IB 0 10U 1U
```

specifies a DC ramp, designated previously as VCE, that sweeps from 0 V to 10 V in 0.25 V steps at the same time as a second current source IB sweeps from 0 to 10 μA in 1 μA steps.

Occasionally, on more complex circuits, with SPICE running a DC analysis, the program may indicate a NO DC CONVERGENCE error. This may indicate that the alignment of component parameters is such that the circuit is unstable and prone to oscillation. If SPICE cannot find a stable operating point, which usually occurs near a threshold of the circuit, then the DC convergence error is indicated. If this occurs, at least three possibilities exist to correct the problem.

1. The circuit is modeled incorrectly and should be rechecked.

2. The circuit is inherently unstable and should be stabilized through some external means.

3. Nothing is wrong and SPICE requires a different sweep format.

It is wise to first assume incorrect modeling. If this assumption proves faulty, then try to give SPICE a different outlook. This can be accomplished in two ways.

1. Assume you are sweeping the voltage from menus -1 V to +1 V in 0.01 V steps. Change the sweep from +1 V to -1 V in 0.01 V steps. This is an identical sweep but reverses the direction of that voltage polarity and will often solve the problem. The reversal does not suggest that the circuit would oscillate in a laboratory environment, it simply indicates that SPICE believes this to be the case.

2. The second possibility is to change the step size. It may be desirable to go to a 0.05 V step or a 0.1 V step, which might avoid the problem but would not surrender analysis accuracy. This is sometime of benefit, because SPICE has a tendency to find a lack of convergence at a device threshold. The threshold represents a finite input voltage to SPICE and if the threshold can be bypassed, with a larger step size, then the stability may go unnoticed.

If neither of these approaches are fruitful, then circuit changes that affect the bias points might be considered. Many versions of SPICE attempt to provide the last node voltages computed if a "no convergence" error occurs. Any data presented to the user as a result of a DC convergence error is probably meaningless and should be viewed with extreme skepticism.

4.8 DC SAMPLE CIRCUIT MODEL DESCRIPTION

The previous discussions, in this section, should lead the reader to recognize that three elements are critical for running a DC analysis. These elements are represented by the following:

```
VXX 1 0 DC 10
.DC VXX 2 5 .01
.PLOT DC V(5) V(6) V(5,6)
.PRINT DC V(5) V(6) V(5,6)
```

The first line is the source of the input, the second is the definition of the source and the sweep parameters, and the third and fourth are the definitions of the output nodes requested.

The analysis of the differential amplifier, shown in Figure 4-1, will focuses on the extraction of the data to answer the following questions:

1. What is the DC gain characteristic between VIN and the voltage at node 5 when the negative 12-V supply is swept from −15-V DC to 0-V DC? The gain will be measured between -0.01 V and +0.01 V.

2. What is the most sensitive resistive component in the circuit relative to output node 5?

3. How does the differential gain measurement, based on a .TF analysis at a VEE value of minus 12-V DC compare to the gain computed between -0.01 V and +0.01 V as a result of a DC analysis?

The differential amplifier circuit will have a DC analysis run on it with a sweep of VIN occurring from −100 mV to +100 mV in 5-mV steps. The output will be the differential amplifier outputs at nodes 5 and 6. The differential amplifier will also be analyzed by looking at the sensitivity of node 5 to each circuit element and associated parameter. To gain the necessary information about the effects of VEE on the circuit gain, VEE will be varied as a secondary source in the .DC statement of the analysis. To gain a correlation between the computed differential gain and the measured differential gain, a .TF statement is included. The circuit description is shown below.

```
 DIFFERENTIAL AMPLIFIER EXAMPLE FOR A DC ANALYSIS
.MODEL D1N914 D (IS=1E-13 RS=16 CJO=2.2PF TT=12NS BV=100
+ IBV=1E-13)
.MODEL Q2N2222A NPN(IS=14.34F VAF=74.03 BF=255.9
+ NE=1.307 ISE=14.34F IKF=.2847 XTB=1.5 BR=6.092 IKR=0
+ RC=1 CJC=7.306P MJC=.3416 FC=.5 CJE=22.01P MJE=.377
+ TR=46.91N TF=411.1P ITF=.6 VTF=1.7 XTF=3 RB=10)
VCC 1 0 12
VEE 2 0 -12
VIN 3 0 DC 0
* CAUSES A SPICE CALCULATION OF DIFFERENTIAL GAIN
.TF V(5) VIN
* CAUSES A SENSITIVITY ANALYSIS TO BE PERFORMED
.SENS V(5)
RC1 1 5 7.75K
RC2 1 6 7.75K
RB1 3 8 50
RB2 7 0 50
Q1 5 8 4 Q2N2222A
Q2 6 7 4 Q2N2222A
Q3 4 9 12 Q2N2222A
R1 9 0 3.2K
R2 11 2 1.5K
```

```
RE 12 2 2.2K
D1 9 10 D1N914
D2 10 11 D1N914
* PROVIDES A SWEEP OF VIN NESTED WITHIN A SWEEP OF VEE
.DC VIN -0.1 0.1 0.005 VEE -15 0 1
* PROVIDES PLOTTED DATA TO COMPARE VEE EFFECTS ON THE CIRCUIT GAIN
.PLOT DC V(5) V(6) V(5,6)
.END
```

4.9 DC ANALYSIS SOLUTIONS

The solutions to running the circuit described in Section 4.8 are shown as follows:

```
 DIFFERENTIAL AMPLIFIER EXAMPLE FOR A DC ANALYSIS

 ****     CIRCUIT DESCRIPTION

*******************************************************************************

.MODEL D1N914 D (IS=1E-13 RS=16 CJO=2.2PF TT=12NS BV=100
+ IBV=1E-13)
.MODEL Q2N2222A NPN(IS=14.34F VAF=74.03 BF=255.9
+ NE=1.307 ISE=14.34F IKF=.2847 XTB=1.5 BR=6.092 IKR=0
+ RC=1 CJC=7.306P MJC=.3416 FC=.5 CJE=22.01P MJE=.377
+ TR=46.91N TF=411.1P ITF=.6 VTF=1.7 XTF=3 RB=10)
VCC 1 0 12
VEE 2 0 -12
VIN 3 0 DC 0
.TF V(5) VIN
.SENS V(5)
RC1 1 5 7.75K
RC2 1 6 7.75K
RB1 3 8 50
RB2 7 0 50
Q1 5 8 4 Q2N2222A
Q2 6 7 4 Q2N2222A
Q3 4 9 12 Q2N2222A
R1 9 0 3.2K
R2 11 2 1.5K
RE 12 2 2.2K
D1 9 10 D1N914
D2 10 11 D1N914
.DC VIN -0.1 0.1 .005 VEE -15 0 1
.PLOT DC V(5) V(6) V(5,6)
.END


 DIFFERENTIAL AMPLIFIER EXAMPLE FOR A DC ANALYSIS

 ****     Diode MODEL PARAMETERS

*******************************************************************************

          D1N914
     IS   100.000000E-15
     BV   100
    IBV   100.000000E-15
     RS    16
     TT    12.000000E-09
    CJO     2.200000E-12
```

```
DIFFERENTIAL AMPLIFIER EXAMPLE FOR A DC ANALYSIS

****     BJT MODEL PARAMETERS

******************************************************************************

          Q2N2222A
          NPN
     IS    14.340000E-15
     BF   255.9
     NF     1
    VAF    74.03
    IKF      .2847
    ISE    14.340000E-15
     NE     1.307
     BR     6.092
     NR     1
     RB    10
    RBM    10
     RC     1
    CJE    22.010000E-12
    MJE      .377
    CJC     7.306000E-12
    MJC      .3416
     TF   411.100000E-12
    XTF     3
    VTF     1.7
    ITF      .6
     TR    46.910000E-09
    XTB     1.5
```

```
DIFFERENTIAL AMPLIFIER EXAMPLE FOR A DC ANALYSIS

****     DC TRANSFER CURVES                  TEMPERATURE =   27.000 DEG C

******************************************************************************

LEGEND:         VEE = -15.0VDC

*: V(5)
+: V(6)
=: V(5,6)

  VIN        V(5)
(*+)---------  -5.0000E+00   0.0000E+00   5.0000E+00   1.0000E+01   1.5000E+01
(=)----------  -2.0000E+01  -1.0000E+01   0.0000E+00   1.0000E+01   2.0000E+01

                       - - - - - - - - - - - - - - - - - - - - - - - - - - -
 -1.000E-01  1.010E+01 .          + .            .            *=           .
 -9.500E-02  9.909E+00 .          + .            .            *=           .
 -9.000E-02  9.713E+00 .          + .            .           *.=           .
 -8.500E-02  9.512E+00 .          + .            .           *=            .
 -8.000E-02  9.306E+00 .          + .            .          * =            .
 -7.500E-02  9.097E+00 .          + .            .          * =            .
 -7.000E-02  8.884E+00 .          + .            .         * =.            .
 -6.500E-02  8.670E+00 .          + .            .         * =.            .
 -6.000E-02  8.456E+00 .          + .            .        *  =.            .
 -5.500E-02  8.242E+00 .          + .            .       *   =.            .
 -5.000E-02  8.030E+00 .          + .            .       *  = .            .
 -4.500E-02  7.819E+00 .          + .            .      *   = .            .
 -4.000E-02  7.612E+00 .          + .            .      *   = .            .
 -3.500E-02  7.406E+00 .          + .            .     *   =  .            .
 -3.000E-02  7.196E+00 .           +.            .     *   =  .            .
 -2.500E-02  6.795E+00 .           +.            .    *   =   .            .
 -2.000E-02  6.141E+00 .            .+           .  *   =     .            .
 -1.500E-02  5.453E+00 .            .  +         .*    =      .            .
 -1.000E-02  4.738E+00 .            .    +      *.   =        .            .
 -5.000E-03  4.005E+00 .            .      +  *  . =          .            .
  0.000E+00  3.262E+00 .            .       X    =            .            .
  5.000E-03  2.519E+00 .            .      *  += .            .            .
  1.000E-02  1.786E+00 .            .    *   =  +.            .            .
  1.500E-02  1.071E+00 .            .  *   =     .+           .            .
  2.000E-02  3.837E-01 .            .*    =      .  +         .            .
  2.500E-02 -2.699E-01 .           *.   =        .    +       .            .
  3.000E-02 -5.494E-01 .           *.  =         .     +      .            .
  3.500E-02 -5.724E-01 .           *.  =         .     +      .            .
  4.000E-02 -5.823E-01 .          * . =          .      +     .            .
  4.500E-02 -5.878E-01 .          * . =          .      +     .            .
  5.000E-02 -5.913E-01 .          * . =          .       +    .            .
  5.500E-02 -5.934E-01 .          * . =          .       +    .            .
  6.000E-02 -5.946E-01 .          * .=           .        +   .            .
  6.500E-02 -5.951E-01 .          * .=           .        +   .            .
  7.000E-02 -5.952E-01 .          * .=           .         +  .            .
  7.500E-02 -5.948E-01 .          * =            .          + .            .
  8.000E-02 -5.941E-01 .          * =            .          + .            .
  8.500E-02 -5.931E-01 .          * =            .           +.            .
  9.000E-02 -5.918E-01 .          * =            .           +.            .
  9.500E-02 -5.903E-01 .          *=.            .            +            .
  1.000E-01 -5.885E-01 .          *=.            .            +            .
```

```
LEGEND:          VEE = -14.0VDC

*: V(5)
+: V(6)
=: V(5,6)

  VIN        V(5)
(*+)---------    -5.0000E+00   0.0000E+00   5.0000E+00   1.0000E+01   1.5000E+01
(=)----------    -2.0000E+01  -1.0000E+01   0.0000E+00   1.0000E+01   2.0000E+01

                       - - - - - - - - - - - - - - - - - - - - - - - - - - -
 -1.000E-01  1.061E+01 .          + .             .            . X          .
 -9.500E-02  1.045E+01 .          + .             .            .X           .
 -9.000E-02  1.029E+01 .          + .             .            .X           .
 -8.500E-02  1.012E+01 .          + .             .            *=           .
 -8.000E-02  9.940E+00 .          + .             .            *=           .
 -7.500E-02  9.758E+00 .          + .             .           *.=           .
 -7.000E-02  9.573E+00 .          + .             .           *=            .
 -6.500E-02  9.384E+00 .          + .             .          * =            .
 -6.000E-02  9.194E+00 .          + .             .          * =            .
 -5.500E-02  9.004E+00 .          + .             .         *  =            .
 -5.000E-02  8.813E+00 .          + .             .         * =.            .
 -4.500E-02  8.622E+00 .          + .             .        *  =.            .
 -4.000E-02  8.431E+00 .          + .             .        *  =.            .
 -3.500E-02  8.220E+00 .           +.             .      *   = .            .
 -3.000E-02  7.732E+00 .            +             .     *   =  .            .
 -2.500E-02  7.156E+00 .            .+            .    *   =   .            .
 -2.000E-02  6.541E+00 .            .  +          .   *  =     .            .
 -1.500E-02  5.893E+00 .            .    +        . *  =       .            .
 -1.000E-02  5.219E+00 .            .     +       .*  =        .            .
 -5.000E-03  4.528E+00 .            .       +    *. =          .            .
  0.000E+00  3.827E+00 .            .         X   =            .            .
  5.000E-03  3.126E+00 .            .       *   =+.            .            .
  1.000E-02  2.435E+00 .            .     *  =    .+           .            .
  1.500E-02  1.761E+00 .            .    *  =     . +          .            .
  2.000E-02  1.114E+00 .            .  *  =       .   +        .            .
  2.500E-02  4.988E-01 .            .*  =         .     +      .            .
  3.000E-02 -7.737E-02 .            *  =          .      +     .            .
  3.500E-02 -5.142E-01 .           *. =           .       +    .            .
  4.000E-02 -5.567E-01 .           *.=            .        +   .            .
  4.500E-02 -5.698E-01 .           *.=            .        +   .            .
  5.000E-02 -5.765E-01 .           *.=            .         +  .            .
  5.500E-02 -5.805E-01 .          * .=            .         +  .            .
  6.000E-02 -5.828E-01 .          * =             .          + .            .
  6.500E-02 -5.840E-01 .          * =             .          + .            .
  7.000E-02 -5.845E-01 .          * =             .           +.            .
  7.500E-02 -5.844E-01 .          * =             .           +.            .
  8.000E-02 -5.838E-01 .          *=.             .            +            .
  8.500E-02 -5.828E-01 .          *=.             .            +            .
  9.000E-02 -5.815E-01 .          *=.             .            .+           .
  9.500E-02 -5.798E-01 .          *=.             .            .+           .
  1.000E-01 -5.779E-01 .          *=.             .            .+           .
```

```
LEGEND:          VEE = -13.0VDC

*: V(5)
+: V(6)
=: V(5,6)

    VIN           V(5)
(*+)---------    -5.0000E+00   0.0000E+00   5.0000E+00   1.0000E+01   1.5000E+01
(=)----------    -2.0000E+01  -1.0000E+01   0.0000E+00   1.0000E+01   2.0000E+01
                       - - - - - - - - - - - - - - - - - - - - - - - - - - - - - - - - - - - - - - - - - - - - - - - - - -
 -1.000E-01  1.105E+01 .                    +   .                        .                        . =*                     .
 -9.500E-02  1.092E+01 .                    +   .                        .                        . X                      .
 -9.000E-02  1.079E+01 .                    +   .                        .                        . X                      .
 -8.500E-02  1.066E+01 .                    +   .                        .                        . X                      .
 -8.000E-02  1.051E+01 .                    +   .                        .                        .*=                      .
 -7.500E-02  1.036E+01 .                    +   .                        .                        .X                       .
 -7.000E-02  1.021E+01 .                    +   .                        .                        .X                       .
 -6.500E-02  1.005E+01 .                    +   .                        .                        *=                       .
 -6.000E-02  9.886E+00 .                    +   .                        .                        *=                       .
 -5.500E-02  9.720E+00 .                    +   .                        .                      *=                         .
 -5.000E-02  9.551E+00 .                    +   .                        .                      *=                         .
 -4.500E-02  9.367E+00 .                       +.                        .                    *  =                         .
 -4.000E-02  9.005E+00 .                       +.                        .                  *  =  .                        .
 -3.500E-02  8.550E+00 .                        .+                       .                *   =   .                        .
 -3.000E-02  8.052E+00 .                        .  +                     .              *   =     .                        .
 -2.500E-02  7.513E+00 .                        .    +                   .            * =         .                        .
 -2.000E-02  6.937E+00 .                        .        +               .        *   =           .                        .
 -1.500E-02  6.330E+00 .                        .          +             .     *  =               .                        .
 -1.000E-02  5.698E+00 .                        .              +         .   *=                   .                        .
 -5.000E-03  5.049E+00 .                        .                  +     *   =                    .                        .
  0.000E+00  4.392E+00 .                        .                    X   =                        .                        .
  5.000E-03  3.734E+00 .                        .                  *=    +                        .                        .
  1.000E-02  3.085E+00 .                        .              *   =     .   +                    .                        .
  1.500E-02  2.453E+00 .                        .          *   =         .     +                  .                        .
  2.000E-02  1.846E+00 .                        .        *=              .        +               .                        .
  2.500E-02  1.271E+00 .                        .    *   =               .            +           .                        .
  3.000E-02  7.326E-01 .                        .  *=                    .              +         .                        .
  3.500E-02  2.348E-01 .                        .*=                      .                +       .                        .
  4.000E-02 -2.201E-01 .                       *.=                       .                  +     .                        .
  4.500E-02 -5.138E-01 .                       *=                        .                    +   .                        .
  5.000E-02 -5.465E-01 .                       *=                        .                       +.                        .
  5.500E-02 -5.580E-01 .                       *=                        .                       +.                        .
  6.000E-02 -5.640E-01 .                       X.                        .                        +                        .
  6.500E-02 -5.672E-01 .                       X.                        .                        +                        .
  7.000E-02 -5.689E-01 .                       X.                        .                        +                        .
  7.500E-02 -5.696E-01 .                       X.                        .                        .+                       .
  8.000E-02 -5.695E-01 .                       X.                        .                        .+                       .
  8.500E-02 -5.687E-01 .                      =*.                        .                        . +                      .
  9.000E-02 -5.675E-01 .                      =*.                        .                        . +                      .
  9.500E-02 -5.659E-01 .                      =*.                        .                        . +                      .
  1.000E-01 -5.639E-01 .                      =*.                        .                        .    +                   .
```

```
 LEGEND:        VEE = -12.0VDC

*: V(5)
+: V(6)
=: V(5,6)

  VIN          V(5)
(*+)---------    -5.0000E+00   0.0000E+00   5.0000E+00   1.0000E+01   1.5000E+01
(=)----------    -2.0000E+01  -1.0000E+01   0.0000E+00   1.0000E+01   2.0000E+01
                      - - - - - - - - - - - - - - - - - - - - - - - - - -
 -1.000E-01  1.140E+01 .         + .            .           .  =*        .
 -9.500E-02  1.132E+01 .         + .            .           .  X         .
 -9.000E-02  1.122E+01 .         + .            .           . =*         .
 -8.500E-02  1.112E+01 .         + .            .           . =*         .
 -8.000E-02  1.101E+01 .         + .            .           . =*         .
 -7.500E-02  1.090E+01 .         + .            .           . X          .
 -7.000E-02  1.077E+01 .         + .            .           . X          .
 -6.500E-02  1.064E+01 .         + .            .           . X          .
 -6.000E-02  1.050E+01 .          +.            .           .X           .
 -5.500E-02  1.027E+01 .          +.            .           .X           .
 -5.000E-02  9.970E+00 .           +            .           X            .
 -4.500E-02  9.631E+00 .           .+           .          X.            .
 -4.000E-02  9.251E+00 .           . +          .         X .            .
 -3.500E-02  8.830E+00 .           .  +         .        X  .            .
 -3.000E-02  8.367E+00 .           .   +        .       X   .            .
 -2.500E-02  7.867E+00 .           .    +       .     *=    .            .
 -2.000E-02  7.331E+00 .           .      +     .     X     .            .
 -1.500E-02  6.766E+00 .           .       +    .    X      .            .
 -1.000E-02  6.176E+00 .           .         +  .  X        .            .
 -5.000E-03  5.571E+00 .           .          + .*=         .            .
  0.000E+00  4.956E+00 .           .            X           .            .
  5.000E-03  4.342E+00 .           .          X .+          .            .
  1.000E-02  3.737E+00 .           .         X  .  +        .            .
  1.500E-02  3.148E+00 .           .       X    .    +      .            .
  2.000E-02  2.582E+00 .           .      X     .     +     .            .
  2.500E-02  2.047E+00 .           .    X       .      +    .            .
  3.000E-02  1.547E+00 .           .   X        .       +   .            .
  3.500E-02  1.085E+00 .           .  X         .        +  .            .
  4.000E-02  6.634E-01 .           . X          .         + .            .
  4.500E-02  2.835E-01 .           .X           .          +.            .
  5.000E-02 -5.533E-02 .           X            .           +            .
  5.500E-02 -3.544E-01 .          X.            .           .+           .
  6.000E-02 -5.043E-01 .          X.            .           .+           .
  6.500E-02 -5.285E-01 .          X.            .           . +          .
  7.000E-02 -5.380E-01 .         =*.            .           . +          .
  7.500E-02 -5.427E-01 .         =*.            .           . +          .
  8.000E-02 -5.450E-01 .         =*.            .           . +          .
  8.500E-02 -5.457E-01 .         =*.            .           .  +         .
  9.000E-02 -5.454E-01 .         =*.            .           .  +         .
  9.500E-02 -5.444E-01 .         =*.            .           .  +         .
  1.000E-01 -5.427E-01 .         =*.            .           .   +        .
```

```
 LEGEND:          VEE = -11.0VDC

*: V(5)
+: V(6)
=: V(5,6)

  VIN          V(5)
(*+)---------     -5.0000E+00   0.0000E+00   5.0000E+00   1.0000E+01   1.5000E+01
(=)----------     -2.0000E+01  -1.0000E+01   0.0000E+00   1.0000E+01   2.0000E+01
                       - - - - - - - - - - - - - - - - - - - - - - - - - -
 -1.000E-01  1.167E+01 .          +.            .           .  =*        .
 -9.500E-02  1.161E+01 .          +.            .           .  =*        .
 -9.000E-02  1.154E+01 .          +.            .           .  =*        .
 -8.500E-02  1.145E+01 .          +.            .           . = *        .
 -8.000E-02  1.134E+01 .          +.            .           . =*         .
 -7.500E-02  1.121E+01 .           +            .           . =*         .
 -7.000E-02  1.105E+01 .           +            .           .= *         .
 -6.500E-02  1.087E+01 .           +            .           .=*          .
 -6.000E-02  1.067E+01 .           .+           .           = *          .
 -5.500E-02  1.043E+01 .           . +          .           =*           .
 -5.000E-02  1.015E+01 .           . +          .          =*            .
 -4.500E-02  9.842E+00 .           .  +         .         = *            .
 -4.000E-02  9.493E+00 .           .   +        .        = *.            .
 -3.500E-02  9.105E+00 .           .    +       .       = * .            .
 -3.000E-02  8.679E+00 .           .     +      .      = *  .            .
 -2.500E-02  8.217E+00 .           .      +     .     =*    .            .
 -2.000E-02  7.722E+00 .           .        +   .    =*     .            .
 -1.500E-02  7.199E+00 .           .         +  .  = *      .            .
 -1.000E-02  6.653E+00 .           .          + . =*        .            .
 -5.000E-03  6.091E+00 .           .            += *        .            .
  0.000E+00  5.522E+00 .           .            =X          .            .
  5.000E-03  4.952E+00 .           .           =*  +        .            .
  1.000E-02  4.391E+00 .           .         =* .   +       .            .
  1.500E-02  3.845E+00 .           .        =*  .     +     .            .
  2.000E-02  3.321E+00 .           .      = *   .      +    .            .
  2.500E-02  2.827E+00 .           .     =*     .       +   .            .
  3.000E-02  2.365E+00 .           .    =*      .        +  .            .
  3.500E-02  1.939E+00 .           .   =*       .         + .            .
  4.000E-02  1.551E+00 .           .  =*        .          +.            .
  4.500E-02  1.202E+00 .           . =*         .           +            .
  5.000E-02  8.916E-01 .           .=*          .           +            .
  5.500E-02  6.176E-01 .           = *          .           .+           .
  6.000E-02  3.781E-01 .           =*           .           . +          .
  6.500E-02  1.705E-01 .          =*            .           . +          .
  7.000E-02 -8.138E-03 .          =*            .           .  +         .
  7.500E-02 -1.609E-01 .         = *            .           .  +         .
  8.000E-02 -2.909E-01 .         =*.            .           .  +         .
  8.500E-02 -3.975E-01 .         =*.            .           .   +        .
  9.000E-02 -4.558E-01 .        = *.            .           .   +        .
  9.500E-02 -4.761E-01 .        = *.            .           .   +        .
  1.000E-01 -4.845E-01 .        = *.            .           .   +        .
```

```
 LEGEND:          VEE = -10.0VDC

*: V(5)
+: V(6)
=: V(5,6)

  VIN          V(5)
(*+)---------    -5.0000E+00   0.0000E+00    5.0000E+00    1.0000E+01    1.5000E+01
(=)----------    -2.0000E+01  -1.0000E+01    0.0000E+00    1.0000E+01    2.0000E+01
                 _ _ _ _ _ _ _ _ _ _ _ _ _ _ _ _ _ _ _ _ _ _ _ _ _ _ _ _ _ _ _ _ _ _ _ _ _ _
 -1.000E-01  1.171E+01 .           .+            .            . = *         .
 -9.500E-02  1.166E+01 .           .+            .            .=  *         .
 -9.000E-02  1.159E+01 .           . +           .            .=  *         .
 -8.500E-02  1.150E+01 .           . +           .            .=  *         .
 -8.000E-02  1.140E+01 .           . +           .            .=  *         .
 -7.500E-02  1.129E+01 .           . +           .            .= *          .
 -7.000E-02  1.115E+01 .           .  +          .            =  *          .
 -6.500E-02  1.099E+01 .           .  +          .            =  *          .
 -6.000E-02  1.080E+01 .           .   +         .           =. *           .
 -5.500E-02  1.058E+01 .           .   +         .           =. *           .
 -5.000E-02  1.033E+01 .           .    +        .          = .*            .
 -4.500E-02  1.005E+01 .           .     +       .          = *             .
 -4.000E-02  9.731E+00 .           .     +       .         = *.             .
 -3.500E-02  9.377E+00 .           .      +      .         = * .            .
 -3.000E-02  8.988E+00 .           .       +     .        = *  .            .
 -2.500E-02  8.564E+00 .           .        +    .      =  *   .            .
 -2.000E-02  8.110E+00 .           .          + .     =  *     .            .
 -1.500E-02  7.630E+00 .           .          +.    =  *       .            .
 -1.000E-02  7.128E+00 .           .           +  =  *         .            .
 -5.000E-03  6.611E+00 .           .           .X  *           .            .
  0.000E+00  6.087E+00 .           .           =  X            .            .
  5.000E-03  5.563E+00 .           .          =.*  +           .            .
  1.000E-02  5.046E+00 .           .        =  *     +         .            .
  1.500E-02  4.544E+00 .           .       =  *.      +        .            .
  2.000E-02  4.064E+00 .           .      =  * .       +       .            .
  2.500E-02  3.610E+00 .           .     = *   .        +      .            .
  3.000E-02  3.187E+00 .           .   =  *    .         +     .            .
  3.500E-02  2.797E+00 .           .  =  *     .          +    .            .
  4.000E-02  2.443E+00 .           .  = *      .           +.               .
  4.500E-02  2.125E+00 .           . =  *      .            +               .
  5.000E-02  1.843E+00 .           . =  *      .            .+              .
  5.500E-02  1.594E+00 .           .=  *       .            . +             .
  6.000E-02  1.377E+00 .           .=  *       .            . +             .
  6.500E-02  1.189E+00 .           =  *        .            .  +            .
  7.000E-02  1.027E+00 .           =  *        .            .  +            .
  7.500E-02  8.892E-01 .          =. *         .            .  +            .
  8.000E-02  7.719E-01 .          =. *         .            .   +           .
  8.500E-02  6.728E-01 .          =. *         .            .   +           .
  9.000E-02  5.893E-01 .          =. *         .            .   +           .
  9.500E-02  5.192E-01 .          =.*          .            .   +           .
  1.000E-01  4.606E-01 .         = .*          .            .   +           .
```

```
 LEGEND:          VEE = -9.0VDC

*: V(5)
+: V(6)
=: V(5,6)

   VIN         V(5)
(*+)---------   -5.0000E+00   0.0000E+00   5.0000E+00   1.0000E+01   1.5000E+01
(=)----------   -2.0000E+01  -1.0000E+01   0.0000E+00   1.0000E+01   2.0000E+01
                       - - - - - - - - - - - - - - - - - - - - - - - - - -
 -1.000E-01  1.175E+01 .           .   +        .           =     *      .
 -9.500E-02  1.169E+01 .           .   +        .           =    *       .
 -9.000E-02  1.163E+01 .           .   +        .           =    *       .
 -8.500E-02  1.156E+01 .           .    +       .           =    *       .
 -8.000E-02  1.147E+01 .           .    +       .           =    *       .
 -7.500E-02  1.136E+01 .           .    +       .          =.    *       .
 -7.000E-02  1.124E+01 .           .    +       .          =.   *        .
 -6.500E-02  1.110E+01 .           .     +      .          =.   *        .
 -6.000E-02  1.093E+01 .           .     +      .         = . *          .
 -5.500E-02  1.073E+01 .           .      +     .         = . *          .
 -5.000E-02  1.051E+01 .           .      +     .        =  .*           .
 -4.500E-02  1.025E+01 .           .       +    .       =   .*           .
 -4.000E-02  9.966E+00 .           .        +   .       =   *            .
 -3.500E-02  9.645E+00 .           .         +  .      =   *.            .
 -3.000E-02  9.293E+00 .           .         +  .     =   * .            .
 -2.500E-02  8.909E+00 .           .          + .    =   *  .            .
 -2.000E-02  8.496E+00 .           .            +   =   *   .            .
 -1.500E-02  8.058E+00 .           .            .+ =   *    .            .
 -1.000E-02  7.601E+00 .           .            . X   *     .            .
 -5.000E-03  7.130E+00 .           .            .= + *      .            .
  0.000E+00  6.653E+00 .           .            =   X       .            .
  5.000E-03  6.175E+00 .           .           =.  * +      .            .
  1.000E-02  5.704E+00 .           .          = . *   +     .            .
  1.500E-02  5.247E+00 .           .        =   .*     +    .            .
  2.000E-02  4.809E+00 .           .       =    *       +   .            .
  2.500E-02  4.397E+00 .           .      =   * .        +  .            .
  3.000E-02  4.013E+00 .           .     =   *  .         + .            .
  3.500E-02  3.660E+00 .           .    =    *  .          +.            .
  4.000E-02  3.340E+00 .           .   =    *   .           +            .
  4.500E-02  3.053E+00 .           .   =   *    .           .+           .
  5.000E-02  2.798E+00 .           .  =   *     .           .+           .
  5.500E-02  2.574E+00 .           . =    *     .           . +          .
  6.000E-02  2.379E+00 .           . =   *      .           . +          .
  6.500E-02  2.210E+00 .           .=    *      .           .  +         .
  7.000E-02  2.066E+00 .           .=   *       .           .  +         .
  7.500E-02  1.942E+00 .           .=   *       .           .   +        .
  8.000E-02  1.837E+00 .           =    *       .           .   +        .
  8.500E-02  1.749E+00 .           =    *       .           .   +        .
  9.000E-02  1.674E+00 .           =   *        .           .   +        .
  9.500E-02  1.612E+00 .           =   *        .           .   +        .
  1.000E-01  1.560E+00 .           =   *        .           .    +       .
```

```
LEGEND:          VEE = -8.0VDC

*: V(5)
+: V(6)
=: V(5,6)

  VIN          V(5)
(*+)---------    -5.0000E+00   0.0000E+00   5.0000E+00   1.0000E+01   1.5000E+01
(=)----------    -2.0000E+01  -1.0000E+01   0.0000E+00   1.0000E+01   2.0000E+01
                      - - - - - - - - - - - - - - - - - - - - - - - - - -
 -1.000E-01  1.178E+01 .            .     +     .           =.   *       .
 -9.500E-02  1.173E+01 .            .     +     .           =.  *        .
 -9.000E-02  1.168E+01 .            .     +     .           =.  *        .
 -8.500E-02  1.161E+01 .            .     +     .         =  .  *        .
 -8.000E-02  1.153E+01 .            .      +    .         =  .  *        .
 -7.500E-02  1.144E+01 .            .      +    .         =  .  *        .
 -7.000E-02  1.133E+01 .            .      +    .         =  . *         .
 -6.500E-02  1.120E+01 .            .      +    .        =   . *         .
 -6.000E-02  1.105E+01 .            .       +   .        =   . *         .
 -5.500E-02  1.088E+01 .            .       +   .        =   .*          .
 -5.000E-02  1.068E+01 .            .        +  .       =    .*          .
 -4.500E-02  1.045E+01 .            .        +  .      =     .*          .
 -4.000E-02  1.020E+01 .            .         + .      =     .*          .
 -3.500E-02  9.910E+00 .            .          +.      =     *           .
 -3.000E-02  9.594E+00 .            .           +     =     *.           .
 -2.500E-02  9.250E+00 .            .           +    =     * .           .
 -2.000E-02  8.879E+00 .            .           .+  =     *  .           .
 -1.500E-02  8.485E+00 .            .           . +=     *   .           .
 -1.000E-02  8.074E+00 .            .           . = +   *    .           .
 -5.000E-03  7.649E+00 .            .           .=   + *     .           .
  0.000E+00  7.219E+00 .            .           =     X      .           .
  5.000E-03  6.788E+00 .            .          =.    * +     .           .
  1.000E-02  6.364E+00 .            .         = .   *    +    .           .
  1.500E-02  5.952E+00 .            .        =  . *      +   .           .
  2.000E-02  5.558E+00 .            .       =   .*        +  .           .
  2.500E-02  5.188E+00 .            .      =    *          + .           .
  3.000E-02  4.843E+00 .            .     =     *           +.           .
  3.500E-02  4.527E+00 .            .    =     *.            +           .
  4.000E-02  4.241E+00 .            .    =    * .            .+          .
  4.500E-02  3.985E+00 .            .    =   *  .            .+          .
  5.000E-02  3.758E+00 .            .   =    *  .            . +         .
  5.500E-02  3.559E+00 .            .  =    *   .            . +         .
  6.000E-02  3.385E+00 .            .  =    *   .            .  +        .
  6.500E-02  3.236E+00 .            .  =   *    .            .  +        .
  7.000E-02  3.107E+00 .            . =    *    .            .  +        .
  7.500E-02  2.998E+00 .            . =    *    .            .   +       .
  8.000E-02  2.906E+00 .            . =    *    .            .   +       .
  8.500E-02  2.827E+00 .            . =   *     .            .   +       .
  9.000E-02  2.762E+00 .            .=    *     .            .   +       .
  9.500E-02  2.707E+00 .            .=    *     .            .   +       .
  1.000E-01  2.661E+00 .            .=    *     .            .    +      .
```

```
LEGEND:          VEE = -7.0VDC

*: V(5)
+: V(6)
=: V(5,6)

   VIN        V(5)
(*+)---------   -5.0000E+00   0.0000E+00   5.0000E+00   1.0000E+01   1.5000E+01
(=)----------   -2.0000E+01  -1.0000E+01   0.0000E+00   1.0000E+01   2.0000E+01
                      - - - - - - - - - - - - - - - - - - - - - - - - - - -
-1.000E-01  1.181E+01 .            .         +  .         =  .    *       .
-9.500E-02  1.177E+01 .            .         +  .         =  .    *       .
-9.000E-02  1.172E+01 .            .         +  .         =  .   *        .
-8.500E-02  1.166E+01 .            .         +  .         =  .   *        .
-8.000E-02  1.159E+01 .            .         +  .         =  .   *        .
-7.500E-02  1.151E+01 .            .          + .         =  .   *        .
-7.000E-02  1.142E+01 .            .          + .        =   .   *        .
-6.500E-02  1.131E+01 .            .          + .        =   .  *         .
-6.000E-02  1.117E+01 .            .          + .        =   .  *         .
-5.500E-02  1.102E+01 .            .           +.       =    .  *         .
-5.000E-02  1.085E+01 .            .           +.       =    . *          .
-4.500E-02  1.065E+01 .            .            +      =     . *          .
-4.000E-02  1.042E+01 .            .            +      =     .*           .
-3.500E-02  1.017E+01 .            .            .+    =      *            .
-3.000E-02  9.892E+00 .            .            . +  =       *            .
-2.500E-02  9.588E+00 .            .            .  + =      *.            .
-2.000E-02  9.259E+00 .            .            .  +=      * .            .
-1.500E-02  8.910E+00 .            .            .  =+     *  .            .
-1.000E-02  8.545E+00 .            .            . =  +   *   .            .
-5.000E-03  8.168E+00 .            .            .=    + *    .            .
 0.000E+00  7.785E+00 .            .            =      X     .            .
 5.000E-03  7.402E+00 .            .           =.     * +    .            .
 1.000E-02  7.025E+00 .            .          = .    *   +   .            .
 1.500E-02  6.660E+00 .            .         =  .   *     +  .            .
 2.000E-02  6.311E+00 .            .        =   .  *       + .            .
 2.500E-02  5.983E+00 .            .       =    .  *        +.            .
 3.000E-02  5.678E+00 .            .       =    . *          +            .
 3.500E-02  5.399E+00 .            .      =     .*           +            .
 4.000E-02  5.147E+00 .            .     =      *            .+           .
 4.500E-02  4.921E+00 .            .     =      *            . +          .
 5.000E-02  4.722E+00 .            .    =      *.            . +          .
 5.500E-02  4.547E+00 .            .    =      *.            .  +         .
 6.000E-02  4.396E+00 .            .   =      * .            .  +         .
 6.500E-02  4.265E+00 .            .   =      * .            .  +         .
 7.000E-02  4.153E+00 .            .   =      * .            .   +        .
 7.500E-02  4.058E+00 .            .  =       * .            .   +        .
 8.000E-02  3.977E+00 .            .  =      *  .            .   +        .
 8.500E-02  3.909E+00 .            .  =      *  .            .   +        .
 9.000E-02  3.852E+00 .            .  =      *  .            .   +        .
 9.500E-02  3.804E+00 .            .  =      *  .            .    +       .
 1.000E-01  3.764E+00 .            .  =      *  .            .    +       .
```

```
LEGEND:          VEE = -6.0VDC

*: V(5)
+: V(6)
=: V(5,6)

  VIN          V(5)
(*+)----------    -5.0000E+00   0.0000E+00   5.0000E+00   1.0000E+01   1.5000E+01
(=)-----------    -2.0000E+01  -1.0000E+01   0.0000E+00   1.0000E+01   2.0000E+01

                         - - - - - - - - - - - - - - - - - - - - - - - - - - - - - - - - - - - - - - - -
  -1.000E-01  1.183E+01  .                   .                   +            =      .      *            .
  -9.500E-02  1.180E+01  .                   .                   +            =      .      *            .
  -9.000E-02  1.176E+01  .                   .                   +            =      .      *            .
  -8.500E-02  1.171E+01  .                   .                   +            =      .    *              .
  -8.000E-02  1.165E+01  .                   .                   +            =      .    *              .
  -7.500E-02  1.158E+01  .                   .                   +           =       .    *              .
  -7.000E-02  1.150E+01  .                   .                   .+          =       .    *              .
  -6.500E-02  1.141E+01  .                   .                   .+          =       .    *              .
  -6.000E-02  1.129E+01  .                   .                   .+          =       .   *               .
  -5.500E-02  1.116E+01  .                   .                   .+        =         .   *               .
  -5.000E-02  1.101E+01  .                   .                   .  +      =         .   *               .
  -4.500E-02  1.084E+01  .                   .                   .  +     =          . *                 .
  -4.000E-02  1.065E+01  .                   .                   .   +    =          . *                 .
  -3.500E-02  1.043E+01  .                   .                   .   +  =            .*                  .
  -3.000E-02  1.019E+01  .                   .                   .     +=            *                   .
  -2.500E-02  9.923E+00  .                   .                   .     =+            *                   .
  -2.000E-02  9.637E+00  .                   .                   .   =  +           *.                   .
  -1.500E-02  9.334E+00  .                   .                   .   =    +       *  .                   .
  -1.000E-02  9.015E+00  .                   .                   .  =      +    *    .                   .
  -5.000E-03  8.686E+00  .                   .                   .=          +  *    .                   .
   0.000E+00  8.353E+00  .                   .                   =             X     .                   .
   5.000E-03  8.019E+00  .                   .                  =.           *  +    .                   .
   1.000E-02  7.690E+00  .                   .                =  .         *    +    .                   .
   1.500E-02  7.372E+00  .                   .              =    .        *         +.                   .
   2.000E-02  7.068E+00  .                   .              =    .       *          +.                   .
   2.500E-02  6.782E+00  .                   .             =     .       *           +                   .
   3.000E-02  6.518E+00  .                   .           =       .     *             +                   .
   3.500E-02  6.276E+00  .                   .           =       .    *              .+                  .
   4.000E-02  6.058E+00  .                   .          =        .    *              .  +                .
   4.500E-02  5.863E+00  .                   .          =        .  *                .  +                .
   5.000E-02  5.691E+00  .                   .        =          .  *                .   +               .
   5.500E-02  5.541E+00  .                   .        =          .*                  .   +               .
   6.000E-02  5.410E+00  .                   .       =           .*                  .   +               .
   6.500E-02  5.298E+00  .                   .       =           .*                  .     +             .
   7.000E-02  5.202E+00  .                   .       =           .*                  .     +             .
   7.500E-02  5.121E+00  .                   .       =           *                   .     +             .
   8.000E-02  5.051E+00  .                   .     =             *                   .     +             .
   8.500E-02  4.993E+00  .                   .     =             *                   .     +             .
   9.000E-02  4.944E+00  .                   .     =             *                   .       +           .
   9.500E-02  4.904E+00  .                   .     =             *                   .       +           .
   1.000E-01  4.869E+00  .                   .     =             *                   .       +           .
```

```
 LEGEND:          VEE = -5.0VDC

*: V(5)
+: V(6)
=: V(5,6)

  VIN          V(5)
(*+)---------    -5.0000E+00   0.0000E+00   5.0000E+00   1.0000E+01   1.5000E+01
(=)----------    -2.0000E+01  -1.0000E+01   0.0000E+00   1.0000E+01   2.0000E+01

                       ---------------------------------------------------
 -1.000E-01  1.186E+01 .           .            .  +    =    .    *      .
 -9.500E-02  1.183E+01 .           .            .  +    =    .    *      .
 -9.000E-02  1.180E+01 .           .            .  +   =     .    *      .
 -8.500E-02  1.176E+01 .           .            .  +   =     .    *      .
 -8.000E-02  1.171E+01 .           .            .  +   =     .   *       .
 -7.500E-02  1.165E+01 .           .            .  +   =     .   *       .
 -7.000E-02  1.159E+01 .           .            .  +   =     .   *       .
 -6.500E-02  1.151E+01 .           .            .  +   =     .   *       .
 -6.000E-02  1.141E+01 .           .            .   + =      .   *       .
 -5.500E-02  1.130E+01 .           .            .   + =      .  *        .
 -5.000E-02  1.118E+01 .           .            .   + =      .  *        .
 -4.500E-02  1.103E+01 .           .            .    X       .  *        .
 -4.000E-02  1.087E+01 .           .            .    X       . *         .
 -3.500E-02  1.068E+01 .           .            .    =+      . *         .
 -3.000E-02  1.048E+01 .           .            .   = +      .*          .
 -2.500E-02  1.026E+01 .           .            .  =   +     .*          .
 -2.000E-02  1.001E+01 .           .            .  =   +     *           .
 -1.500E-02  9.756E+00 .           .            . =     +   *.           .
 -1.000E-02  9.485E+00 .           .            .=       +  *.           .
 -5.000E-03  9.205E+00 .           .            .=       + * .           .
  0.000E+00  8.921E+00 .           .            =         X  .           .
  5.000E-03  8.637E+00 .           .           =.        * + .           .
  1.000E-02  8.358E+00 .           .           =.        *  +.           .
  1.500E-02  8.087E+00 .           .          = .       *   +.           .
  2.000E-02  7.829E+00 .           .         =  .      *     +           .
  2.500E-02  7.587E+00 .           .         =  .      *     .+          .
  3.000E-02  7.363E+00 .           .        =   .     *      .+          .
  3.500E-02  7.159E+00 .           .       =    .     *      . +         .
  4.000E-02  6.975E+00 .           .       =    .    *       . +         .
  4.500E-02  6.811E+00 .           .       =    .    *       .  +        .
  5.000E-02  6.666E+00 .           .      =     .   *        .  +        .
  5.500E-02  6.540E+00 .           .      =     .   *        .  +        .
  6.000E-02  6.430E+00 .           .      =     .   *        .   +       .
  6.500E-02  6.337E+00 .           .     =      .  *         .   +       .
  7.000E-02  6.256E+00 .           .     =      .  *         .   +       .
  7.500E-02  6.188E+00 .           .     =      .  *         .   +       .
  8.000E-02  6.130E+00 .           .     =      .  *         .   +       .
  8.500E-02  6.082E+00 .           .     =      .  *         .    +      .
  9.000E-02  6.041E+00 .           .     =      .  *         .    +      .
  9.500E-02  6.007E+00 .           .    =       .  *         .    +      .
  1.000E-01  5.979E+00 .           .    =       .  *         .    +      .
```

```
LEGEND:          VEE = -4.0VDC

*: V(5)
+: V(6)
=: V(5,6)

  VIN          V(5)
(*+)---------    -5.0000E+00   0.0000E+00   5.0000E+00   1.0000E+01   1.5000E+01
(=)----------    -2.0000E+01  -1.0000E+01   0.0000E+00   1.0000E+01   2.0000E+01
                            - - - - - - - - - - - - - - - - - - - - - - - - - - - - - -
 -1.000E-01  1.189E+01 .          .          .   +=      .   *      .
 -9.500E-02  1.187E+01 .          .          .    X      .   *      .
 -9.000E-02  1.184E+01 .          .          .    X      .   *      .
 -8.500E-02  1.181E+01 .          .          .    X      .   *      .
 -8.000E-02  1.177E+01 .          .          .    X      .   *      .
 -7.500E-02  1.172E+01 .          .          .    X      .  *       .
 -7.000E-02  1.167E+01 .          .          .    X      .  *       .
 -6.500E-02  1.160E+01 .          .          .   =+      .  *       .
 -6.000E-02  1.153E+01 .          .          .   =+      .  *       .
 -5.500E-02  1.144E+01 .          .          .   = +     .  *       .
 -5.000E-02  1.134E+01 .          .          .   = +     . *        .
 -4.500E-02  1.122E+01 .          .          .  =  +     . *        .
 -4.000E-02  1.109E+01 .          .          .  =   +    . *        .
 -3.500E-02  1.094E+01 .          .          .  =   +    .*         .
 -3.000E-02  1.077E+01 .          .          . =    +    .*         .
 -2.500E-02  1.059E+01 .          .          . =     +   .*         .
 -2.000E-02  1.039E+01 .          .          .=      +   .*         .
 -1.500E-02  1.018E+01 .          .          .=       +  *          .
 -1.000E-02  9.955E+00 .          .          .=       +  *          .
 -5.000E-03  9.726E+00 .          .          .=        +*.          .
  0.000E+00  9.492E+00 .          .          =          X.          .
  5.000E-03  9.259E+00 .          .         =.         *+.          .
  1.000E-02  9.030E+00 .          .         =.        *  +          .
  1.500E-02  8.808E+00 .          .        = .        *  +          .
  2.000E-02  8.597E+00 .          .        = .       *   .+         .
  2.500E-02  8.399E+00 .          .       =  .       *   . +        .
  3.000E-02  8.216E+00 .          .       =  .      *    . +        .
  3.500E-02  8.049E+00 .          .      =   .      *    . +        .
  4.000E-02  7.899E+00 .          .      =   .      *    .  +       .
  4.500E-02  7.766E+00 .          .      =   .     *     .  +       .
  5.000E-02  7.648E+00 .          .     =    .     *     .  +       .
  5.500E-02  7.546E+00 .          .     =    .     *     .   +      .
  6.000E-02  7.458E+00 .          .     =    .    *      .   +      .
  6.500E-02  7.382E+00 .          .     =    .    *      .   +      .
  7.000E-02  7.317E+00 .          .    =     .    *      .   +      .
  7.500E-02  7.262E+00 .          .    =     .    *      .   +      .
  8.000E-02  7.216E+00 .          .    =     .    *      .    +     .
  8.500E-02  7.177E+00 .          .    =     .    *      .    +     .
  9.000E-02  7.144E+00 .          .    =     .    *      .    +     .
  9.500E-02  7.117E+00 .          .    =     .    *      .    +     .
  1.000E-01  7.094E+00 .          .    =     .   *       .    +     .
```

```
 LEGEND:          VEE = -3.0VDC

*: V(5)
+: V(6)
=: V(5,6)

  VIN          V(5)
(*+)---------    -5.0000E+00   0.0000E+00   5.0000E+00   1.0000E+01   1.5000E+01
(=)----------    -2.0000E+01  -1.0000E+01   0.0000E+00   1.0000E+01   2.0000E+01
                         - - - - - - - - - - - - - - - - - - - - - - - - - -
  -1.000E-01  1.192E+01 .           .            .    = +    .    *       .
  -9.500E-02  1.190E+01 .           .            .    = +    .    *       .
  -9.000E-02  1.188E+01 .           .            .    = +    .    *       .
  -8.500E-02  1.185E+01 .           .            .    =  +   .    *       .
  -8.000E-02  1.182E+01 .           .            .    =  +   .    *       .
  -7.500E-02  1.179E+01 .           .            .   =   +   .    *       .
  -7.000E-02  1.175E+01 .           .            .   =   +   .    *       .
  -6.500E-02  1.170E+01 .           .            .   =   +   .   *        .
  -6.000E-02  1.164E+01 .           .            .   =   +   .   *        .
  -5.500E-02  1.157E+01 .           .            .   =   +   .   *        .
  -5.000E-02  1.149E+01 .           .            .   =   +   .   *        .
  -4.500E-02  1.140E+01 .           .            .  =     +  .   *        .
  -4.000E-02  1.130E+01 .           .            .  =     +  .  *         .
  -3.500E-02  1.119E+01 .           .            .  =     +  .  *         .
  -3.000E-02  1.106E+01 .           .            .  =      + .  *         .
  -2.500E-02  1.092E+01 .           .            . =       + . *          .
  -2.000E-02  1.076E+01 .           .            . =       + . *          .
  -1.500E-02  1.060E+01 .           .            .=         +. *          .
  -1.000E-02  1.043E+01 .           .            .=         +.*           .
  -5.000E-03  1.025E+01 .           .            =           +*           .
   0.000E+00  1.007E+01 .           .            =           X            .
   5.000E-03  9.887E+00 .           .            =           *+           .
   1.000E-02  9.709E+00 .           .           =.          *.+           .
   1.500E-02  9.537E+00 .           .           =.          *. +          .
   2.000E-02  9.373E+00 .           .          = .         * . +          .
   2.500E-02  9.220E+00 .           .          = .         * . +          .
   3.000E-02  9.079E+00 .           .         =  .         * .  +         .
   3.500E-02  8.950E+00 .           .         =  .        *  .  +         .
   4.000E-02  8.835E+00 .           .         =  .        *  .  +         .
   4.500E-02  8.733E+00 .           .         =  .        *  .   +        .
   5.000E-02  8.643E+00 .           .        =   .       *   .   +        .
   5.500E-02  8.564E+00 .           .        =   .       *   .   +        .
   6.000E-02  8.497E+00 .           .        =   .       *   .   +        .
   6.500E-02  8.439E+00 .           .        =   .       *   .   +        .
   7.000E-02  8.389E+00 .           .        =   .       *   .    +       .
   7.500E-02  8.347E+00 .           .        =   .       *   .    +       .
   8.000E-02  8.312E+00 .           .       =    .       *   .    +       .
   8.500E-02  8.282E+00 .           .       =    .       *   .    +       .
   9.000E-02  8.257E+00 .           .       =    .      *    .    +       .
   9.500E-02  8.237E+00 .           .       =    .      *    .    +       .
   1.000E-01  8.219E+00 .           .       =    .      *    .    +       .
```

```
LEGEND:          VEE = -2.0VDC

*: V(5)
+: V(6)
=: V(5,6)

  VIN          V(5)
(*+)---------   -5.0000E+00   0.0000E+00   5.0000E+00   1.0000E+01   1.5000E+01
(=)----------   -2.0000E+01  -1.0000E+01   0.0000E+00   1.0000E+01   2.0000E+01

                         - - - - - - - - - - - - - - - - - - - - - - - - - - -
  -1.000E-01  1.194E+01 .            .            .  =       + .    *       .
  -9.500E-02  1.193E+01 .            .            .  =       + .    *       .
  -9.000E-02  1.192E+01 .            .            .  =       + .    *       .
  -8.500E-02  1.190E+01 .            .            .  =       + .    *       .
  -8.000E-02  1.188E+01 .            .            .  =        +.    *       .
  -7.500E-02  1.186E+01 .            .            .  =        +.    *       .
  -7.000E-02  1.183E+01 .            .            .  =        +.    *       .
  -6.500E-02  1.179E+01 .            .            .  =        +.    *       .
  -6.000E-02  1.175E+01 .            .            .  =        +.    *       .
  -5.500E-02  1.171E+01 .            .            .  =        +.   *        .
  -5.000E-02  1.165E+01 .            .            .  =        +.   *        .
  -4.500E-02  1.159E+01 .            .            . =         +.   *        .
  -4.000E-02  1.152E+01 .            .            . =         +.   *        .
  -3.500E-02  1.144E+01 .            .            . =          +   *        .
  -3.000E-02  1.135E+01 .            .            . =          +   *        .
  -2.500E-02  1.125E+01 .            .            . =          +  *         .
  -2.000E-02  1.114E+01 .            .            .=           +  *         .
  -1.500E-02  1.103E+01 .            .            .=           .+ *         .
  -1.000E-02  1.091E+01 .            .            .=           .+*          .
  -5.000E-03  1.079E+01 .            .            =            .+*          .
   0.000E+00  1.066E+01 .            .            =            . X          .
   5.000E-03  1.053E+01 .            .            =            .*+          .
   1.000E-02  1.041E+01 .            .           =.            .*+          .
   1.500E-02  1.029E+01 .            .           =.            .* +         .
   2.000E-02  1.017E+01 .            .           =.            *  +         .
   2.500E-02  1.007E+01 .            .          = .            *  +         .
   3.000E-02  9.967E+00 .            .          = .            *   +        .
   3.500E-02  9.877E+00 .            .          = .            *   +        .
   4.000E-02  9.797E+00 .            .          = .           *.   +        .
   4.500E-02  9.726E+00 .            .          = .           *.   +        .
   5.000E-02  9.664E+00 .            .         =  .           *.   +        .
   5.500E-02  9.610E+00 .            .         =  .           *.   +        .
   6.000E-02  9.563E+00 .            .         =  .           *.    +       .
   6.500E-02  9.523E+00 .            .         =  .           *.    +       .
   7.000E-02  9.489E+00 .            .         =  .           *.    +       .
   7.500E-02  9.460E+00 .            .         =  .           *.    +       .
   8.000E-02  9.436E+00 .            .         =  .           *.    +       .
   8.500E-02  9.416E+00 .            .         =  .          * .    +       .
   9.000E-02  9.399E+00 .            .         =  .          * .    +       .
   9.500E-02  9.384E+00 .            .         =  .          * .    +       .
   1.000E-01  9.373E+00 .            .         =  .          * .    +       .
```

```
LEGEND:          VEE = -1.0VDC

*: V(5)
+: V(6)
=: V(5,6)

  VIN           V(5)
(*+)---------   -5.0000E+00   0.0000E+00   5.0000E+00   1.0000E+01   1.5000E+01
(=)----------   -2.0000E+01  -1.0000E+01   0.0000E+00   1.0000E+01   2.0000E+01

                        - - - - - - - - - - - - - - - - - - - - - - - - - - - - - - - - - - - - - - - - - - - - - - - - - - -
 -1.000E-01  1.198E+01 .                        .                        .=                       .  +     *               .
 -9.500E-02  1.197E+01 .                        .                        .=                       .  +     *               .
 -9.000E-02  1.196E+01 .                        .                        .=                       .  +     *               .
 -8.500E-02  1.196E+01 .                        .                        .=                       .  +     *               .
 -8.000E-02  1.195E+01 .                        .                        .=                       .  +     *               .
 -7.500E-02  1.194E+01 .                        .                        .=                       .  +     *               .
 -7.000E-02  1.192E+01 .                        .                        .=                       .  +     *               .
 -6.500E-02  1.191E+01 .                        .                        .=                       .  +     *               .
 -6.000E-02  1.189E+01 .                        .                        .=                       .  +     *               .
 -5.500E-02  1.187E+01 .                        .                        .=                       .  +     *               .
 -5.000E-02  1.185E+01 .                        .                        .=                       .  +     *               .
 -4.500E-02  1.182E+01 .                        .                        .=                       .    +   *               .
 -4.000E-02  1.179E+01 .                        .                        .=                       .    +   *               .
 -3.500E-02  1.175E+01 .                        .                        .=                       .    +   *               .
 -3.000E-02  1.171E+01 .                        .                        .=                       .      +*                .
 -2.500E-02  1.166E+01 .                        .                        .=                       .      +*                .
 -2.000E-02  1.162E+01 .                        .                        .=                       .      +*                .
 -1.500E-02  1.156E+01 .                        .                        =                        .      +*                .
 -1.000E-02  1.151E+01 .                        .                        =                        .      +*                .
 -5.000E-03  1.145E+01 .                        .                        =                        .      +*                .
  0.000E+00  1.139E+01 .                        .                        =                        .      X                 .
  5.000E-03  1.133E+01 .                        .                        =                        .      *+                .
  1.000E-02  1.127E+01 .                        .                        =                        .      *+                .
  1.500E-02  1.121E+01 .                        .                        =                        .      *+                .
  2.000E-02  1.115E+01 .                        .                       =.                        .      *+                .
  2.500E-02  1.110E+01 .                        .                       =.                        .      *+                .
  3.000E-02  1.105E+01 .                        .                       =.                        .      *+                .
  3.500E-02  1.101E+01 .                        .                       =.                        .    *   +               .
  4.000E-02  1.097E+01 .                        .                       =.                        .    *   +               .
  4.500E-02  1.094E+01 .                        .                       =.                        .   *    +               .
  5.000E-02  1.091E+01 .                        .                       =.                        .   *    +               .
  5.500E-02  1.088E+01 .                        .                       =.                        .   *    +               .
  6.000E-02  1.086E+01 .                        .                       =.                        .   *    +               .
  6.500E-02  1.084E+01 .                        .                       =.                        .   *    +               .
  7.000E-02  1.082E+01 .                        .                       =.                        .   *    +               .
  7.500E-02  1.080E+01 .                        .                       =.                        .   *    +               .
  8.000E-02  1.079E+01 .                        .                       =.                        .   *    +               .
  8.500E-02  1.078E+01 .                        .                      = .                        .   *    +               .
  9.000E-02  1.077E+01 .                        .                      = .                        .   *    +               .
  9.500E-02  1.077E+01 .                        .                      = .                        .   *    +               .
  1.000E-01  1.076E+01 .                        .                      = .                        .   *    +               .
```

```
LEGEND:          VEE = 0.0VDC

*: V(5)
+: V(6)
=: V(5,6)

  VIN          V(5)
(*+)---------   -5.0000E+00   0.0000E+00   5.0000E+00   1.0000E+01   1.5000E+01
(=)----------   -2.0000E+01  -1.0000E+01   0.0000E+00   1.0000E+01   2.0000E+01
                 - - - - - - - - - - - - - - - - - - - - - - - - - - - - - - -
 -1.000E-01  1.200E+01 .           .           =           .   X       .
 -9.500E-02  1.200E+01 .           .           =           .   X       .
 -9.000E-02  1.200E+01 .           .           =           .   X       .
 -8.500E-02  1.200E+01 .           .           =           .   X       .
 -8.000E-02  1.200E+01 .           .           =           .   X       .
 -7.500E-02  1.200E+01 .           .           =           .   X       .
 -7.000E-02  1.200E+01 .           .           =           .   X       .
 -6.500E-02  1.200E+01 .           .           =           .   X       .
 -6.000E-02  1.200E+01 .           .           =           .   X       .
 -5.500E-02  1.200E+01 .           .           =           .   X       .
 -5.000E-02  1.200E+01 .           .           =           .   X       .
 -4.500E-02  1.200E+01 .           .           =           .   X       .
 -4.000E-02  1.200E+01 .           .           =           .   X       .
 -3.500E-02  1.200E+01 .           .           =           .   X       .
 -3.000E-02  1.200E+01 .           .           =           .   X       .
 -2.500E-02  1.200E+01 .           .           =           .   X       .
 -2.000E-02  1.200E+01 .           .           =           .   X       .
 -1.500E-02  1.200E+01 .           .           =           .   X       .
 -1.000E-02  1.200E+01 .           .           =           .   X       .
 -5.000E-03  1.200E+01 .           .           =           .   X       .
  0.000E+00  1.200E+01 .           .           =           .   X       .
  5.000E-03  1.200E+01 .           .           =           .   X       .
  1.000E-02  1.200E+01 .           .           =           .   X       .
  1.500E-02  1.200E+01 .           .           =           .   X       .
  2.000E-02  1.200E+01 .           .           =           .   X       .
  2.500E-02  1.200E+01 .           .           =           .   X       .
  3.000E-02  1.200E+01 .           .           =           .   X       .
  3.500E-02  1.200E+01 .           .           =           .   X       .
  4.000E-02  1.200E+01 .           .           =           .   X       .
  4.500E-02  1.200E+01 .           .           =           .   X       .
  5.000E-02  1.200E+01 .           .           =           .   X       .
  5.500E-02  1.200E+01 .           .           =           .   X       .
  6.000E-02  1.200E+01 .           .           =           .   X       .
  6.500E-02  1.200E+01 .           .           =           .   X       .
  7.000E-02  1.200E+01 .           .           =           .   X       .
  7.500E-02  1.200E+01 .           .           =           .   X       .
  8.000E-02  1.200E+01 .           .           =           .   X       .
  8.500E-02  1.200E+01 .           .           =           .   X       .
  9.000E-02  1.200E+01 .           .           =           .   X       .
  9.500E-02  1.200E+01 .           .           =           .   X       .
  1.000E-01  1.200E+01 .           .           =           .   X       .
                 - - - - - - - - - - - - - - - - - - - - - - - - - - - - - - -
```

```
DIFFERENTIAL AMPLIFIER EXAMPLE FOR A DC ANALYSIS

****     SMALL SIGNAL BIAS SOLUTION       TEMPERATURE =   27.000 DEG C

******************************************************************************

NODE   VOLTAGE     NODE   VOLTAGE     NODE   VOLTAGE     NODE   VOLTAGE


(    1)   12.0000  (    2)  -12.0000  (    3)    0.0000  (    4)    -.6421

(    5)    4.9565  (    6)    4.9565  (    7)-292.7E-06  (    8)-292.7E-06

(    9)   -7.2921  (   10)   -7.9451  (   11)   -8.5981  (   12)   -7.9515

    VOLTAGE SOURCE CURRENTS
    NAME         CURRENT

    VCC         -1.818E-03
    VEE          4.108E-03
    VIN         -5.855E-06

    TOTAL POWER DISSIPATION   7.11E-02  WATTS

****     SMALL-SIGNAL CHARACTERISTICS

      V(5)/VIN = -1.231E+02

      INPUT RESISTANCE AT VIN =  9.944E+03

      OUTPUT RESISTANCE AT V(5) =  7.431E+03
```

```
DIFFERENTIAL AMPLIFIER EXAMPLE FOR A DC ANALYSIS

****     DC SENSITIVITY ANALYSIS          TEMPERATURE =   27.000 DEG C

******************************************************************************

DC SENSITIVITIES OF OUTPUT V(5)

          ELEMENT          ELEMENT          ELEMENT        NORMALIZED
           NAME             VALUE         SENSITIVITY     SENSITIVITY
                                         (VOLTS/UNIT) (VOLTS/PERCENT)

           RC1            7.750E+03        -8.714E-04      -6.753E-02
           RC2            7.750E+03        -3.697E-05      -2.865E-03
           RB1            5.000E+01         7.207E-04       3.603E-04
           RB2            5.000E+01        -7.207E-04      -3.603E-04
           R1             3.200E+03         1.285E-03       4.111E-02
           R2             1.500E+03        -2.632E-03      -3.947E-02
           RE             2.200E+03         3.175E-03       6.985E-02
           VCC            1.200E+01         9.995E-01       1.199E-01
           VEE           -1.200E+01         5.650E-01      -6.780E-02
           VIN            0.000E+00        -1.231E+02       0.000E+00
```

```
D1
SERIES RESISTANCE
          RS         1.600E+01      -2.632E-03      -4.210E-04
INTRINSIC PARAMETERS
          IS         1.000E-13       3.001E+11       3.001E-04
          N          1.000E+00      -7.156E-01      -7.156E-03
D2
SERIES RESISTANCE
          RS         1.600E+01      -2.632E-03      -4.210E-04
INTRINSIC PARAMETERS
          IS         1.000E-13       3.001E+11       3.001E-04
          N          1.000E+00      -7.156E-01      -7.156E-03
Q1
          RB         1.000E+01       7.207E-04       7.207E-05
          RC         1.000E+00       3.744E-05       3.744E-07
          RE         0.000E+00       0.000E+00       0.000E+00
          BF         2.559E+02      -1.466E-04      -3.752E-04
          ISE        1.434E-14       1.971E+12       2.827E-04
          BR         6.092E+00       5.725E-12       3.488E-13
          ISC        0.000E+00       0.000E+00       0.000E+00
          IS         1.434E-14      -2.236E+14      -3.207E-02
          NE         1.307E+00      -4.106E-01      -5.366E-03
          NC         2.000E+00       0.000E+00       0.000E+00
          IKF        2.847E-01      -3.399E-02      -9.677E-05
          IKR        0.000E+00       0.000E+00       0.000E+00
          VAF        7.403E+01       2.758E-03       2.042E-03
          VAR        0.000E+00       0.000E+00       0.000E+00
Q2
          RB         1.000E+01      -7.207E-04      -7.207E-05
          RC         1.000E+00      -3.697E-05      -3.697E-07
          RE         0.000E+00       0.000E+00       0.000E+00
          BF         2.559E+02       4.612E-05       1.180E-04
          ISE        1.434E-14      -6.202E+11      -8.894E-05
          BR         6.092E+00      -2.732E-12      -1.664E-13
          ISC        0.000E+00       0.000E+00       0.000E+00
          IS         1.434E-14       2.226E+14       3.192E-02
          NE         1.307E+00       1.292E-01       1.688E-03
          NC         2.000E+00       0.000E+00       0.000E+00
          IKF        2.847E-01       3.357E-02       9.556E-05
          IKR        0.000E+00       0.000E+00       0.000E+00
          VAF        7.403E+01      -2.723E-03      -2.016E-03
          VAR        0.000E+00       0.000E+00       0.000E+00
Q3
          RB         1.000E+01       1.868E-05       1.868E-06
          RC         1.000E+00       2.278E-06       2.278E-08
          RE         0.000E+00       0.000E+00       0.000E+00
          BF         2.559E+02      -1.448E-04      -3.706E-04
          ISE        1.434E-14       1.660E+12       2.380E-04
          BR         6.092E+00       2.192E-12       1.335E-13
          ISC        0.000E+00       0.000E+00       0.000E+00
          IS         1.434E-14      -4.382E+12      -6.283E-04
          NE         1.307E+00      -3.552E-01      -4.642E-03
          NC         2.000E+00       0.000E+00       0.000E+00
          IKF        2.847E-01      -2.057E-03      -5.856E-06
          IKR        0.000E+00       0.000E+00       0.000E+00
          VAF        7.403E+01       1.118E-04       8.280E-05
          VAR        0.000E+00       0.000E+00       0.000E+00
```

4.10 EXTRACTION OF DC ANALYSIS DATA

The interpretation of the results, may on the surface, seem overpowering. To assist in this task, a high resolution plot of the DC data is provided in Figures 4-3 and 4-4. This plot is a composite of the data generated from the .PLOT DC statement. Before reviewing specifics of the analysis data, a familiarization with the output is the first order of business.

Looking at the analysis, the first section following the circuit description is a listing of diode model parameters and BJT model parameters. These sections, of the analysis, are simply a listing of all solid state parameters passed to SPICE in the .MODEL descriptions of the D1N914 and Q2N2222A devices. The DC transfer curves appear next and were generated due to the .PLOT DC command. All three requested outputs are plotted as a function of the VIN. SPICE performs an analysis for each value of VEE shown at the start of each analysis. Looking at the first analysis where VEE is -15-V DC, the first two columns are VIN and V(5) where V(5) represents the first output node requested from the analysis. The plot is in text character format with various symbols used to designate each output. The * symbol is V(5), the + symbol is V(6) and the = symbol is V(5,6). The reader is directed to the start of the plot where VIN is -1.000E-01 (-0.1 V). The plot shows the existence of all three graphic symbols. As the plot continues (VIN = 0.000E+00), only two symbols exist with the symbols for the + and * being replaced with an X. This indicates that the plots are realizing an intersection point based on the two different scales shown. Care must be taken to assure an understanding of what symbols are making up the X to avoid misinterpretation of the plotted data.

The family of curves, showing the effects at node 5 with a variation of VIN and VEE, is shown in high-resolution graphics form in Figure 4-5. Each curve represents a different value of VEE. The leftmost curve shows VEE at a value of -15-V DC with each subsequent curve being a drop in VEE of 1-V DC. The reader is directed to notice that the amplifier is almost perfectly balanced with the zero crossing of 0-V DC at the input. The question was previously raised about the maximum swing the input could realize without affecting the linearity characteristics of the output. A review of the plotted data will reveal that the linearity degrades as the value of VEE approaches 0-V DC. One possible conclusion is that maximum gain and optimal linearity occurs when VEE is -15VDC.

If the goal is to maintain some optimal gain with a maximally linear input range, then the fourth curve (VEE = minus 12-V DC) might be the choice. Assuming this is the desired outcome, the individual plot of V(5) versus VIN at VEE = -12-V DC is shown in Figure 4-2. For completeness, a plot of V(5), V(6), and V(5,6) is provided in Figure 4-4. Careful study of Figure 4-3 suggests that good overall linearity occurs between +40 mV and -40 mV at the input. This input range permits the maximum linear output swing.

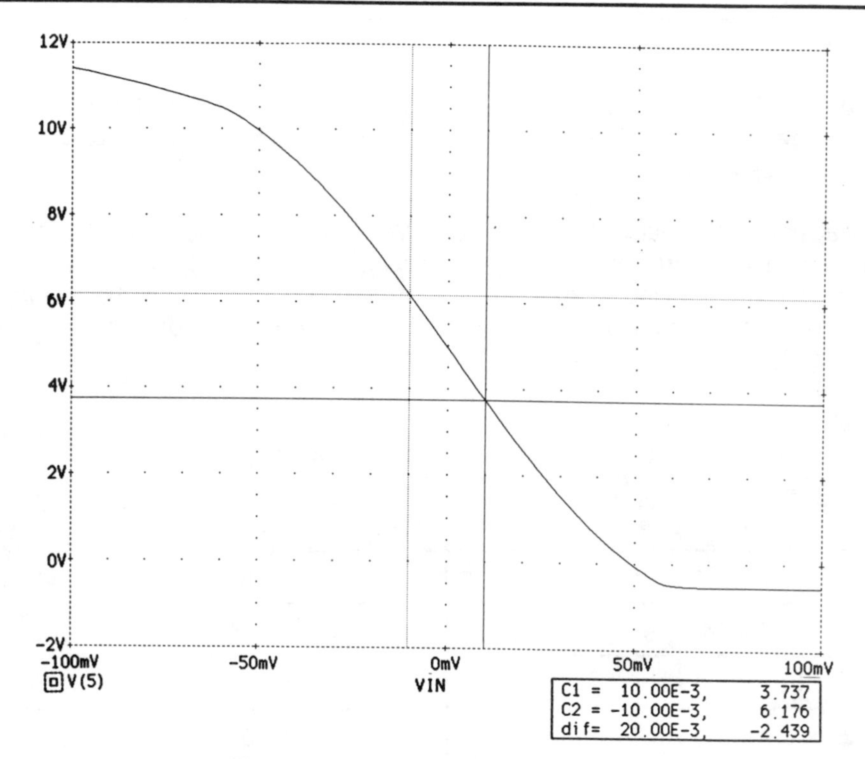

Figure 4-2. Plot of VOUT versus VIN.

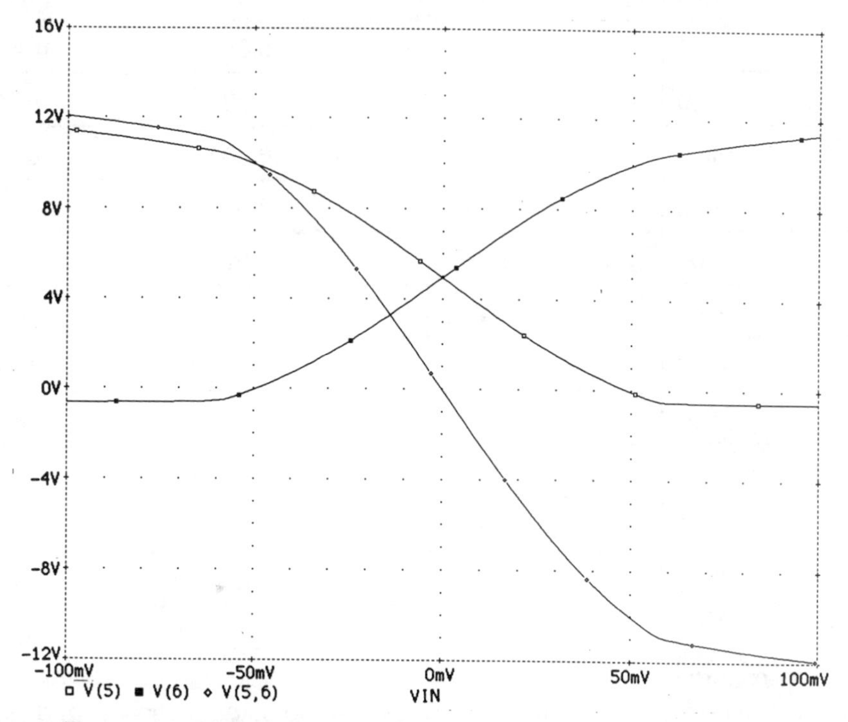

Figure 4-3. Plot of VOUT versus VIN for V(5), V(6), and V(5,6).

This 80 mV input signal will cause an output swing, looking at the differential output data of V(5,6), of approximately +8.5 V to -8.5 V. The total swing is approximately 17 V.

A calculation of the gain, at each value of VEE, can be made by extracting the data from each separate plot. The gain calculation is arbitrarily chosen based on a -0.01-V to +0.01-V input range. To assist in the data reduction, the two input points are identified, on each plot, by underlining the data. The data, having been removed from the plots, is shown in Table 4-8.

Table 4-8. DC Analysis Data Summary

VEE Voltage	Vout with Vin = -10 mV	Vout with Vin = 10 mV	Differential gain
-15.00	4.738	1.786	-147.60
-14.00	5.219	2.435	-139.20
-13.00	5.698	3.085	-130.65
-12.00	6.176	3.737	-121.95
-11.00	6.653	4.391	-113.10
-10.00	7.128	5.046	-104.10
-9.00	7.601	5.704	-94.85
-8.00	8.074	6.364	-85.50
-7.00	8.545	7.025	-76.00
-6.00	9.015	7.690	-66.25
-5.00	9.485	8.358	-56.35
-4.00	9.955	9.030	-46.25
-3.00	10.430	9.709	-36.05
-2.00	10.910	10.410	-25.00
-1.00	11.510	11.270	-12.00
0.00	12.000	12.000	0.00

The differential gain of the circuit at a selected VEE voltage of -12-V DC is -121.95. The negative sign simply indicates that the output is an inverted signal from the input. The gain correlates well to the results of the .TF analysis, which follows the .PLOT DC results. In reviewing the data shown in section 4-9, Small-signal characteristics the differential gain [V(5)/VIN] is shown to be -123.1. The discrepancy lies in the linearity of the waveform. If perfect linearity existed over the computed 20 mV range, then the number would correlate perfectly. More gain and output swing might be realized by a shift in the DC output operating point of the amplifier. This might be realized by an output voltage value of more than the 4.9565 V predicted by the operating point analysis shown in the small-signal characteristics portion of the analysis.

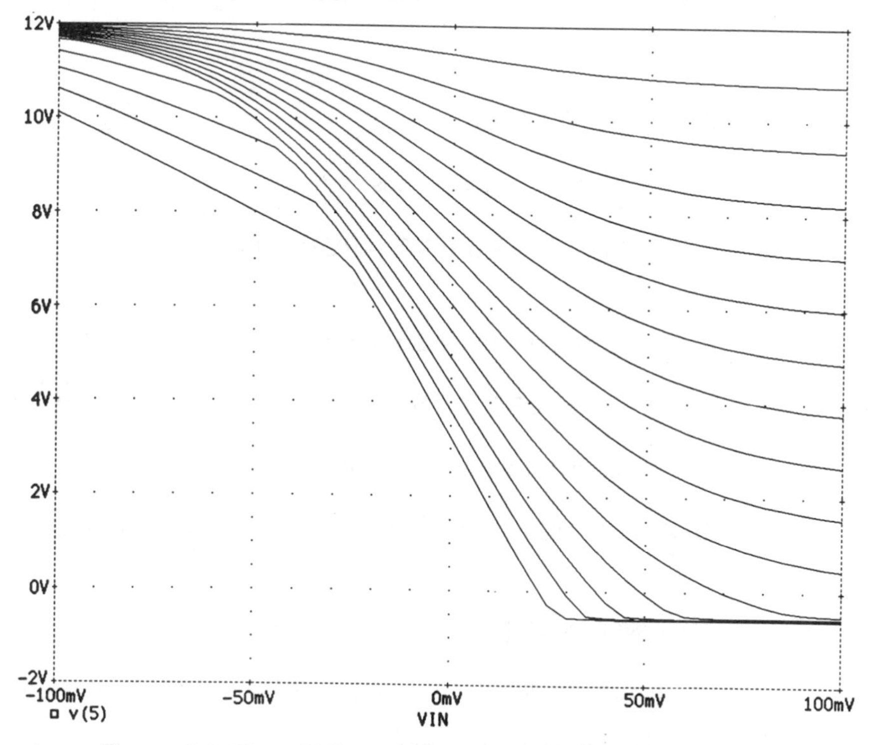

Figure 4-4. Plot of VIN versus VOUT with Variation in VEE

The small-signal bias solution is a precursor to the sensitivity analysis requested with the .SENS statement. Note that this is an actual operating point analysis and not merely a calculation of a DC operating point. Because no .OP statement occurred in the input file, the conclusion must be that the sensitivity analysis produced the operating point analysis. The final data presented is the DC sensitivity analysis. As indicated earlier in this section, all affects of all components and their parameters, on the requested output, are shown. For example, looking at only the resistors in the circuit and excluding any resistance associated with the solid state devices, the single resistor with the greatest effect on the voltage at node 5 is RE. RE will cause the output voltage at node 5 to change 3.175 mV/ohm or a resistance change of 3.175E-03 V/unit. The most sensitive of the two power supplies (VCC and VEE) is the VCC supply, which will cause a 0.9995-V change at node 5 for every 1-V change at VCC. As a point of interest, the reader is directed to review the volts/unit value for

VIN. The value shown is −123.1 V/unit. If we assume a 1-V change in VIN, then this value is normalized and is, in fact, the differential gain of the circuit. The reader is again cautioned that the sensitivity values shown for each component and parameter are only valid at the specified bias conditions. Any change in parameters or component values may cause a significant change on all other component sensitivities.

4.11 SUMMARY OF THE DC ANALYSIS COMMANDS

1. Sensitivity

   .SENS OV1 [OV2 ...]

2. Diode circuit interconnect description

   DXXXXXXX NA NC MNAME [AREA] [OFF] [IC=VD]

3. Diode model parameter description

   .MODEL MNAME D [PNAME1=PVAL1 PNAME2=PVAL2 ...]

4. Transistor circuit interconnect description

   QXXXXXXX NC NB NE [NS] MNAME [AREA] [OFF] [IC=VBE, VCE]

5. Transistor model parameter description

   .MODEL MNAME NPN [PNAME1=PVAL1 PNAME2=PVAL2 ...]

   .MODEL MNAME PNP [PNAME1=PVAL1 PNAME2=PVAL2 ...]

6. JFET circuit interconnect description

   JXXXXXXX ND NG NS MNAME [AREA] [OFF] [IC=VDS,VGS]

7. JFET model parameter description

   .MODEL MNAME NJF [PNAME1=PVAL1 PNAME2=PVAL2 ...]

   .MODEL MNAME PJF [PNAME1=PVAL1 PNAME2=PVAL2 ...]

8. MOSFET circuit interconnect description

   MXXXXXXX ND NG NS NB MNAME [L=VAL] [W=VAL]
   + [AD=VAL] [AS=VAL] [PD=VAL] [PS=VAL]

+ [NRD=VAL] [NRS=VAL] [OFF] [IC=VDS,VGS,VBS]

9. MOSFET model parameter description

.MODEL MNAME NMOS [PNAME1=PVAL1 PNAME2=PVAL2 ..]

.MODEL MNAME PMOS [PNAME1=PVAL1 PNAME2=PVAL2 ..]

10. DC voltage source description

VXXXXXXX N+ N- DC DCVALUE AC ACVALUE ACPHASE

11. DC current source description

IXXXXXXX N+ N- DC DCVALUE AC ACVALUE ACPHASE

12. Print command

.PRINT DC OV1 [OV2 ... OV8]

13. Plot command

.PLOT DC OV1 [OV2 ... OV8]

14. DC sweep range description

.DC SRC1 VALSTRT1 VALSTOP1 VALINCR1 [SRC2 START2
STOP2 INCR2]

5

AC AND DC ANALYSIS

The scope of this section is to present the reader with the following concepts:

1. Dependent voltage-controlled voltage source
2. Dependent voltage-controlled current source
3. Dependent current-controlled voltage source
4. Dependent current-controlled current source
5. Subcircuit definitions
6. Combining the DC and AC analyses

The section introduces the dependent voltage/current source concept to provide a better understanding of their applications in conjunction with subcircuit definitions. Subcircuit definitions most often use such dependent sources to realize simplified model definition.

5.1 DEPENDENT SOURCES

SPICE provides, the user, with the ability to define and control four different dependent sources (see items 1 to 4 above). A dependent source, as the name implies, has an output that is dependent on some input. For example, a DC amplifier may represent a dependent source. If the amplifier has a gain of 1000 and if 10-mV DC is applied to the input, the amplifier will attain a 10-V DC level.

SPICE allows circuits to contain linear dependent sources characterized by the following four equations:

I = G*V, V = E*V, I = F*I, V = H*I

where *G, E, F,* and *H* are constants representing transconductance, voltage gain, current gain, and resistance, respectively. In a simpler sense, the constants are purely multipliers.

5.1.1 LINEAR VOLTAGE-CONTROLLED CURRENT SOURCES

The voltage-controlled current source (VCCS) is designated by G followed by up to seven numbers. The VCCS, like all dependent voltage and current sources, is a four-lead device with a multiplier constant. The general form is shown by

GXXXXXXX N+ N- NC+ NC- VALUE

where N+ and N- are positive and negative output nodes, NC+ and NC- are the positive and negative controlling input nodes, and VALUE is the transconductance in mhos (multiplier). The basic characteristics associated with the VCCS are as follows:

Input impedance	-	infinite
Output impedance	-	infinite
Bandwidth	-	infinite
Losses	-	none

Assume, for example, that it is desirable to specify a VCCS designated G1 (e.g., see Figure 5-1). The current outputs are nodes 2 and 0 and the voltage control inputs are nodes 5 and 0. The transconductance in mhos is 0.1E-03. While transconductance is the technical term for the multiplier, since transconductance is defined as 1/R and 1/R * V = I, the value is actually gain. Assume the control voltage, at any given moment in time, is 10 V. The output current is (0.1 mmho * 10 V) = 0.1E-3 * 10 V = 1 mA. The control nodes may be anything and might typically be the nodes of a component whose voltage is being sensed and converted to current for another application within the circuit. The above example could be described, using a SPICE statement as follows:

G1 2 0 5 0 0.1MMHO (NOTE MMHO = E-3 = 1/R)

The schematic of this example is shown in Figure 5-1. In this example, if we applied 10 V at node 5 and placed a 1000 Ω resistor from node 2 to ground, the voltage at node 2, as a result of the current source, would be -1.0 V. This negative voltage is due to the convention of the current flow in the current source. The reader is again reminded that the current flow, in SPICE, is from the N- node to the N+ node. If a positive voltage were required at node 2, then -10 V could be applied to node 5 or output nodes 2 and 0 could be reversed. Either approach will provide the same result. Again, it is noteworthy to mention that all nodes must have a path to ground. If the output nodes had been other than 2 and 0, a DC path must have existed from one of the two output nodes and ground.

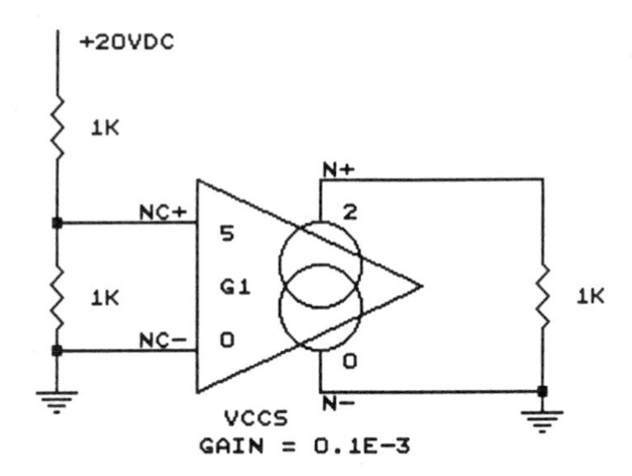

Figure 5-1. Voltage-controlled current source example.

5.1.2 LINEAR VOLTAGE-CONTROLLED VOLTAGE SOURCES

The voltage-controlled voltage source (VCVS) is designated by E followed by up to seven numbers. The VCVS, like all dependent voltage and current sources, is a four-lead device. The VCVS can be characterized as simply a voltage amplifier. The general form is

EXXXXXXX N+ N- NC+ NC- VALUE

where N+, N-, NC+, NC-, and VALUE are defined as above.

The basic characteristics associated with the VCVS are as follows:

Input impedance	-	infinite
Output impedance	-	zero
Bandwidth	-	infinite
Losses	-	none

An example of a VCVS might be a source designated E1, whose voltage outputs are nodes 2 and 3, and whose voltage control inputs are nodes 14 and 1 (e.g., see Figure 5-2). Assume the control voltage, at any given moment in time, is 10 V. The VCVS has a voltage gain of 2.0. The voltage at nodes 2 and 3 is (2.0 * 10 V) = 20 V. The VCVS is, therefore, a voltage amplifier. The description of the above defined VCVS is shown below.

```
E1 2 3 14 1 2.0
```

In this example, if we applied 10 V at nodes 14 to 1 and metered the voltage from nodes 2 to 3, the voltage at node 2, with respect to node 3, as a result of the voltage source, would be +20 V.

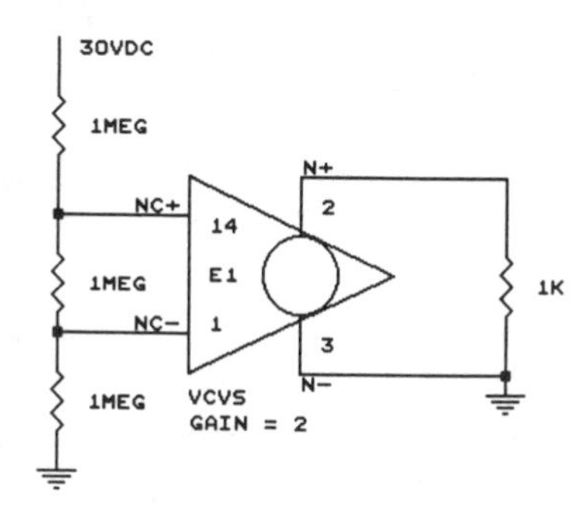

Figure 5-2. Voltage-controlled voltage source example.

5.1.3 LINEAR CURRENT-CONTROLLED CURRENT SOURCES

The current-controlled current source (CCCS) is designated by F followed by up to seven numbers. The CCCS, like all dependent voltage and current sources, is a four-lead device with a multiplier constant. The general form is

FXXXXXXX N+ N- VNAM VALUE

where N+ is the positive node (current flows from - to + terminal), N- is the negative node (current flows from + to - terminal), VNAM is the name of the voltage source whose function is that of an ammeter sensing current flow, and VALUE is the current gain (multiplier). The basic characteristics associated with the CCCS are as follows:

Input impedance	-	zero
Output impedance	-	infinite
Bandwidth	-	infinite
Losses	-	none

The example shown is a current source designated F1 with a current-sensing voltage source VSENS (e.g., see Figure 5-3). This current-sensing voltage source is an ammeter. The current is assumed to be flowing to the positive node and from the negative node of VSENS. The current source output nodes are 13 and 5 and the CCCS has a current gain of 5. Assume that the current flowing from the positive to the negative node of VSENS is 1 mA. The resulting current, flowing into node 13 and out of node 5, is 5 * 1 mA = 5 mA. If a 100 Ω resistor were connected between nodes 13 and 5, a voltage of -0.5 V would exist at node 13 with respect to 5.

The CCCS could be defined as follows:

F1 13 5 VSENS 5

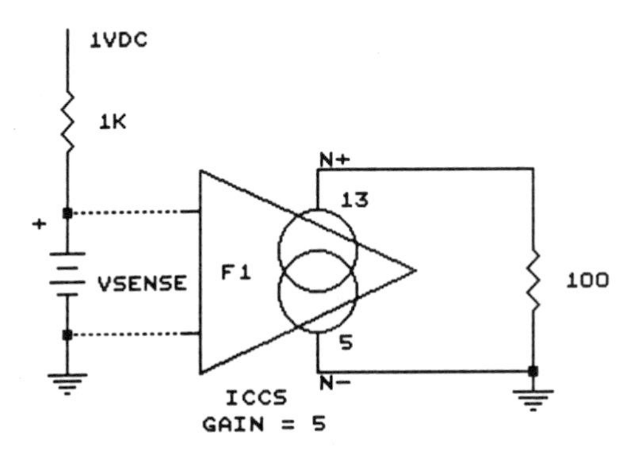

Figure 5-3. Current-controlled current source example.

5.1.4 LINEAR CURRENT-CONTROLLED VOLTAGE SOURCES

The current-controlled voltage source (CCVS) is designated by H followed by up to seven numbers. The CCVS, like all dependent voltage and current sources, is a four-lead device with a multiplier constant. The general form is

HXXXXXXX N+ N- VNAM VALUE

where N+ positive and N- are the negative nodes, VNAM is the name of the voltage source whose function is that of an ammeter sensing current flow, and VALUE is the resistance in ohms (multiplier). The basic characteristics associated with the CCVS are as follows:

Input impedance	-	zero
Output impedance	-	zero
Bandwidth	-	infinite
Losses	-	none

Assume you are defining a voltage source designated HX with a controlling voltage source VZ and a transresistance (gain) of 500 Ω (as in Figure 5-4). Assume the current is flowing into the positive node of VZ, through VZ, and out the negative node. If the current through VZ is 1 mA and the gain is 500 (500 Ω), then the voltage at node 5 with respect to node 17 is 500 * 1 mA = 500 mV. The CCVS could be defined as follows:

```
HX 5 17 VZ 0.5K
```

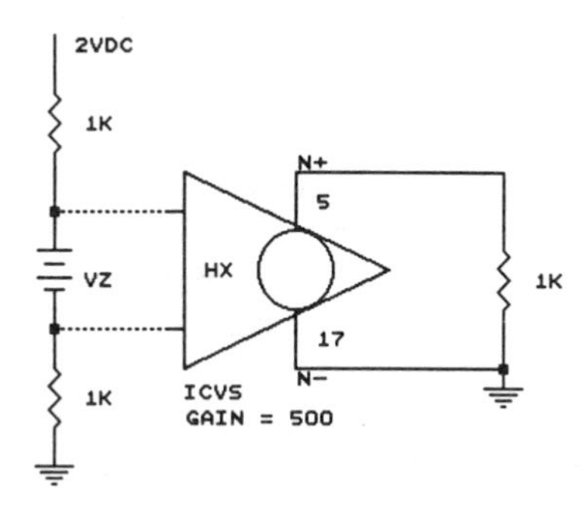

Figure 5-4. Current-controlled voltage source example.

5.1.5 DEPENDENT SOURCE APPLICATIONS

A possible application of dependent current sources is shown in Figure 5-5. The figure describes an optical isolator (OP4N25). Node 1 is the anode of the sensor light emitting diode (LED) with node 2 being the cathode. The LED is modeled as a diode in series with an ammeter VM. VM has a specified voltage of 0 V. The current flowing through VM, and subsequently D1, is sensed by VM and converted to an output voltage using a CCVS, designated H1. The delay time, rise time, and fall time characteristics of the OP4N25 are established by the use of R1 and C1. The resulting voltage across C1 is amplified using a VCCS. The VCCS is designated G1 and directly supplies current to the base of Q1. Q1 represents the output of the optical isolator and presents three outputs consisting of the collector, base, and emitter. The base is brought out, as in many optical isolator circuits, to allow for the presence of external base biasing and is driven by a high-impedance current source.

The detail associated with the various value determinations is outside the scope of this text but the general approach will be presented. The model parameters for the OP4N25 optical isolator (4N25) are developed from a composite of multiple source vendor data sheets. The LED parameters, of interest, are N, RS, and IS. The value of N, for integrated circuits is normally 1. For discrete devices, such as diodes, the value is closer to 2. The value of RS is derived by looking at the forward bias characteristics of a typical LED and computing the resistance over the normal operating region. A value of 0.75 is chosen for RS. The value of IS is again extracted from typical LED data and is determined based on the ID and VD in the forward bias region where the plot of the logarithm of ID versus VD is a linear relationship. In this instance, 25E-13 is the computed value.

The LED dynamic characteristics are next estimated from data sheets for CJO = 40 pF. The rise time (Tr) of the 4N25 is found to be 1 μs and is attributed predominately to the LED. Using the equation Tr = 2.3 * R1 * C1 and assuming that C1 = 1000 pF, the resulting calculation of R1 yields a resistance of approximately 450 Ω. Having now defined the DC and dynamic parameters of the LED, the focus can be shifted to the phototransistor parameters. From the data sheets, the current gain through the 4N25 is approximately 0.5. If the VCCS is assumed to have a gain of 1 and the beta of the output transistor is assumed to be 150, then the gain value of the CCVS is 3.33E-3. To verify the accuracy of the calculations, assume that VM has 10 mA flowing through it. The voltage out of the CCVS is 10E-3 * 3.33E-3 or 33.3E-6. From a static DC standpoint, this voltage is applied to the input of the VCCS and yields a current of 33.3E-6 * 1 or 33.3 μA. Because the collector current of Q1 is approximately the base current multiplied by beta, the collector current becomes 33.3 μA * 150 or 5 mA. If 10 mA was originally applied to the input and 5 mA resulted at the output, a current gain of 0.5 has been realized. The LED dark current specification for ID is 8 nA. the value of IS, for Q1, can be estimated as ID/beta or 3.33E-11. The values of NF, BF, IKF, and VAF are assumed to be 1.3, 150, 0.1, and 100, respectively.

The phototransistor (Q1) dynamic parameters are estimated based on the manufacturers specifications with a fall time (Tf) of 2 μs, a saturated switching storage time (Ts) of 4 μs, and a saturated fall time (Tfs) of 8 μs. BR effects the storage time and is assumed to be a value of 10. The following equations are then used to compute TF, TR, CJC, and CJE.

$$TF = (Tf)/(14.45 * beta) = 0.923E-9$$

$$TR = (Ts/BR)/[\ln (BF/10)] = 148E-9$$

$$CJC = (2 * Tfs)/(2.237E3 * BF) = 47.4E-12$$

$$CJE = (3.5 * CJC) = 167E-12$$

The parameters BR, CJE, NF, and VAF are estimates that might be refined by measurement. The resulting SPICE model circuit description listed as follows.

```
VM 1 12 DC 0
D1 12 2 LED
.MODEL LED D(IS=25E-13 RS=0.75 CJO=35P N=2)
R1 10 11 450
C1 11 0 1000PF
H1 10 0 VM 3.33E-3
G1 5 6 11 0 1.0
Q1 5 6 4 QNPN
.MODEL QNPN NPN(IS=3.33E-11 NF=1.35 CJC=47.4P CJE=167P TF=.923N
+ TR=148N BF=150 BR=10 IKF=0.1 VAF=100)
```

The advantage of using the dependent sources is that they allow simple modeling to occur. The optical isolator circuit model defines a set of realistic performance characteristics and provides the DC isolation characteristics expected of an optical isolator device. As mentioned previously, the dependent sources find significant usage in operational amplifier models and related linear circuits. An example of a use of dependent sources is shown in this LM741 operational amplifier subcircuit, discussed in the next section.

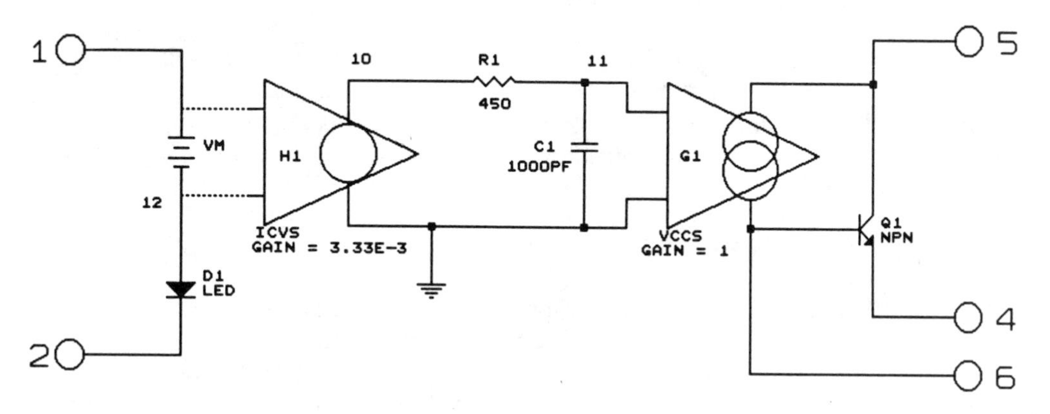

Figure 5-5. Optical isolator model.

```
.SUBCKT LM741 10 20 30 40 50
* + input=10; - input=20; output=30; VCC=40; VEE=50
.MODEL D2 D(IS=10E-15 N=2.118)
.MODEL D3 D(IS=9.71E-15 N=2.118)
.MODEL D1 D(IS=6E-15 N=.1 XTI=.1)
.MODEL D4 D(IS=1E-14 N=1)
VP 10 11 0
VN 20 21 0
DP 11 1 D2
DN 21 1 D3
GO 3 0 2 0 161
RD 3 0 60
DH 3 5 D1
DL 6 3 D1
VH 40 5 .7
VL 6 50 2.2
D1 3 9 D4
D2 9 3 D4
FA 20 0 VN .276
I1 1 0 140.5N
C1 1 0 33.3FF IC=-.863V
FP 4 0 VP 162.2
FN 0 4 VN 161
GC 0 4 1 0 6.6N
RT 4 0 1
CT 4 0 80N
G2 0 2 4 0 1
R2 2 0 100E3
CC 2 3 30P EX 9 0 POLY(2) 3 0 3 30 0 1 -1.64
RO 3 30 40
.ENDS LM741
```

Notice that the LM741 model has described the operational amplifier and all characteristics using only diode models and dependent sources. No transistor models have been used. The absence of transistors allows for (1) simplification of the model, (2) simpler performance characteristic definition, and (3) greater speed of analysis.

5.2 SUBCIRCUIT DEFINITION

A subcircuit, in SPICE, can be viewed as a computer subroutine. In a computer subroutine, certain parameters are passed to the subroutine, in the form of variables. The subroutine performs the required calculations and returns certain computed variables following its execution. In SPICE, if a circuit is to be used repeatedly in some application, two options exist. The first option is to define the circuit each time it is used. The second option is to define a subcircuit. Assume that the LM741 model, shown above, were to be used six times in a complex circuit. The choices would be (1) define each element, in the circuit, six times or (2) define a subcircuit and call it six times with a call statement. The LM741 has 29 statements that define the actual circuit (less the .SUBCKT and .ENDS statements). If the circuit, being simulated, uses six LM741s, then some 174 statements would be necessary. If the LM741 is treated as a subcircuit, only 37 statements would accomplish the same task. The optimal approach is the use of subcircuit calls. This section describes the procedures to define a subcircuit and the commands necessary to call the subcircuit.

5.2.1 SUBCIRCUIT CREATION

A subcircuit is simply a description of the desired circuit with the first statement being a .SUBCKT command and the last being a .ENDS command. Assume that the optical isolator, discussed previously, were to be made into a subcircuit. The circuit description is as follows:

```
VM 1 12 DC 0
D1 12 2 LED
.MODEL LED D(IS=25E-13 RS=0.75 CJO=35P N=2)
R1 10 11 450
C1 11 0 1000PF
H1 10 0 VM 3.33E-3
G1 5 6 11 0 1.0
Q1 5 6 4 QNPN
.MODEL QNPN NPN(IS=3.33E-11 NF=1.35 CJC=47.4P CJE=167P TF=.923N
+ TR=148N BF=150 BR=10 IKF=0.1 VAF=100)
```

The circuit can be turned into a subcircuit by placing a .SUBCKT statement prior to the first statement of the circuit. The general form of this statement is shown as follows.

.SUBCKT subnam N1 [N2 N3 ...]

To best see the process of subcircuit creation, assume the optical isolator is a model of a 4N25. Assume that the name of the subcircuit is OP4N25. Any name could be used as long as the name starts with an alphabetical character. Thus, the name could have been OPTISO. It is next necessary to define the nodes within the subcircuit that are of interest for external interface. In this case, assume the input nodes (1 and 2) and output nodes (4 and 5) are of interest. Notice that node 6 (base connection) is not of interest to us. The nodes of interest must be defined as part of the subcircuit statement. The sequence of nodes is not important within the subcircuit but is significant for external interfacing to the subcircuit. The resulting statement, inserted at the front of the subcircuit would be

.SUBCKT OP4N25 1 2 4 5

To provide a mental reminder of node functions, a comment statement is normally placed after the .SUBCKT statement. Assume the following comment is placed after the .SUBCKT statement.

* ANODE INPUT = 1; CATHODE INPUT = 2; EMITTER
* OUTPUT = 4; COLLECTOR OUTPUT = 5

The last statement in a subcircuit is a .ENDS statement with the following general form.

.ENDS [SUBNAM]

In this example the statement would be as follows:

.ENDS OP4N25

The resulting subcircuit would be described as follows:

```
.SUBCKT OP4N25 1 2 4 5
* ANODE INPUT = 1; CATHODE INPUT = 2; EMITTER
* OUTPUT = 4; COLLECTOR OUTPUT = 5
VM 1 12 DC 0
D1 12 2 LED
.MODEL LED D(IS=25E-13 RS=0.75 CJO=35P N=2)
R1 10 11 450
C1 11 0 1000PF
H1 10 0 VM 3.33E-3
G1 5 6 11 0 1.0
Q1 5 6 4 QNPN
.MODEL QNPN NPN(IS=3.33E-11 NF=1.35 CJC=47.4P CJE=167P TF=.923N
+ TR=148N BF=150 BR=10 IKF=0.1 VAF=100)
.ENDS OP4N25
```

To summarize, a subcircuit that consists of SPICE elements can be defined and referenced in a fashion similar to device models. The subcircuit is defined in the input file by a grouping of element statements; the SPICE program automatically inserts the group of elements wherever the subcircuit is referenced. There is no limit on the size or complexity of subcircuits, and subcircuits may contain and call other subcircuits.

A subcircuit definition begins with a .SUBCKT statement. SUBNAM is the subcircuit name, and N1, N2, ... are nodes of external interest within the subcircuit. Zero node may not be used as part of the .SUBCKT statement. The group of element statements, which immediately follow the .SUBCKT statement, defines the internal connections of the subcircuit. The last statement in a subcircuit definition is the .ENDS statement. Control statements may not appear within a subcircuit definition. However, subcircuit definitions may contain anything else, including other subcircuit definitions, device models, and subcircuit calls. Note that any device models or subcircuit definitions included as part of a subcircuit definition are strictly local (i.e., such models and definitions are not known outside the subcircuit definition). Also, any element nodes not included on the .SUBCKT statement are strictly local and are not known external to the subcircuit. The exception is node 0 (ground) which is always global. Note in the subcircuit example, that no ground node has been defined. As a result, a DC path to ground must be provided external to the subcircuit.

5.2.2 SUBCIRCUIT CALLS

A subcircuit, once defined, is treated as a circuit element. The first character to describe the element is X. The X designation may be followed with up to seven alphanumeric characters. The external node connections are passed, as part of the subcircuit call, along with the name of the subcircuit. The general form is

XYYYYYYY N1 [N2 N3 ...] SUBNAM

The sequence of node connections must match exactly with the sequence of nodes defined within the subcircuit. If four nodes are defined as part of the .SUBCKT statement, then four nodes must be defined using the subcircuit call.

For example, assume we want to connect up two OP4N25s in the configuration shown in Figure 5-6. The circuit shows an interconnection of two L4N25 subcircuits designated X1 and X2. The two subcircuits have external interface elements consisting of resistors and voltage sources. The interfaces to the subcircuits are shown accordingly.

The circuit might be described as follows:

```
OPTICAL ISOLATOR CIRCUIT
VX 1 0 DC 1
R1 1 2 1K
R2 4 3 1K
R3 4 5 1K
VCC 4 0 DC 5
X1 2 0 90 3 OP4N25
X2 3 90 90 5 OP4N25
```

```
* THIS IS A RESISTOR TO CONNECT 2 DIFFERENT GRDS
RG 90 0 1T
.SUBCKT OP4N25 1 2 4 5
* ANODE INPUT = 1; CATHODE INPUT = 2; EMITTER
* OUTPUT = 4; COLLECTOR OUTPUT = 5
VM 1 12 DC 0
D1 12 2 LED
.MODEL LED D(IS=25E-13 RS=0.75 CJO=35P N=2)
R1 10 11 450
C1 11 0 1000PF
H1 10 0 VM 3.33E-3
G1 5 6 11 0 1.0
Q1 5 6 4 QNPN
.MODEL QNPN NPN(IS=3.33E-11 NF=1.35 CJC=47.4P CJE=167P TF=.923N
+ TR=148N BF=150 BR=10 IKF=0.1 VAF=100)
.ENDS OP4N25
.DC VX 0 20 .1
.PLOT DC V(5)
.END
```

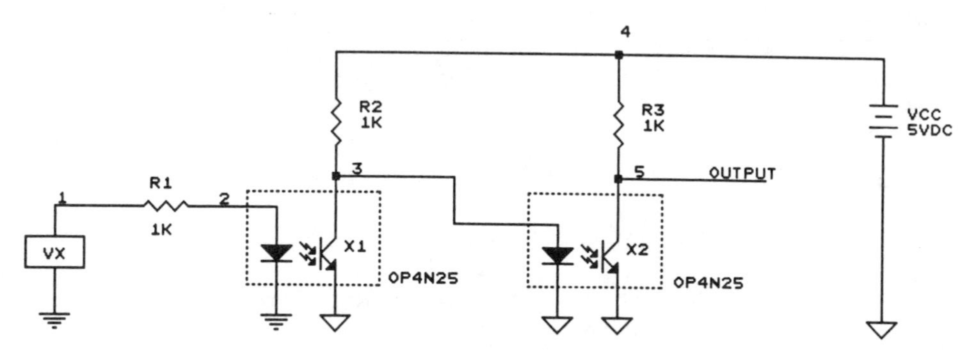

Figure 5-6. Cascaded optical isolator model.

Notice that the following has occurred:

1. The node designations for X1 and X2 are in the exact sequence as defined in the subcircuit with the node sequence being anode input, cathode input, emitter output, and collector output.

2. A ground (node 0) has been defined external to the subcircuit since the subcircuit contained no internal ground node in this example.

3. The node numbers, external to the subcircuit, may be the same as used internal to the subcircuit as they have no knowledge of each other and represent a mapping.

4. Component names external to the subcircuit have no knowledge of internal subcircuit component names.

5. The subcircuit may be treated as a "black box" from an external standpoint. The subcircuit represents, in this instance, a function with four connections.

6. A resistor, designated RG, has been place between the two grounds. The resistor permits SPICE to run and satisfies the requirement for each node having a DC path to ground. The value is set to 1T Ω (1E12). This effectively simulates the DC resistance of air.

5.3 AC/DC EXAMPLE CIRCUIT

The OP4N25 optical isolator circuit, discussed earlier in this section, is shown in Figure 5-7. The circuit represents a test circuit to verify the DC and AC performance characteristics of the OP4N25. The OP4N25, having been previously modeled as a subcircuit, will be subjected to an analysis to determine the following information:

1. The input resistance, output resistance, and current gain

2. The voltages at each node

3. The plot of output current versus input current, looking for current gain characteristics

4. The plot of the AC current gain, phase shift, and 3-dB point on the output (Note: the value of the 3-dB point is valid for current gain/loss as well as voltage gain/loss; although the input normally is set to unity, this is not necessary to determine the 3-dB point at the output node/source).

5. The sensitivity of the output current to each component in the circuit

To realize these results, it will become necessary to treat the optical isolator as a two-port (2 inputs and 2 outputs) current device. Voltage readings are only applicable as they apply to a specified application of the circuit. The input diode is purely a current driven device and the output collector is a current sink device. If the user knows the current gain/loss of an optical isolator, then the appropriate input drive circuit, output load resistance (RL), and output voltage source (VCC) can be selected to realize the desired output voltages.

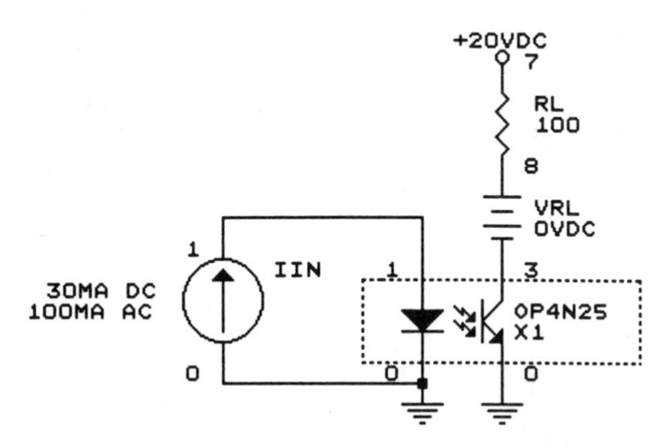

Figure 5-7. Optical isolator parameter verification circuit.

To treat the optical isolator as a current device suggests that the optical isolator should be driven with a current source and metered on the output for current sinking capability. A review of Figure 5-7 shows an input current source (IIN) in series with the anode of the input diode. The output circuit consists of a (1) 20-V DC power supply (VCC) in series with a (2) 100-Ω load resistor (RL) in series with an (3) ammeter (VRL), which is (4) connected to the collector of the output. This will allow the input current to be swept, over a specified range, and the effects monitored at the output. Although not a practical application of the optical isolator, the circuit does represent a functional test circuit to determine device performance. This procedure provides the following advantages.

1. Verifies that the modeling of the circuit is consistent with manufacturers specifications.

2. Provides information about the circuit that is not provided by the manufacturer (e.g., AC frequency characteristics).

3. Provides information to assure the desired interface circuitry will produce the desired output result in an actual circuit application.

5.4 SAMPLE AC/DC CIRCUIT INPUT FILE

The resulting circuit, which is shown in Figure 5-7, is described as follows:

```
OPTICAL ISOLATOR CIRCUIT
.SUBCKT OP4N25 1 2 4 5
* ANODE INPUT = 1; CATHODE INPUT = 2; EMITTER
* OUTPUT = 4; COLLECTOR OUTPUT = 5
VM 1 12 DC 0
D1 12 2 LED
.MODEL LED D(IS=25E-13 RS=0.75 CJO=35P N=2)
R1 10 11 450
C1 11 0 1000PF
H1 10 0 VM 3.33E-3
```

```
G1 5 6 11 0 1.0
Q1 5 6 4 QNPN
.MODEL QNPN NPN(IS=3.33E-11 NF=1.35 CJC=47.4P CJE=167P TF=.923N
+ TR=148N BF=150 BR=10 IKF=0.1 VAF=100)
.ENDS OP4N25
X1 1 0 0 3 OP4N25
IIN 0 1 DC .03 AC .1
RL 7 8 100
VCC 7 0 20
VRL 8 3 0
.DC IIN 0 .1 .002
.PLOT DC V(3) I(VRL)
.PRINT DC V(3) V(1) I(VRL)
.AC DEC 25 5000 500K
.PLOT AC IM(VRL) IDB(VRL) IP(VRL) VM(3) VP(3)
.PRINT AC IM(VRL) IDB(VRL) IP(VRL) VM(3) VP(3) VDB(3)
.OP
.TF I(VRL) IIN
.OPTIONS NODE LIST LIMPTS=5001 ITL5=0
.SENS I(VRL)
.END
```

The circuit description has the following statements and characteristics that should be of interest to the reader.

1. The circuit uses the subcircuit OP4N25, which was described previously. Note that, because the base lead of the OP4N25 is not of interest, it was not included as a connection external to the subcircuit.

2. The subcircuit is labeled X1 and has the cathode of the input diode (LED) and the emitter of the output transistor connected to ground.

3. The circuit is driven by a current source associated with both a DC and an AC component. The DC component is 30 mA and is used to determine the DC operating point information. The AC input consists of a 100 mA AC. It should be noted that the AC component is not normalized to 1 A because of the unrealistic linear region that such a current would cause the part to operate in.

4. VRL serves as the output ammeter as discussed previously.

5. .DC and .AC statements cause both a DC and an AC analysis to occur. The DC statement sweeps IIN from 0 to 100 mA and plots the current through VRL I(VRL) versus the input current IIN. The AC sweep is from 500 Hz to 500 KHz and plots the magnitude, phase, and DB gain/loss of the current in the output.

6. In addition, the .OP, .TF, and .SENS commands are used. In this application of the .TF command, the current gain of the circuit will be computed. The .SENS command is included to permit the determination of which component has the greatest effect on the output current sink capability of the optical isolator.

5.5 SAMPLE CIRCUIT ANALYSIS AND RESULT INTERPRETATION

The following results are obtained after running the circuit described in Section 5.4. The extraction of the desired data is discussed at the end of this section.

```
OPTICAL ISOLATOR CIRCUIT

 ****     CIRCUIT DESCRIPTION

******************************************************************************

.SUBCKT OP4N25 1 2 4 5
* ANODE INPUT = 1; CATHODE INPUT = 2; EMITTER
* OUTPUT = 4; COLLECTOR OUTPUT = 5
VM 1 12 DC 0
D1 12 2 LED
.MODEL LED D(IS=25E-13 RS=0.75 CJO=35P N=2)
R1 10 11 450
C1 11 0 1000PF
H1 10 0 VM 3.33E-3
G1 5 6 11 0 1.0
Q1 5 6 4 QNPN
.MODEL QNPN NPN(IS=3.33E-11 NF=1.35 CJC=47.4P CJE=167P TF=.923N
+ TR=148N BF=150 BR=10 IKF=0.1 VAF=100)
.ENDS OP4N25
X1 1 0 0 3 OP4N25
IIN 0 1 DC .03 AC .1
RL 7 8 100
VCC 7 0 20
VRL 8 3 0
.DC IIN 0 .1 .002
.PLOT DC V(3) I(VRL)
.PRINT DC V(3) V(1) I(VRL)
.AC DEC 25 5000 500K
.PLOT AC IM(VRL) IDB(VRL) IP(VRL) VM(3) VP(3)
.PRINT AC IM(VRL) IDB(VRL) IP(VRL) VM(3) VP(3) VDB(3)
.OP
.TF I(VRL) IIN
.OPTIONS NODE LIST LIMPTS=5001 ITL5=0
.SENS I(VRL)
.END
```

```
OPTICAL ISOLATOR CIRCUIT

****     ELEMENT NODE TABLE

******************************************************************************
0              IIN       VCC       X1.C1     X1.D1     X1.H1
               X1.Q1

1              IIN       X1.VM

3              VRL       X1.G1     X1.Q1

7              RL        VCC

8              RL        VRL

X1.6           X1.G1     X1.Q1

X1.10          X1.H1     X1.R1

X1.11          X1.C1     X1.R1

X1.12          X1.D1     X1.VM
```

```
OPTICAL ISOLATOR CIRCUIT

****     Diode MODEL PARAMETERS

******************************************************************************
            X1.LED
      IS     2.500000E-12
       N     2
      RS      .75
     CJO    35.000000E-12
```

```
OPTICAL ISOLATOR CIRCUIT

****     BJT MODEL PARAMETERS

******************************************************************************
            X1.QNPN
            NPN

      IS    33.300000E-12
      BF    150
      NF    1.35
     VAF    100
     IKF    0.1
      BR    10
      NR    1
     CJE    167.000000E-12
     CJC    47.400000E-12
      TF    923.000000E-12
      TR    148.000000E-09
```

```
OPTICAL ISOLATOR CIRCUIT

 ****     CIRCUIT ELEMENT SUMMARY

******************************************************************************

**** RESISTORS

NAME                 NODES      MODEL      VALUE        TC1        TC2        TCE

RL                7       8                1.00E+02
X1.R1         X1.10   X1.11                4.50E+02

**** CAPACITORS

NAME                 NODES      MODEL      VALUE   In. Cond.      TC1        TC2

X1.C1         X1.11       0                1.00E-09  0.00E+00

**** VOLTAGE-CONTROLLED CURRENT SOURCES

NAME              +       - FUNCTION      GAIN    CONTROLLING NODES

X1.G1             3    X1.6 LINEAR     1.00E+00  (X1.11, 0)

**** CURRENT-CONTROLLED CURRENT SOURCES

NAME              +       - FUNCTION      GAIN    CONTROLLING SOURCES

X1.H1         X1.10       0 LINEAR     3.33E-03 X1.VM

**** INDEPENDENT SOURCES

NAME                 NODES     DC VALUE  AC VALUE  AC PHASE

VCC               7       0  2.00E+01  0.00E+00  0.00E+00  degrees

VRL               8       3  0.00E+00  0.00E+00  0.00E+00  degrees

X1.VM             1   X1.12  0.00E+00  0.00E+00  0.00E+00  degrees

IIN               0       1  3.00E-02  1.00E-01  0.00E+00  degrees


**** DIODES

NAME              +       -  MODEL        AREA

X1.D1         X1.12       0 X1.LED      1.00E+00

**** BIPOLAR JUNCTION TRANSISTORS

NAME              C        B         E         S MODEL          AREA

X1.Q1             3     X1.6         0         0 X1.QNPN      1.00E+00
```

```
OPTICAL ISOLATOR CIRCUIT

****     DC TRANSFER CURVES                 TEMPERATURE =   27.000 DEG C

******************************************************************************
```

IIN	V(3)	V(1)	I(VRL)
0.000E+00	2.000E+01	1.549E-17	1.026E-09
2.000E-03	1.988E+01	1.062E+00	1.187E-03
4.000E-03	1.977E+01	1.099E+00	2.348E-03
6.000E-03	1.965E+01	1.122E+00	3.486E-03
8.000E-03	1.954E+01	1.138E+00	4.601E-03
1.000E-02	1.943E+01	1.151E+00	5.696E-03
1.200E-02	1.932E+01	1.162E+00	6.771E-03
1.400E-02	1.922E+01	1.172E+00	7.826E-03
1.600E-02	1.911E+01	1.180E+00	8.864E-03
1.800E-02	1.901E+01	1.188E+00	9.884E-03
2.000E-02	1.891E+01	1.195E+00	1.089E-02
2.200E-02	1.881E+01	1.201E+00	1.188E-02
2.400E-02	1.872E+01	1.207E+00	1.285E-02
2.600E-02	1.862E+01	1.213E+00	1.381E-02
2.800E-02	1.853E+01	1.218E+00	1.475E-02
3.000E-02	1.843E+01	1.223E+00	1.568E-02
3.200E-02	1.834E+01	1.228E+00	1.660E-02
3.400E-02	1.825E+01	1.233E+00	1.750E-02
3.600E-02	1.816E+01	1.237E+00	1.840E-02
3.800E-02	1.807E+01	1.241E+00	1.928E-02
4.000E-02	1.799E+01	1.245E+00	2.015E-02
4.200E-02	1.790E+01	1.249E+00	2.101E-02
4.400E-02	1.781E+01	1.253E+00	2.186E-02
4.600E-02	1.773E+01	1.257E+00	2.270E-02
4.800E-02	1.765E+01	1.261E+00	2.353E-02
5.000E-02	1.757E+01	1.264E+00	2.435E-02
5.200E-02	1.748E+01	1.268E+00	2.516E-02
5.400E-02	1.740E+01	1.271E+00	2.596E-02
5.600E-02	1.732E+01	1.275E+00	2.675E-02
5.800E-02	1.725E+01	1.278E+00	2.754E-02
6.000E-02	1.717E+01	1.281E+00	2.831E-02
6.200E-02	1.709E+01	1.285E+00	2.908E-02
6.400E-02	1.702E+01	1.288E+00	2.984E-02
6.600E-02	1.694E+01	1.291E+00	3.059E-02
6.800E-02	1.687E+01	1.294E+00	3.134E-02
7.000E-02	1.679E+01	1.297E+00	3.208E-02
7.200E-02	1.672E+01	1.300E+00	3.281E-02
7.400E-02	1.665E+01	1.303E+00	3.353E-02
7.600E-02	1.657E+01	1.306E+00	3.425E-02
7.800E-02	1.650E+01	1.308E+00	3.496E-02
8.000E-02	1.643E+01	1.311E+00	3.567E-02
8.200E-02	1.636E+01	1.314E+00	3.637E-02
8.400E-02	1.629E+01	1.317E+00	3.706E-02
8.600E-02	1.623E+01	1.320E+00	3.775E-02
8.800E-02	1.616E+01	1.322E+00	3.843E-02
9.000E-02	1.609E+01	1.325E+00	3.910E-02
9.200E-02	1.602E+01	1.327E+00	3.977E-02
9.400E-02	1.596E+01	1.330E+00	4.044E-02
9.600E-02	1.589E+01	1.333E+00	4.110E-02
9.800E-02	1.582E+01	1.335E+00	4.175E-02
1.000E-01	1.576E+01	1.338E+00	4.240E-02

```
 OPTICAL ISOLATOR CIRCUIT

 ****     DC TRANSFER CURVES                  TEMPERATURE =   27.000 DEG C
******************************************************************************
 LEGEND:
*: V(3)
+: I(VRL)

  IIN        V(3)
(*)----------      1.4000E+01    1.6000E+01    1.8000E+01    2.0000E+01    2.2000E+01
(+)----------      0.0000E+00    2.0000E-02    4.0000E-02    6.0000E-02    8.0000E-02

                      - - - - - - - - - - - - - - - - - - - - - - - - - - - - - - - -
  0.000E+00 2.000E+01 +            .            .            *            .
  2.000E-03 1.988E+01 .+           .            .           *.            .
  4.000E-03 1.977E+01 . +          .            .          * .            .
  6.000E-03 1.965E+01 . +          .            .          * .            .
  8.000E-03 1.954E+01 .  +         .            .         *  .            .
  1.000E-02 1.943E+01 .   +        .            .        *   .            .
  1.200E-02 1.932E+01 .   +        .            .        *   .            .
  1.400E-02 1.922E+01 .    +       .            .       *    .            .
  1.600E-02 1.911E+01 .     +      .            .      *     .            .
  1.800E-02 1.901E+01 .     +      .            .      *     .            .
  2.000E-02 1.891E+01 .      +     .            .     *      .            .
  2.200E-02 1.881E+01 .       +    .            .    *       .            .
  2.400E-02 1.872E+01 .       +    .            .    *       .            .
  2.600E-02 1.862E+01 .        +   .            .   *        .            .
  2.800E-02 1.853E+01 .         +  .            .  *         .            .
  3.000E-02 1.843E+01 .         +  .            .  *         .            .
  3.200E-02 1.834E+01 .          + .            . *          .            .
  3.400E-02 1.825E+01 .          + .            . *          .            .
  3.600E-02 1.816E+01 .           +.            .*           .            .
  3.800E-02 1.807E+01 .            +            *            .            .
  4.000E-02 1.799E+01 .            +            *            .            .
  4.200E-02 1.790E+01 .            .+          *.            .            .
  4.400E-02 1.781E+01 .            .+          *.            .            .
  4.600E-02 1.773E+01 .            . +        * .            .            .
  4.800E-02 1.765E+01 .            . +        * .            .            .
  5.000E-02 1.757E+01 .            .  +      *  .            .            .
  5.200E-02 1.748E+01 .            .  +      *  .            .            .
  5.400E-02 1.740E+01 .            .   +    *   .            .            .
  5.600E-02 1.732E+01 .            .   +    *   .            .            .
  5.800E-02 1.725E+01 .            .    +  *    .            .            .
  6.000E-02 1.717E+01 .            .    +  *    .            .            .
  6.200E-02 1.709E+01 .            .     +*     .            .            .
  6.400E-02 1.702E+01 .            .     +*     .            .            .
  6.600E-02 1.694E+01 .            .     *+     .            .            .
  6.800E-02 1.687E+01 .            .     *+     .            .            .
  7.000E-02 1.679E+01 .            .    *  +    .            .            .
  7.200E-02 1.672E+01 .            .    *  +    .            .            .
  7.400E-02 1.665E+01 .            .   *    +   .            .            .
  7.600E-02 1.657E+01 .            .   *    +   .            .            .
  7.800E-02 1.650E+01 .            .  *      +  .            .            .
  8.000E-02 1.643E+01 .            .  *      +  .            .            .
  8.200E-02 1.636E+01 .            . *        + .            .            .
  8.400E-02 1.629E+01 .            . *        + .            .            .
  8.600E-02 1.623E+01 .            .*          +.            .            .
  8.800E-02 1.616E+01 .            .*          +.            .            .
  9.000E-02 1.609E+01 .            .*          +.            .            .
  9.200E-02 1.602E+01 .            *            +            .            .
  9.400E-02 1.596E+01 .            *            +            .            .
  9.600E-02 1.589E+01 .           *.            .+           .            .
  9.800E-02 1.582E+01 .           *.            .+           .            .
  1.000E-01 1.576E+01 .          * .            . +          .            .
                      - - - - - - - - - - - - - - - - - - - - - - - - - - - - - - - -
```

```
 OPTICAL ISOLATOR CIRCUIT

 ****      SMALL SIGNAL BIAS SOLUTION       TEMPERATURE =   27.000 DEG C
******************************************************************************
 NODE   VOLTAGE     NODE   VOLTAGE     NODE   VOLTAGE     NODE   VOLTAGE

(    1)    1.2230  (    3)   18.4320  (    7)   20.0000  (    8)   18.4320
( X1.6)     .6957  (X1.10) 99.90E-06  (X1.11) 99.90E-06  (X1.12)    1.2230

    VOLTAGE SOURCE CURRENTS
    NAME         CURRENT

    VCC         -1.568E-02
    VRL          1.568E-02
    X1.VM        3.000E-02

    TOTAL POWER DISSIPATION   3.14E-01  WATTS
```

```
 OPTICAL ISOLATOR CIRCUIT

 ****      OPERATING POINT INFORMATION      TEMPERATURE =   27.000 DEG C
******************************************************************************
**** VOLTAGE-CONTROLLED CURRENT SOURCES

NAME         X1.G1
I-SOURCE     9.990E-05

**** CURRENT-CONTROLLED VOLTAGE SOURCES

NAME         X1.H1
V-SOURCE     9.990E-05
I-SOURCE    -9.021E-24

**** DIODES

NAME         X1.D1
MODEL        X1.LED
ID           3.00E-02
VD           1.22E+00
REQ          1.72E+00
CAP          8.42E-11

**** BIPOLAR JUNCTION TRANSISTORS

NAME         X1.Q1
MODEL        X1.QNPN
IB           9.99E-05
IC           1.56E-02
VBE          6.96E-01
VBC         -1.77E+01
VCE          1.84E+01
BETADC       1.56E+02
GM           3.99E-01
RPI          3.50E+02
RX           0.00E+00
RO           7.56E+03
CBE          6.38E-10
CBC          1.65E-11
CBX          0.00E+00
CJS          0.00E+00
BETAAC       1.40E+02
FT           9.71E+07
```

```
****     SMALL-SIGNAL CHARACTERISTICS

      I(VRL)/IIN =  4.621E-01
      INPUT RESISTANCE AT IIN =  2.474E+00
      OUTPUT RESISTANCE AT I(VRL) =  7.656E+03
```

```
 OPTICAL ISOLATOR CIRCUIT

 ****     DC SENSITIVITY ANALYSIS          TEMPERATURE =   27.000 DEG C
******************************************************************************
DC SENSITIVITIES OF OUTPUT I(VRL)

         ELEMENT         ELEMENT         ELEMENT        NORMALIZED
          NAME            VALUE        SENSITIVITY      SENSITIVITY
                                      (AMPS/UNIT)    (AMPS/PERCENT)
          RL            1.000E+02      -2.048E-06      -2.048E-06
          X1.R1         4.500E+02       0.000E+00       0.000E+00
          VCC           2.000E+01       1.306E-04       2.612E-05
          VRL           0.000E+00      -1.306E-04       0.000E+00
          X1.VM         0.000E+00      -4.621E-13       0.000E+00
          IIN           3.000E-02       4.621E-01       1.386E-04
X1.D1
SERIES RESISTANCE
          RS            7.500E-01      -1.386E-14      -1.040E-16
INTRINSIC PARAMETERS
          IS            2.500E-12       9.561E-03       2.390E-16
          N             2.000E+00      -2.774E-13      -5.547E-15
X1.Q1
          RB            0.000E+00       0.000E+00       0.000E+00
          RC            0.000E+00       0.000E+00       0.000E+00
          RE            0.000E+00       0.000E+00       0.000E+00
          BF            1.500E+02       9.176E-05       1.376E-04
          ISE           0.000E+00       0.000E+00       0.000E+00
          BR            1.000E+01      -4.621E-11      -4.621E-12
          ISC           0.000E+00       0.000E+00       0.000E+00
          IS            3.330E-11       1.370E+05       4.561E-08
          NE            1.500E+00       0.000E+00       0.000E+00
          NC            2.000E+00       0.000E+00       0.000E+00
          IKF           1.000E-01       1.609E-02       1.609E-05
          IKR           0.000E+00       0.000E+00       0.000E+00
          VAF           1.000E+02      -2.317E-05      -2.317E-05
          VAR           0.000E+00       0.000E+00       0.000E+00
```

```
OPTICAL ISOLATOR CIRCUIT

****     AC ANALYSIS                          TEMPERATURE =   27.000 DEG C

******************************************************************************
```

FREQ	IM(VRL)	IDB(VRL)	IP(VRL)	VM(3)	VP(3)
5.000E+03	4.620E-02	-2.671E+01	-1.627E+00	4.620E+00	1.784E+02
5.482E+03	4.620E-02	-2.671E+01	-1.784E+00	4.620E+00	1.782E+02
6.011E+03	4.619E-02	-2.671E+01	-1.956E+00	4.619E+00	1.780E+02
6.591E+03	4.619E-02	-2.671E+01	-2.144E+00	4.619E+00	1.779E+02
7.227E+03	4.619E-02	-2.671E+01	-2.351E+00	4.619E+00	1.776E+02
7.924E+03	4.618E-02	-2.671E+01	-2.578E+00	4.618E+00	1.774E+02
8.689E+03	4.618E-02	-2.671E+01	-2.827E+00	4.618E+00	1.772E+02
9.527E+03	4.617E-02	-2.671E+01	-3.099E+00	4.617E+00	1.769E+02
1.045E+04	4.617E-02	-2.671E+01	-3.398E+00	4.617E+00	1.766E+02
1.145E+04	4.616E-02	-2.671E+01	-3.726E+00	4.616E+00	1.763E+02
1.256E+04	4.615E-02	-2.672E+01	-4.085E+00	4.615E+00	1.759E+02
1.377E+04	4.614E-02	-2.672E+01	-4.479E+00	4.614E+00	1.755E+02
1.510E+04	4.612E-02	-2.672E+01	-4.910E+00	4.612E+00	1.751E+02
1.656E+04	4.611E-02	-2.673E+01	-5.383E+00	4.611E+00	1.746E+02
1.815E+04	4.608E-02	-2.673E+01	-5.902E+00	4.608E+00	1.741E+02
1.991E+04	4.606E-02	-2.673E+01	-6.470E+00	4.606E+00	1.735E+02
2.183E+04	4.603E-02	-2.674E+01	-7.093E+00	4.603E+00	1.729E+02
2.393E+04	4.599E-02	-2.675E+01	-7.775E+00	4.599E+00	1.722E+02
2.624E+04	4.595E-02	-2.675E+01	-8.522E+00	4.595E+00	1.715E+02
2.877E+04	4.590E-02	-2.676E+01	-9.341E+00	4.590E+00	1.707E+02
3.155E+04	4.584E-02	-2.678E+01	-1.024E+01	4.584E+00	1.698E+02
3.459E+04	4.577E-02	-2.679E+01	-1.122E+01	4.577E+00	1.688E+02
3.793E+04	4.568E-02	-2.681E+01	-1.229E+01	4.568E+00	1.677E+02
4.159E+04	4.557E-02	-2.683E+01	-1.347E+01	4.557E+00	1.665E+02
4.560E+04	4.544E-02	-2.685E+01	-1.476E+01	4.544E+00	1.652E+02
5.000E+04	4.529E-02	-2.688E+01	-1.616E+01	4.529E+00	1.638E+02
5.482E+04	4.511E-02	-2.691E+01	-1.770E+01	4.511E+00	1.623E+02
6.011E+04	4.490E-02	-2.696E+01	-1.937E+01	4.490E+00	1.606E+02
6.591E+04	4.464E-02	-2.701E+01	-2.120E+01	4.464E+00	1.588E+02
7.227E+04	4.434E-02	-2.706E+01	-2.319E+01	4.434E+00	1.568E+02
7.924E+04	4.398E-02	-2.714E+01	-2.536E+01	4.398E+00	1.546E+02
8.689E+04	4.355E-02	-2.722E+01	-2.772E+01	4.355E+00	1.523E+02
9.527E+04	4.305E-02	-2.732E+01	-3.027E+01	4.305E+00	1.497E+02
1.045E+05	4.246E-02	-2.744E+01	-3.304E+01	4.246E+00	1.470E+02
1.145E+05	4.178E-02	-2.758E+01	-3.603E+01	4.178E+00	1.440E+02
1.256E+05	4.099E-02	-2.775E+01	-3.925E+01	4.099E+00	1.407E+02
1.377E+05	4.007E-02	-2.794E+01	-4.271E+01	4.007E+00	1.373E+02
1.510E+05	3.902E-02	-2.817E+01	-4.641E+01	3.902E+00	1.336E+02
1.656E+05	3.783E-02	-2.844E+01	-5.035E+01	3.783E+00	1.297E+02
1.815E+05	3.649E-02	-2.876E+01	-5.453E+01	3.649E+00	1.255E+02
1.991E+05	3.501E-02	-2.912E+01	-5.894E+01	3.501E+00	1.211E+02
2.183E+05	3.337E-02	-2.953E+01	-6.356E+01	3.337E+00	1.164E+02
2.393E+05	3.159E-02	-3.001E+01	-6.837E+01	3.159E+00	1.116E+02
2.624E+05	2.969E-02	-3.055E+01	-7.335E+01	2.969E+00	1.066E+02
2.877E+05	2.769E-02	-3.115E+01	-7.846E+01	2.769E+00	1.015E+02
3.155E+05	2.562E-02	-3.183E+01	-8.367E+01	2.562E+00	9.633E+01
3.459E+05	2.350E-02	-3.258E+01	-8.892E+01	2.350E+00	9.108E+01
3.793E+05	2.137E-02	-3.340E+01	-9.419E+01	2.137E+00	8.581E+01
4.159E+05	1.928E-02	-3.430E+01	-9.941E+01	1.928E+00	8.059E+01
4.560E+05	1.725E-02	-3.527E+01	-1.046E+02	1.725E+00	7.544E+01
5.000E+05	1.530E-02	-3.630E+01	-1.096E+02	1.530E+00	7.041E+01

```
OPTICAL ISOLATOR CIRCUIT

****     AC ANALYSIS                         TEMPERATURE =   27.000 DEG C

******************************************************************************

 FREQ         VDB(3)

  5.000E+03   1.329E+01
  5.482E+03   1.329E+01
  6.011E+03   1.329E+01
  6.591E+03   1.329E+01
  7.227E+03   1.329E+01
  7.924E+03   1.329E+01
  8.689E+03   1.329E+01
  9.527E+03   1.329E+01
  1.045E+04   1.329E+01
  1.145E+04   1.329E+01
  1.256E+04   1.328E+01
  1.377E+04   1.328E+01
  1.510E+04   1.328E+01
  1.656E+04   1.328E+01
  1.815E+04   1.327E+01
  1.991E+04   1.327E+01
  2.183E+04   1.326E+01
  2.393E+04   1.325E+01
  2.624E+04   1.325E+01
  2.877E+04   1.324E+01
  3.155E+04   1.322E+01
  3.459E+04   1.321E+01
  3.793E+04   1.319E+01
  4.159E+04   1.317E+01
  4.560E+04   1.315E+01
  5.000E+04   1.312E+01
  5.482E+04   1.309E+01
  6.011E+04   1.304E+01
  6.591E+04   1.299E+01
  7.227E+04   1.294E+01
  7.924E+04   1.286E+01
  8.689E+04   1.278E+01
  9.527E+04   1.268E+01
  1.045E+05   1.256E+01
  1.145E+05   1.242E+01
  1.256E+05   1.225E+01
  1.377E+05   1.206E+01
  1.510E+05   1.183E+01
  1.656E+05   1.156E+01
  1.815E+05   1.124E+01
  1.991E+05   1.088E+01
  2.183E+05   1.047E+01
  2.393E+05   9.992E+00
  2.624E+05   9.453E+00
  2.877E+05   8.847E+00
  3.155E+05   8.170E+00
  3.459E+05   7.421E+00
  3.793E+05   6.597E+00
  4.159E+05   5.701E+00
  4.560E+05   4.733E+00
  5.000E+05   3.696E+00
```

```
OPTICAL ISOLATOR CIRCUIT
****     AC ANALYSIS                          TEMPERATURE =   27.000 DEG C
******************************************************************************
LEGEND: *: IM(VRL)       +: IDB(VRL)       =: IP(VRL)       $: VM(3)         0: VP(3)

  FREQ         IM(VRL)
(*)----------    1.0000E-02   1.0000E-01   1.0000E+00   1.0000E+01   1.0000E+02
(+)----------   -4.0000E+01  -3.5000E+01  -3.0000E+01  -2.5000E+01  -2.0000E+01
(=)----------   -1.5000E+02  -1.0000E+02  -5.0000E+01  -7.1054E-15   5.0000E+01
($)----------    1.0000E+00   1.0000E+01   1.0000E+02   1.0000E+03   1.0000E+04
(0)----------    5.0000E+01   1.0000E+02   1.5000E+02   2.0000E+02   2.5000E+02
                       - - - - - - - - - - - - - - - - - - - - - - - - - - -
  5.000E+03  4.620E-02 .        X   .            .      0 +   =            .
  5.482E+03  4.620E-02 .        X   .            .      0 +   =            .
  6.011E+03  4.619E-02 .        X   .            .      0 +  =.            .
  6.591E+03  4.619E-02 .        X   .            .      0 +  =.            .
  7.227E+03  4.619E-02 .        X   .            .      0 +  =.            .
  7.924E+03  4.618E-02 .        X   .            .      0 +  =.            .
  8.689E+03  4.618E-02 .        X   .            .      0 +  =.            .
  9.527E+03  4.617E-02 .        X   .            .      0 +  =.            .
  1.045E+04  4.617E-02 .        X   .            .      0 +  =.            .
  1.145E+04  4.616E-02 .        X   .            .      0 +  =.            .
  1.256E+04  4.615E-02 .        X   .            .      0 +  =.            .
  1.377E+04  4.614E-02 .        X   .            .      0 +  =.            .
  1.510E+04  4.612E-02 .        X   .            .      0 +  =.            .
  1.656E+04  4.611E-02 .        X   .            .     0  +  =.            .
  1.815E+04  4.608E-02 .        X   .            .     0  + = .            .
  1.991E+04  4.606E-02 .        X   .            .     0 +  = .            .
  2.183E+04  4.603E-02 .        X   .            .     0 +  = .            .
  2.393E+04  4.599E-02 .        X   .            .     0 +  = .            .
  2.624E+04  4.595E-02 .        X   .            .     0 +  = .            .
  2.877E+04  4.590E-02 .        X   .            .    0  +  = .            .
  3.155E+04  4.584E-02 .        X   .            .    0  + =  .            .
  3.459E+04  4.577E-02 .        X   .            .    0  + =  .            .
  3.793E+04  4.568E-02 .        X   .            .    0  + =  .            .
  4.159E+04  4.557E-02 .        X   .            .   0   +=   .            .
  4.560E+04  4.544E-02 .        X   .            .   0   +=   .            .
  5.000E+04  4.529E-02 .        X   .            .   0   +=   .            .
  5.482E+04  4.511E-02 .        X   .            .  0    X    .            .
  6.011E+04  4.490E-02 .        X   .            .  0    X    .            .
  6.591E+04  4.464E-02 .        X   .            . 0   =+     .            .
  7.227E+04  4.434E-02 .        X   .            . 0   =+     .            .
  7.924E+04  4.398E-02 .        X   .            .0   =+      .            .
  8.689E+04  4.355E-02 .        X   .            .0   =+      .            .
  9.527E+04  4.305E-02 .        X   .            0    = +     .            .
  1.045E+05  4.246E-02 .        X   .           0.   =  +     .            .
  1.145E+05  4.178E-02 .        X   .          0 .   = +      .            .
  1.256E+05  4.099E-02 .       X    .          0 .  =  +      .            .
  1.377E+05  4.007E-02 .       X    .         0  . =  +       .            .
  1.510E+05  3.902E-02 .       X    .        0   .=   +       .            .
  1.656E+05  3.783E-02 .       X    .       0    =   +        .            .
  1.815E+05  3.649E-02 .       X    .      0    =.  +         .            .
  1.991E+05  3.501E-02 .       X    .    0     = . +          .            .
  2.183E+05  3.337E-02 .      X     .   0    =   .+           .            .
  2.393E+05  3.159E-02 .      X     .  0    =    +            .            .
  2.624E+05  2.969E-02 .      X     . 0    =    +.            .            .
  2.877E+05  2.769E-02 .     X      0     =   +  .            .            .
  3.155E+05  2.562E-02 .     X     0.   =   +    .            .            .
  3.459E+05  2.350E-02 .    X     0 .  =  +      .            .            .
  3.793E+05  2.137E-02 .    X   0   . = +        .            .            .
  4.159E+05  1.928E-02 .   X   0    = +          .            .            .
  4.560E+05  1.725E-02 .   X  0    X.            .            .            .
  5.000E+05  1.530E-02 .  X 0    += .            .            .            .
```

5.5.1 ELEMENT NODE TABLE

The first part of the analysis, following the input file listing is the element node table. The analysis of a node table was previously discussed in Section 3.0, but now has new data. The format remains the same, however, in the case of node 1, the reader is directed to the existence of a X1.VM component. This component is the voltage source VM contained within the subcircuit X1. The internal nodes are also shown. For example, node 6 within subcircuit X1 (X1.6) has the subcircuit components G1 (X1.G1) and Q1 (X1.Q1) connected to it. Following the node listing, is a listing of each solid state model and the parameters defined for use within that model.

5.5.2 CIRCUIT ELEMENT SUMMARY

The circuit element summary, as discussed in Section 3, is a listing of every circuit component, its value, node connections, and optional parameter definitions. For example, the current source IIN is connected between nodes 0 and 9, has a DC current value of 30 mA, and an AC current value of 100 mA with zero phase shift at the source.

5.5.3 DC TRANSFER CURVES

The DC transfer curves are shown next. The DC transfer curves are first shown as data, followed by plots. A high-resolution plot of IIN versus I(VRL) is shown in Figure 5-8 for purposes of clarity. The DC transfer curves plot of I(VRL) versus IIN is of interest. The plot shows a relatively nonlinear characteristic between the input current source and the output current sensor. If the internal components to the subcircuit are linear, then why should such a nonlinear plot exist? At first, the reader might be inclined to assume that the load resistor RL was causing the output-to-current limit. However, note that the collector voltage is 15.76-V DC when the input current is 100 mA, which indicates that the output transistor of the optical isolator cannot absorb the required amount of current. Ideally, the output voltage should be 20 V (IIN * gain * 100 ohms) or 15-V DC. The problem would seem to lie with the current handling ability of the output transistor rather than the current limiting capabilities of the load resistor. An inspection of the load resistor reveals a 100 Ω resistor driven by a 20-V DC (VCC) source. Simple math indicates that, if the transistor was saturating (VCE = 0V), then the available current through RL is 20 V/100 Ω = 200 mA. The problem is either the base drive to the transistor or the parameters controlling the high current gain of the device. The non linearity of the IIN versus I(VRL) curve can be seen in Figure 5-8. A plot of the actual response and the ideal linear response are shown together. An inspection of the internal optical isolator circuitry shows a linear base current as a function of input current. The problem lies in the transistor parameters and is most likely associated with the beta of the transistor. This does not suggest that the transistor is modeled incorrectly, but rather that this is the actual transistor characteristic that could be expected with an actual device. The reader is reminded that beta is not a linear function over a large range of base currents. What might be expected is a beta, which is low at low base currents, peaks at some midrange base current, and drops again at high base currents.

5.5.4 SMALL SIGNAL BIAS SOLUTION

This data is of little use to the realization of the stated analysis goals. Of interest is the circuit power consumption of 314 mW, which is significant for this type of device. The OP4N25 consumes the majority of the power with the load resistor RL consuming only 24.59 mW of the total power. The major consumption results because the output collector (node 3) is at 18.432 V with a current flow of 15.68 mA accounting for 298.41 mW of the power consumption. This is, again, because the output transistors inability to absorb the desired current.

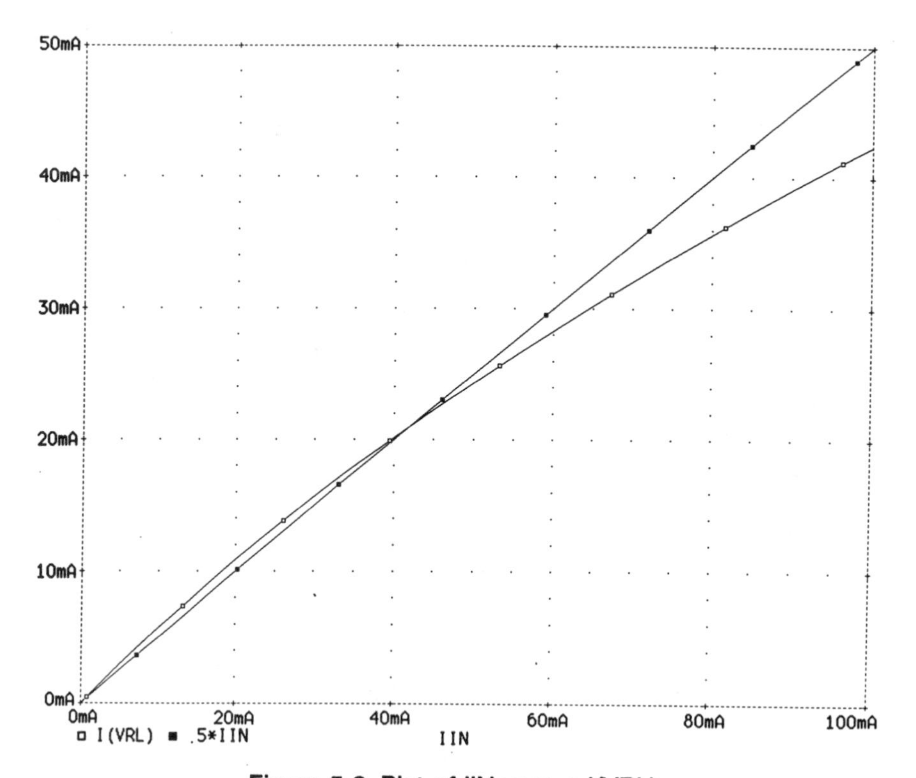

Figure 5-8. Plot of IIN versus I(VRL).

5.5.5 OPERATING POINT INFORMATION

Following the small signal bias solution, produced as a result of the .OP statement, is additional information on each dependent voltage/current source and solid state device in the circuit. Notice that the transistor X1.Q1, signifying transistor Q1 in subcircuit X1, has a base current of 99.9 μA and a collector current of 15.6 mA with an operating point BETADC of 156. Because the original specification of the transistor was for a beta of 150, this value is consistent with the expected beta.

As part of the operating point information, the small-signal characteristics are displayed. This data is produced by the .TF command. The data shows, at the operating point computed, a current gain of 0.4621, between IIN and I(VRL). Additionally, with 30 mA flowing into the anode of the optical isolator, a resistance of only 2.474 Ω is seen by the current source. The output collector resistance is 7.656 kΩ, reflecting the output current limiting that exists.

5.5.6 DC SENSITIVITY ANALYSIS

The DC sensitivity analysis is presented next. This analysis was a result of the .SENS statement in the input file. A review of the data reveals that the forward beta (BF) has a significant affect (normalized sensitivity A/ %) on the output. Notice also that a lesser affect is realized by the IKF parameter (corner for the forward beta high current rolloff). The value for BF is 13.76 mA/percent change in beta. If the beta was increased 100 percent, then the analysis suggests a 137.6 mA increase in the optical isolator collector current.

5.5.7 AC ANALYSIS

To aid in this discussion, high-resolution plots of IM(VRL), IP(VRL), and IDB(VRL) are provided in Figures 5-9, 5-10, and 5-11. A review of Figure 5-9 reveals that the output current amplitude [IM(VRL)] is not constant across frequency. The resulting output current (IM(VRL), as shown in the AC analysis section, is 46.2 mA at 500 Hz. The output remains approximately at this amplitude until the input frequency reaches approximately 16 KHz, where the output current amplitude decreases sharply. The 46.2 mA current is the current produced at 500 Hz when a 100-mA AC input current is applied by IIN. The attenuation at 500 Hz is shown to be -26.71 dB. This is computed by SPICE in the following way.

$$IDB = 20 * \text{Log}\ (46.2\ \text{mA})$$

The reader is directed to notice that the decibel value is not a calculation of gain, but rather a pure calculation of the output current magnitude. The 3-dB frequency of the output can be read directly from the IDB(VRL) data or computed based on IM(VRL). First, using the IDB, notice that the IDB attenuation at 500 Hz is -26.71 dB. If the 3 dB frequency was being sought, the frequency at which the IDB falls 3 more decibels must be determined. If the decibel value is essentially constant from 500 Hz to 16 KHz, then this must assumed to be the decibel value at or near DC.

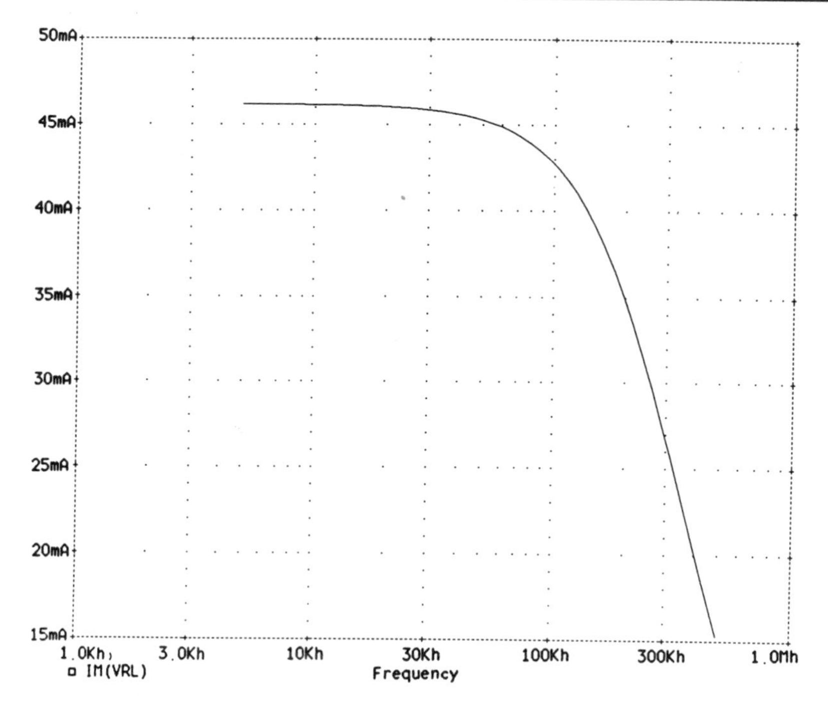

Figure 5-9. IM(VRL) versus frequency.

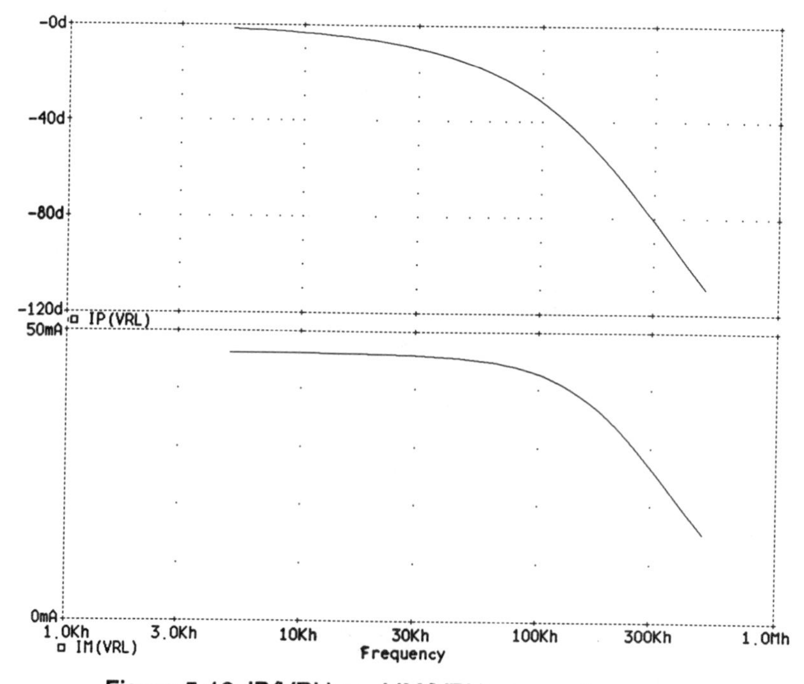

Figure 5-10. IP(VRL) and IM(VRL) versus frequency.

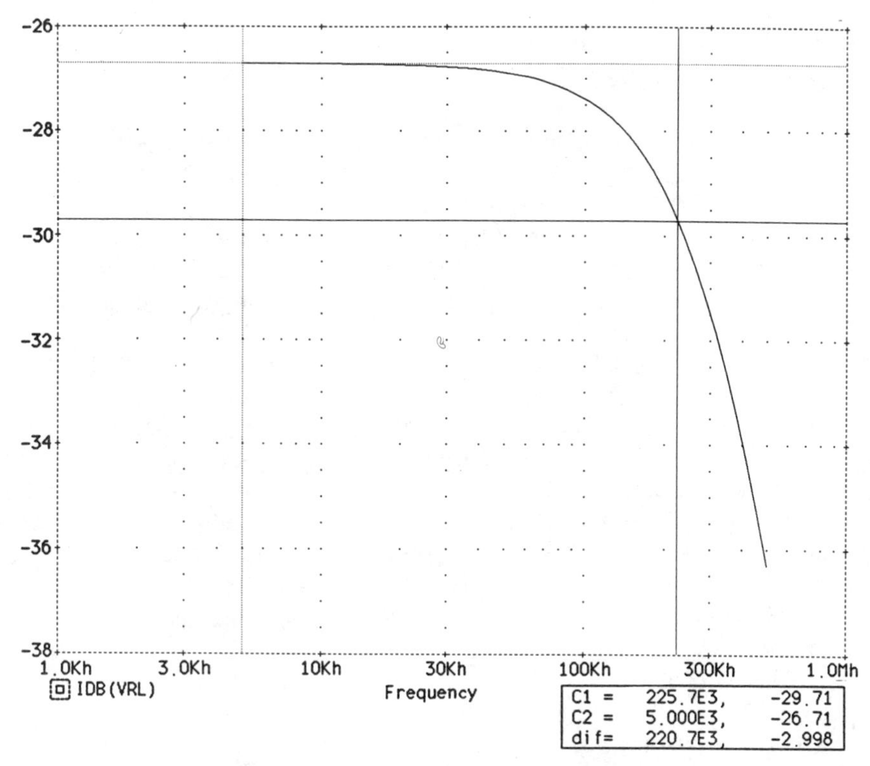

Figure 5-11. IDB(VRL) versus frequency.

If we subtract 3 dB from the -26.71 dB, the frequency at which the output is -29.71 dB is the 3-dB frequency. That decibel value occurs between 218.3 KHz and 239.3 KHz. Assuming a relatively linear relationship between computed points, the predicted 3-dB frequency is computed as

(239.3 KHz - 218.3 KHz)/(30.01 dB - 29.53 dB) = 43.75 KHz/dB

The desired decibel was -29.71 and the decibel at 218.3 KHz was -29.53 dB. The difference is 0.18 dB. Multiplying 0.18 dB * 43.75 KHz/dB = 7.875 KHz. Therefore, the actual 3-dB frequency is 218.3 KHz + 7.875 KHz = 226.175 KHz.

A similar calculation can be made using the fact that the 3-dB point of the output current occurs at 0.7071067812 * IM(VRL) at 500 Hz. Computing this value predicts a 3-dB amplitude of 32.67 mA. A review of the printed data reveals that this amplitude exists between 239.3 KHz and 218.3 KHz. These two frequencies are the same frequencies encountered in the decibel calculations above. The reader is directed to Figure 5-11. The figure is a high-resolution plot of IDB(VRL) versus frequency. The plot shows a cursor intersection at the 3-dB point on the output current AC waveform. The measured value, of the 3-dB point on the plot, is 225.7 KHz. The difference between the computed value, based on a linear assumption, and the measured value is 475 Hz or a 0.21-percent error. This small error indicated that linear extrapolation techniques may be used over small frequency intervals with little loss of accuracy.

The last two items of interest are the input-to-output current phase relationship and the slope of the output after the 3-dB point has occurred. The output phase, as seen in Figure 5-10, is in phase with the input as might be expected. In reviewing the AC analysis data, the phase of the voltage at node 3 V(3), is 180 degrees out of phase with the current. This should appear reasonable to the reader, as voltage should decrease as current increases, implying a 180 degree phase difference.

The slope of the IDB(VRL) roll-off characteristic is not directly obtainable from this data. To properly predict slope, the decibel change per decade should be graphically determined. Because the final frequency of the analysis is 500 KHz and half that frequency is 250 KHz, the decibel-per-octave variation can be mathematically extracted. The result, from the data, is approximately a 6 dB/octave slope. One decade contains approximately 3.325 octaves; the decibel-per-decade slope is 6 * 3.325 or approximately 20 dB/decade which is the expected roll-off slope for a single RC filter. This roll-off is created, within the optical isolator subcircuit, by C1 and R1 and the output transistor Q1.

5.6 SUMMARY

In this chapter we have looked at (1) the definition and usage of dependent sources, (2) the creation of subcircuits, and (3) the implementation of a combined DC and AC analysis using current sources. A current source analysis was selected to acquaint the reader with current sources and current sinks as opposed to the traditional voltage sources and voltage nodes. A summary of the SPICE commands presented in this section is included.

1. Dependent voltage controlled voltage source

   EXXXXXXX N+ N- NC+ NC- VALUE

2. Dependent voltage controlled current source

   GXXXXXXX N+ N- NC+ NC- VALUE

3. Dependent current controlled voltage source

   HXXXXXXX N+ N- VNAM VALUE

4. Dependent current controlled current source

   FXXXXXXX N+ N- VNAM VALUE

5. Subcircuit definitions

```
.SUBCKT subnam N1 [N2 N3 ...]
****** subcircuit data ******
.ENDS [subnam]
```

6. Subcircuit calls

```
XYYYYYYY N1 [N2 N3 ...] SUBNAM
```

TRANSIENT ANALYSIS

Transient analysis, which was introduced in section 1.3.4 is time dependent compared to the AC analysis, which is frequency dependent. Stated another way, the transient analysis deals with a single frequency and the AC analysis deals with a range of frequencies. Because the transient analysis deals with a single frequency, the effects of that frequency, on the circuit, can be seen as a function of time. On an oscilloscope, the horizonal display is in time. The same is true for the transient analysis. As such, the plot of transient analysis data resembles an oscilloscope display.

DC and AC analyses require three basic elements to function:

1. An input source to stimulate the circuit being evaluated.

2. A definition of a sweep range to bound the analysis.

3. A definition of the output voltage or current to be evaluated.

The transient analysis requires elements 2 and 3, but it may not require an input source to stimulate the circuit. Most dynamic circuits have an output which is derived from an input. One exception to this lies with the oscillator. An oscillator requires no stimulus source, as the circuit is self-stimulating and thus produces an output independent of an input. In this chapter, the various stimuli that can be produced with a transient analysis are described and the use of each is demonstrated.

6.1 TRANSIENT ANALYSIS VOLTAGE AND CURRENT SOURCES

Before looking at the stimuli, it is necessary to expand the scope of the voltage and current source definition. Previously, the reader has been exposed to the definition of a DC source, an AC source, and the combination of the two. The general form of the expanded source is shown below.

```
VXXXXXXX N+ N- [DC] VALUE [ACMAG [ACPHASE]]
                     [PULSE V1 V2 [TD [TR [TF [PW [PER]]]]]]
          or         [SIN VO VA [FREQ [TD [THETA]]]]
          or         [EXP V1 V2 [TD1 [T1 [TD2 [T2]]]]]
          or         [PWL T1 V1 T2 V2 ... TN VN]
          or         [SFFM VO VA FREQ [MDI [FS]]]

IXXXXXXX N+ N- [DC] VALUE [ACMAG [ACPHASE]]
                     [PULSE I1 I2 [TD [TR [TF [PW [PER]]]]]]
          or         [SIN IO IA [FREQ [TD [THETA]]]]
          or         [EXP I1 I2 [TD1 [T1 [TD2 [T2]]]]]
          or         [PWL T1 I1 T2 I2 ... TN IN]
          or         [SFFM IO IA FREQ [MDI [FS]]]
```

where again N+ and N- are the positive and negative nodes, DC specifies a DC component of the source and AC specifies an AC component of the source, ACMAG is the AC magnitude value of the source, ACPHASE is the AC phase value of the source, PULSE is the pulse function of transient analysis, SIN is the sin function of transient analysis, EXP is the exponential function of transient analysis, PWL is the piece-wise linear function of the transient analysis, and SFFM is the single frequency FM function of the transient analysis.

Before discussing examples of the various applications of the transient functions, a discussion of the specific parameter definitions of each function is appropriate. The following sections will discuss each of the five waveform functions in detail and an example will demonstrate graphically the effect each parameter has on the waveform creation.

6.1.1 PULSE FUNCTION

The PULSE function, as the name implies, allows the user to define a pulse and it's characteristics. Any pulse has seven basic characteristics.

1. Initial amplitude

2. Pulsed amplitude

3. Delay to the start of the pulse

4. Rise time

5. Fall time

6. Pulse width

7. Period (1/period = frequency)

The pulse definition capability of SPICE fully supports these seven characteristics. The general form of the PULSE function is shown below. V1 and V2 are mandatory parameters with the balance being optional. If optional parameters are desired, all parameters must be input in the sequence designated.

PULSE (VI VP TD TR TF PW PER)

The parameters, default values, and associated parameter units are shown in Table 6-1. The relationship of each of the parameters to the generated waveform is shown in Figure 6-1.

Table 6-1. PULSE Definition Parameters

Parameters	Default Values	Units
VI (initial value)		V or A
VP (pulsed value)		V or A
TD (delay time)	0.0	sec
TR (rise time)	TSTEP	sec
TF (fall time)	TSTEP	sec
PW (pulse width)	TSTOP	sec
PER (period)	TSTOP	sec

A single pulse, so specified, is described by the following table:

Time	Value
0	V1
TD	V1
TD+TR	V2
TD+TR+PW	V2
TD+TR+PW+TF	V1
TSTOP	V1

Intermediate points are determined by linear interpolation.

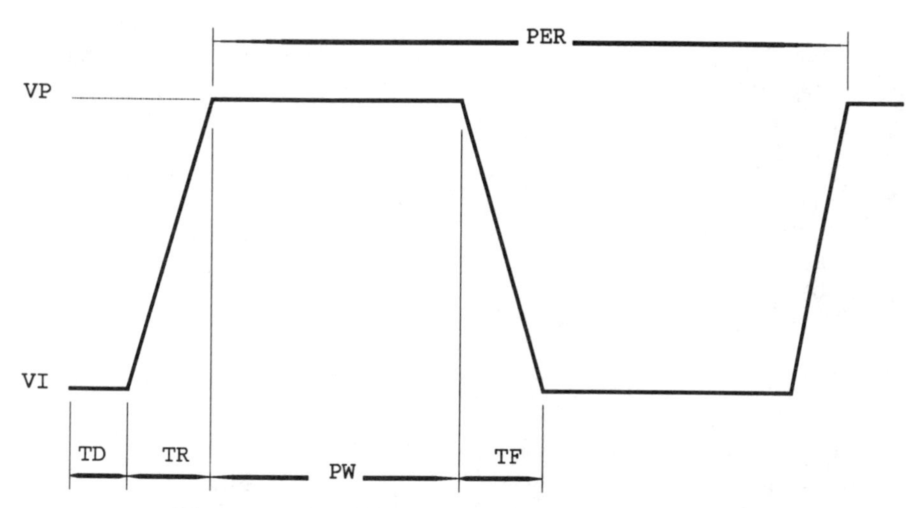

Figure 6-1. PULSE function parameter definitions.

Assume, for example, that a pulse with the following characteristics were required to stimulate a circuit.

Initial value	=	0 V
Pulsed value	=	5 V
Startup delay	=	1 μsec
Rise time	=	0.5 μsec
Fall time	=	1 μsec
Pulse width	=	2 μsec
Period	=	7 μsec

Such a pulse is assumed to be associated with an independent voltage source named VP that is connected between node 1 and ground. The pulse could be describe as follows:

```
VP 1 0 PULSE(0 5 1U 1U 2U 5U 10U)
```

A graphic representation of the pulse appears in Figure 6-2. Note that the pulse starts at time 0 μsec, with a 0 V amplitude and retains this amplitude for 1 μsec. The starting amplitude is a result of the value of VI and the delay from 0 μsec to 1 μsec is a result of TD. Under most circumstances, the value of TD will be set to 0. Why should TD be used at all? The most frequent application of TD is in applying several waveforms to a circuit. Assume you were modeling a circuit that required verification of the performance of a digital-to-analog converter. If the digital-to-analog converter was a 8-bit device, then eight separate waveforms could be developed with specified time offsets which simulate a true binary progression. The affect would be the simulation of an eight binary counter to the digital-to-analog input, which would ideally provide a linear staircase analog output. Other uses might be on a circuit that contained two one-shots, in series, triggered by a single start pulse. By knowing the time relationship from the start pulse to each of the one-shot outputs, a delay factor (TD) and three separate PULSE functions could be used to simulate the start pulse and the two one-shot outputs.

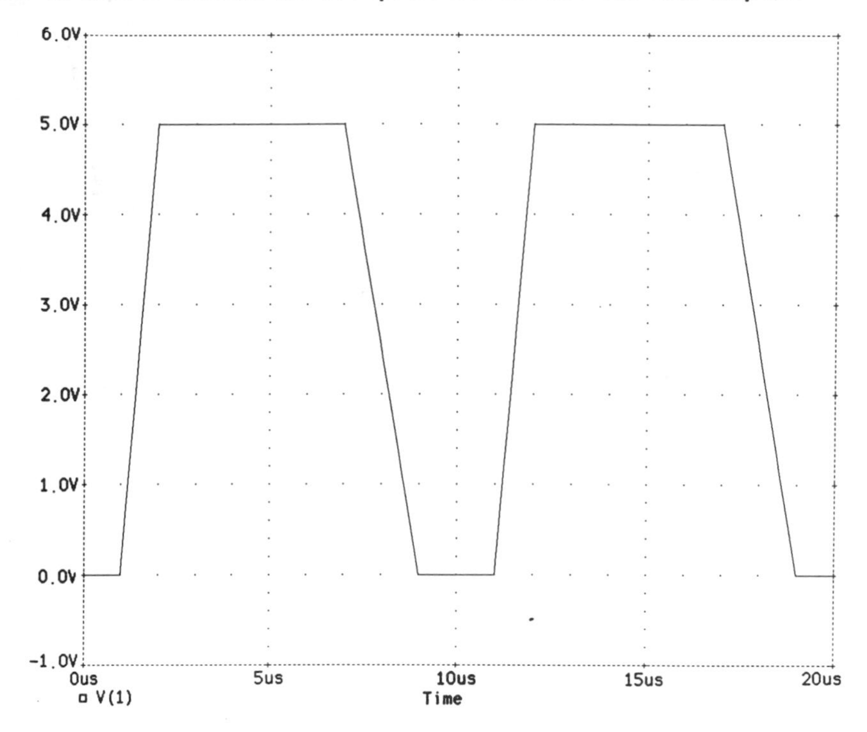

Figure 6-2. PULSE(0 5 1U 1U 2U 5U 10U) example.

Continuing, at 1 μsec, the pulse rises to 5 V in 1 μsec. The 1 μsec rise time is a result of TR and the 5 V is a result of VP. The pulse remains at 5 V from 2 μsec to 7 μsec. This 5 μsec time at 5 V is a result of the PW parameter. The pulse falls from 5 V to 0 V between 7 μsec and 9 μsec. This 2 μsec fall time is a result of the TF parameter. The pulse stays at 0 V until 11 μsec when it again rises. The time from the end of the delay (1 μsec) and the next full cycle is (11 μsec - 1 μsec) 10 μsec. The 10 μsec is defined by the PER parameter.

Several features of the PULSE function are not obvious but are significant:

1. The period of the pulse is independent of any other parameters assuming that the summation of TR, TF, and PW is not greater than PER.

2. The pulse width is defined differently in SPICE than the conventional pulse width definition (50 percent point of the pulse). Because of this, unless the TR and TF are infinitely fast or special provisions made, a 50 percent duty cycle will never be achieved.

3. TR and TF should use a minimum value of PW/1000 or 0.1 nsec, whichever is less. This allows a fast pulse definition to occur without severely hampering the transient analysis. A value of 0 is allowable for TR and TF but is not recommended.

4. The pulse is typically repetitious and represents a single frequency (frequency = 1/period). If a single pulse is desired, starting at VI and rising permanently to VP, then the pulse width should be set to 100 and the period to 101. Without units, SPICE assumes that 100 represents a pulse width of 100 sec.

5. If a pulse is connected, with the nodes reversed, the value of VP is inverted, TR becomes TF, and TF becomes TR. Additionally, PW becomes PERIOD - (TR + TF + PW). The reader should spend some time on this concept to become convinced.

6. VI and VP may be any value. There is nothing to prevent VI from being greater than VP.

The PULSE function can also describe a triangular pulse or a sawtooth. A triangular waveform could be described as follows:

```
PULSE(-1 1 0 2U 2U 1P 4U)
```

This definition will produce a triangular pulse which is symmetrical about 0-V DC with a 2-V peak-to-peak amplitude. The summation of the rise and fall times is equal to the period. The reader is directed to notice that the value of the pulse width is not zero but rather a very small value. This is essential to assure that proper waveform generation is achieved. The graphical result of this pulse definition is shown in Figure 6-3.

A sawtooth pulse is a special form of a triangular pulse. A sawtooth pulse could be generated with the following definition.

```
PULSE(-1 0 0 2U 1P 1P 2U)
```

The resulting waveform is shown in Figure 6-4. The reader should again notice that the value of the pulse width and fall time are small but enter into the overall waveform generation mathematics within SPICE. The reader should also notice that the sawtooth never reaches a value of 0 V because of the existence of a period that is 2 μsec instead of 2.000002 μsec. Care should be exercised in specifying specialized waveforms using the pulse function.

The general form of the PULSE function, as shown in Section 6.1, requires only the specification of VI and VP. If no other parameters are specified, the pulse will start at VI, rise to VP in the <u>step size</u> of the transient analysis (see Section 6.2), and stays at the VP value for the duration of the analysis.

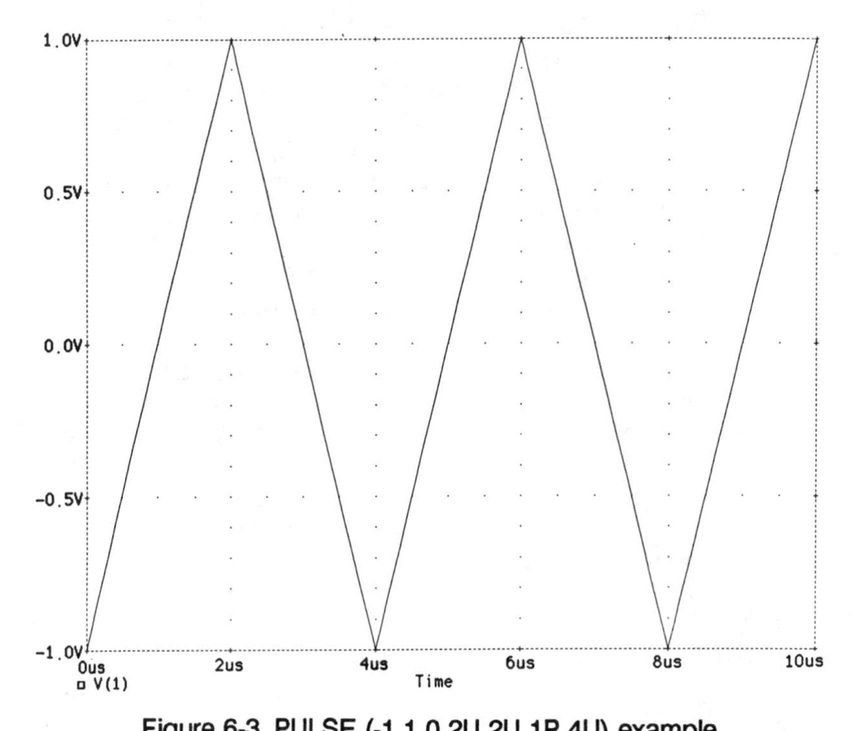

Figure 6-3. PULSE (-1 1 0 2U 2U 1P 4U) example.

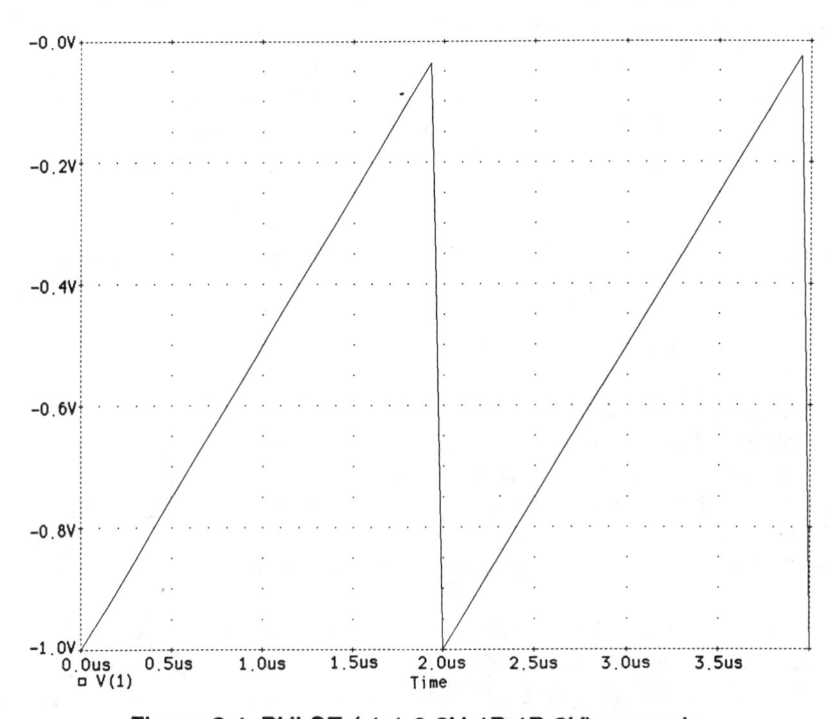

Figure 6-4. PULSE (-1 1 0 2U 1P 1P 2U) example.

A possible example of this application would be a pulse definition as follows:

VIN 1 0 PULSE(-2 5)

The result of this pulse definition is shown in Figure 6-5. The step size, in this example, was 0.1 μsec. Note that the rise time is 0.1 μsec and the pulse becomes a step function that starts at -2.0 V and rises to 5 V in 0.1 μsec and remains at 5 V for the duration of the time.

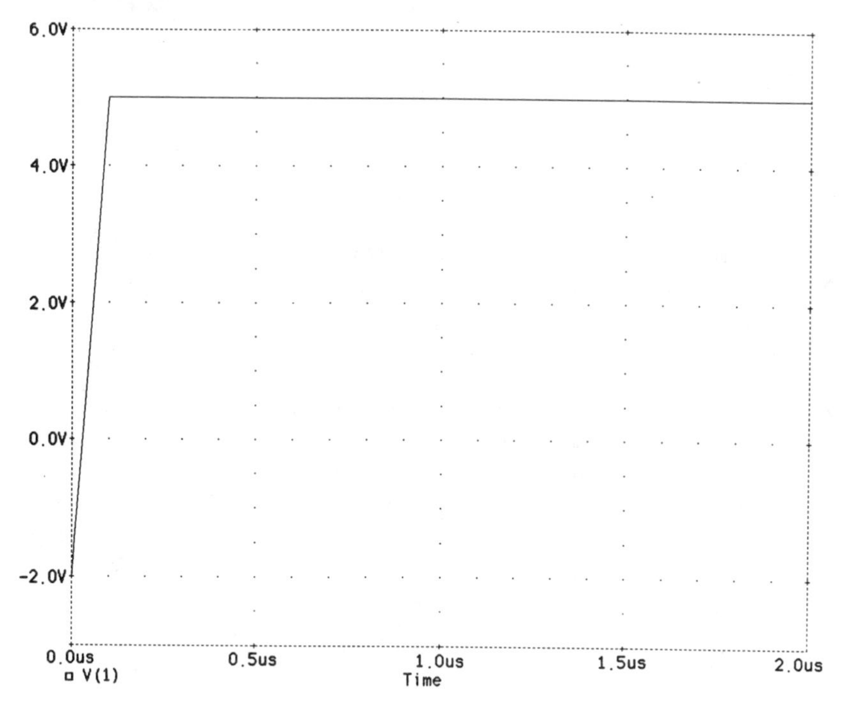

Figure 6-5. PULSE(-2 5) example.

In summary, the special characteristics of the PULSE function are as follows:

1. The function represents a single-frequency square wave under normal use.

2. All PW values must be a positive (nonzero).

3. TR and TF may have any nonnegative value, but, for purposes of analysis speed, the values should be as large as possible.

4. The only required parameters for a pulse definition are VI and VP.

6.1.2 SIN FUNCTION

The SIN function represents a fixed-frequency sinusoidal wave. A sinusoidal wave has the following definable characteristics:

1. A DC offset
2. A peak amplitude (peak-to-peak symmetric about the DC offset)
3. A startup delay
4. A specified frequency
5. An optional exponential rise or decay in the envelope amplitude

These parameters, like the PULSE function parameters, are definable in SPICE. The general form of the SIN function is as follows:

SIN (VO VA FREQ TD THETA)

The parameters, default values, and associated parameter units are shown in Table 6-2. The shape of the waveform is described by the following table:

TIME	VALUE
0 to TD	VO
TD to TSTOP	VO + VA*exp $^{-(\text{time}-\text{TD}) * \text{THETA}}$ *sin[2*π*FREQ*(time–TD)]

Table 6-2. SIN Definition Parameters

Parameters	Default Value	Units
VO (offset)		V or A
VA (amplitude)		V or A
FREQ (frequency)	1/TSTOP	Hz
TD (delay)	0.0	sec
THETA (damping factor)	0.0	1/sec

The sinusoidal wave parameters are shown in Figure 6-6. The sinusoidal waveform consists of a DC offset value of VO. Typically, the DC offset of a sinusoidal waveform is 0-V DC (ground) with the waveform having a symmetric amplitude above and below ground. The value of VO can be any value desired and may serve to simulate the DC bias condition required on the input of a direct coupled amplifier. The parameter VA is the peak amplitude of the waveform above and below the value of VO. Because of this relationship, the peak-to-peak amplitude is twice the value of VA. The positive peak amplitude is therefore VO + VA and the negative peak amplitude is VO - VA. The parameter FREQ is the frequency of the sinusoidal waveform. The start of the waveform may be delay by using a value, other than zero, for TD. The existence of a TD parameter delays the start of the sinusoidal waveform or, more simply, provides a phase shift of the sine wave. The frequency is not affected by the value of TD. The last parameter is THETA whose value creates an exponential decay time constant to control the amplitude of the sinusoidal waveform. An example of this will be shown in the following.

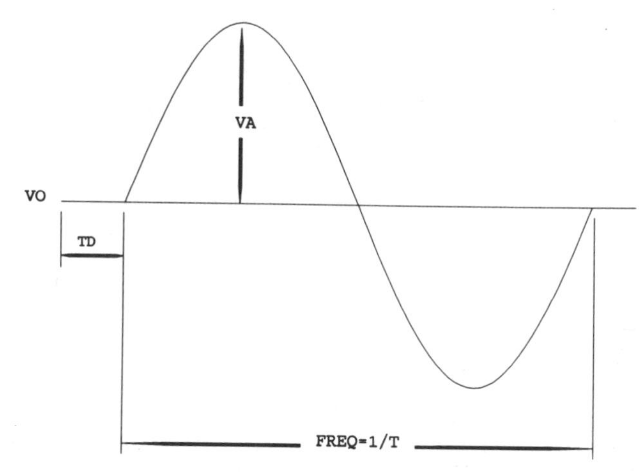

Figure 6-6. SIN function parameter definition.

The parameters VO and VA may be any value. The remaining parameters should have nonnegative values. The example of a sine wave definition is shown in Figure 6.1-2. Assume, for example, that we desire to define a sinusoidal wave with a source name VS, which is connected from node 2 to ground. The sinusoidal wave has a 2-V DC offset with a 2-V peak amplitude, a frequency of 1 KHz, and a 100 μsec delay before startup. Such a SIN function could be described as follows.

```
VS 2 0 SIN(2 2 1K .1M 0)
```

Figure 6-7 depicts the results of such a waveform description. In Figure 6.1 and 6.2, the sinusoidal wave starts at 0 μsec at a 2-V DC level (offset). Typically the value of VO is 0 V as it is normally desired to have a sinusoidal wave symmetrical about 0 V. The 2-V DC level is maintained for 100 μsec. Following the 100 μsec delay, the sinusoidal wave starts to rise.

The 100 μsec is the delay time TD and is normally set to 0. The TD value may be used when interfacing two or more waveforms to a circuit. Assume the user desired to produce a sine/cosine function. One sinusoidal wave could be defined with a TD of 0, which would be the sinusoidal wave. The second could be defined with a TD = (1/frequency) * (¼). Since a cosine function is 90 degrees out of phase, 90 degrees is ¼ of 360 degrees or, in this example, 1.25 μsec. This sine/cosine function is valuable in checking out a variety of phase-sensitive circuits. Such a circuit might be a PSK (phase shift keying) circuit for a computer modem.

At the end of TD (1 μsec), the sinusoidal wave raises to 4 V (2-V DC offset + 2 V peak) and then drops to 0 V (2-V DC offset - 2 V peak). The reader should note that the peak-to-peak amplitude is always 2 * VA and the RMS value is 0.707 * VA. The cycle of the sinusoidal wave begins at 0.1 msec and ends at 1.1 msec. The period is, therefore, 1 ms. The frequency is 1/period = 1/1 msec = 1 KHz, which is the specified frequency.

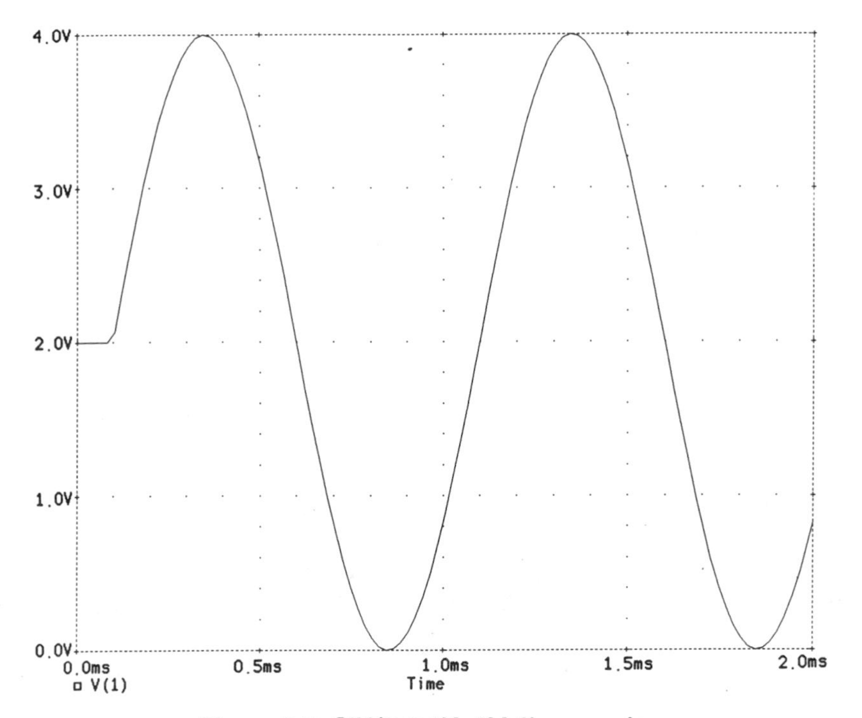

Figure 6-7. SIN(2 2 1K .1M 0) example.

The THETA value is rarely used in most analysis circuits. A review of the equations shown above, that describe the sinusoidal wave, reveal that a positive value of THETA will create an exponentially decaying waveform and a negative value will create an exponentially increasing waveform. The frequency remains constant and only the amplitude is affected. The use of a negative value of THETA produces unusual amplitude results as depicted by the equation. Normally THETA is set to a value of 0. If it is desired to simulate, for example, the decaying waveform of a parallel LC tank circuit, then the use of THETA is appropriate instead of actually modeling the LC circuit. In SPICE, if a circuit stimulus can be accurately modeled using a waveform function, then no advantage is gained by creating the stimulus by physical circuit modeling. An example of the use of THETA could be demonstrated with the following definition of a sinusoidal waveform.

SIN(0 1 1K 0 300)

The result of this sinusoidal waveform definition is shown in Figure 6-8. The peak amplitude of the waveform drops 63.2 percent (one time constant) in approximately 3.33 msec. The value of THETA is the reciprocal of one time constant or 300.

A sinusoidal wave may start toward the negative direction by reversing the node connections or may realize a 180-degrees phase shift by setting TD = (1/frequency) * 0.5. The specification of FREQUENCY, TD, and THETA are optional. If TD and THETA are unspecified then their values will default to 0.

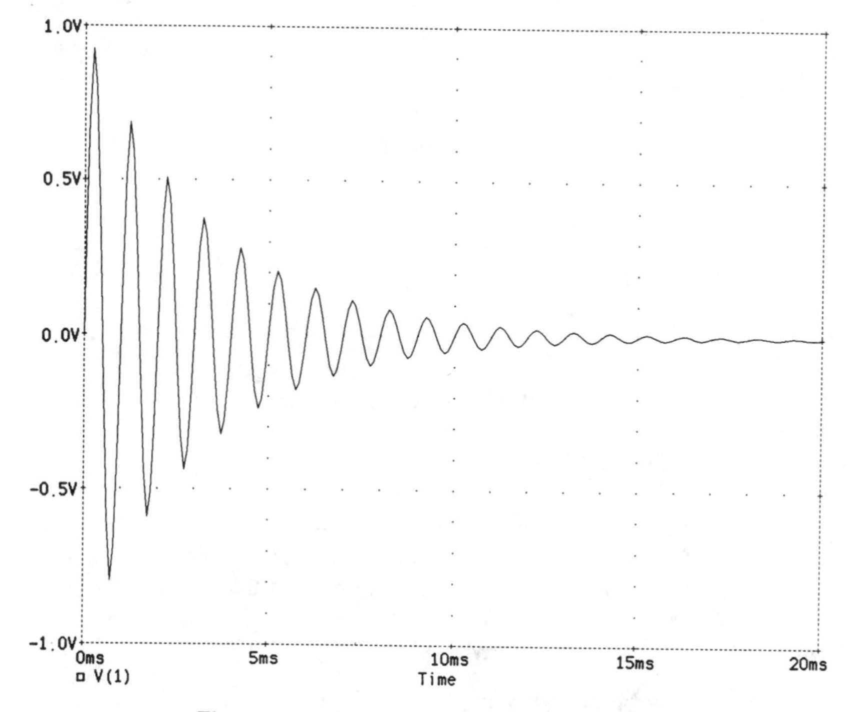

Figure 6-8. SIN(0 1 1K 0 300) example.

If FREQUENCY, TD, and THETA are all unspecified and allowed to default then the sinusoidal wave produced will have a period that is the duration of the analysis (see TSTOP in Section 6.2). An example of this could be as follows.

SIN(0 1)

The resulting waveform is shown in Figure 6-9. The period is shown to be 2 msec which is the duration of the analysis. This form could prove useful when multiple cycles of a sinusoidal waveform are not required for the analysis such as the case of a direct coupled amplifier circuit with a direct coupled output.

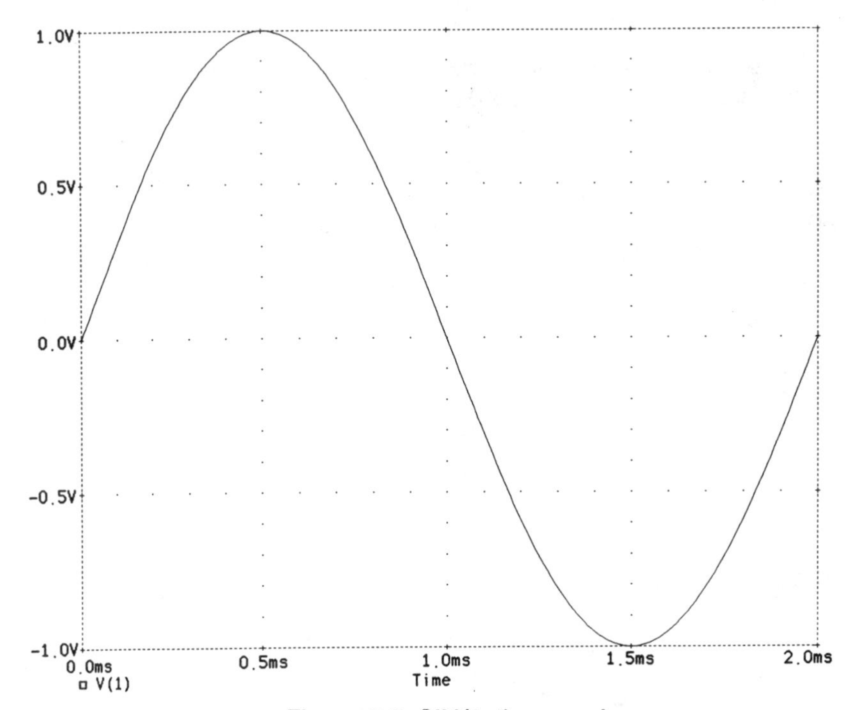

Figure 6-9. SIN(0 1) example.

6.1.3 EXPONENTIAL FUNCTION

The exponential function provides, to the user, a waveform that represents an RC charge/discharge voltage function or an RL charge/discharge current function. The waveform is useful in simulating RC affects without the need to define a circuit to achieve the desired waveform. The general form of the exponential function is as follows with the parameters shown in Table 6-3.

EXP (V1 V2 TD1 T1 TD2 T2)

The equations, associated with the waveform generation, are discussed below. The shape of the waveform is described by the following table:

TIME	VALUE
0 to TD1	V1
TD1 to TD2	$V1+(V2-V1)*(1-exp^{-(time-TD1)/T1})$
TD2 to TSTOP	$V1+(V2-V1)*(1-exp^{-(time-TD1)/T1})$ $+(V1-V2)*(1-exp^{-(time-TD2)/T2})$

Table 6-3. EXP Definition Parameters

Parameters	Default Values	Units
V1 (initial value)		V or A
V2 (pulsed value)		V or A
TD1 (rise delay time)	0.0	sec
T1 (rise time constant)	TSTEP	sec
TD2 (fall delay time)	TD1+TSTEP	sec
T2 (fall time constant)	TSTEP	sec

The above equation is of the standard RC time constant VA * $e^{-T/RC}$, where T is the difference between the constants TD1 or TD2 and any specified time. RC is the time constant for T1 or T2. The exponential waveform is shown in Figure 6-10. The parameters that make up the DC portion of the waveform definition are V1 and V2. V1 is the starting voltage and V2 is the ending voltage, assuming that the number of time constants associated with the rising edge of the waveform are allowed to approach infinity. The value of TD1 is the time the waveform begins to rise from V1 toward V2 and T1 is the RC time constant associated with that rise. TD2 is the point in time when the waveform starts to decay at a rate defined by the RC time constant T2. The reader is cautioned that the value of T1 defines the rise time of the exponential waveform. If sufficient time constants, as defined by T1, are not allowed to occur due to the time value of TD2 compared to T1, the waveform will never achieve the value of V2. This introduces some level of complexity in achieving the desired waveform without some iterative trials. By rearranging the above equations, a value of T1 can be computed based on certain criteria. Assume that TD1, TD2, V1, and V2 are known and the voltage (VY) desired at time TD2 is also known. The following equation can be used to compute the value of T1 to meet VY voltage level.

$$T1 = (TD1-TD2)/[\ln((V2-VY)/(V2-V1))]$$

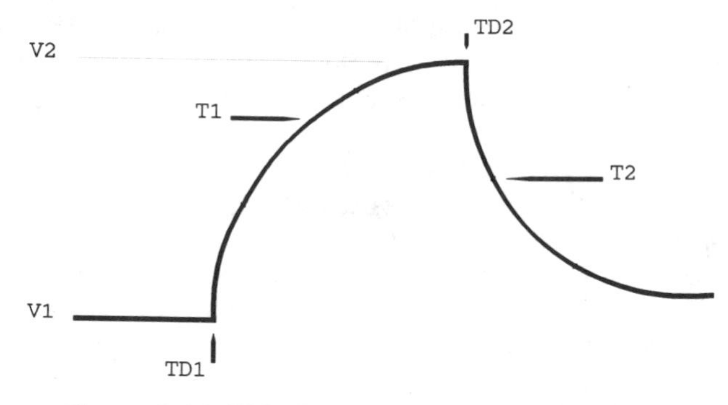

Figure 6-10. EXP Function Parameter Definition

An example of an exponential waveform definition is shown in Figure 6-11. The waveform shown has been designated as a voltage source with the name of VEXP and is connected between node 3 and ground. The waveform has a starting voltage V1 of 0 V, a final voltage of 5 V (V2), a 1 μsec delay (TD), a rise time RC time constant of 0.7 μsec (T1), a fall time start of 2 μsec (TD2), and a fall time RC time constant of 0.3 μsec (T2). The waveform described could be specified as follows:

VEXP 3 0 EXP(0 5 1U .7U 2U .3U)

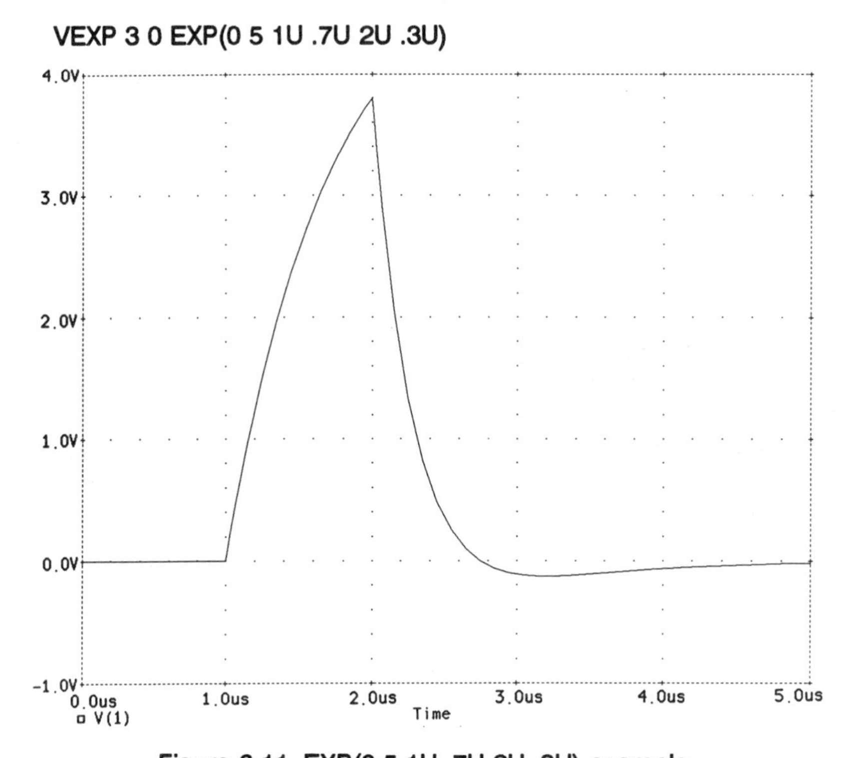

Figure 6-11. EXP(0 5 1U .7U 2U .3U) example.

The waveform of Figure 6-11 rises as expected but, because of the value of TD2, fails to reach a 5 V level. The waveform also achieves a voltage of approximately V1 in the vicinity of 3 μsec. This negative voltage is consistent with the equations shown previously. This should provide further motivation for the reader to carefully evaluate the desired waveform response based on equations used by SPICE.

The specification of V1 and V2 is mandatory with the balance optional. If only T1 and T2 are specified, the value of TD1 is 0, the value of T1, T2, and TD2 are the analysis step size (see Section 6.2). Therefore the waveform would rise from V1 to V2 * 63.21 percent (one time constant) and then return to T1 in one time constant. If the step size were 1 μsec, using the values of V1 and V2 of the previous example, the waveform definition could be as follows.

VEXP 3 0 EXP(0 5 0 1U 1U 1U)

The same results could be achieved with the following waveform definition and are shown in Figure 6-12.

VEXP 3 0 EXP(0 5)

A final observation is that exponential waveforms, in SPICE, are <u>not</u> cyclic. Only one waveform cycle will be produced per voltage source per analysis. If more than one are desired, then multiple voltage sources, with carefully spaced TD1s, could be resistively summed together.

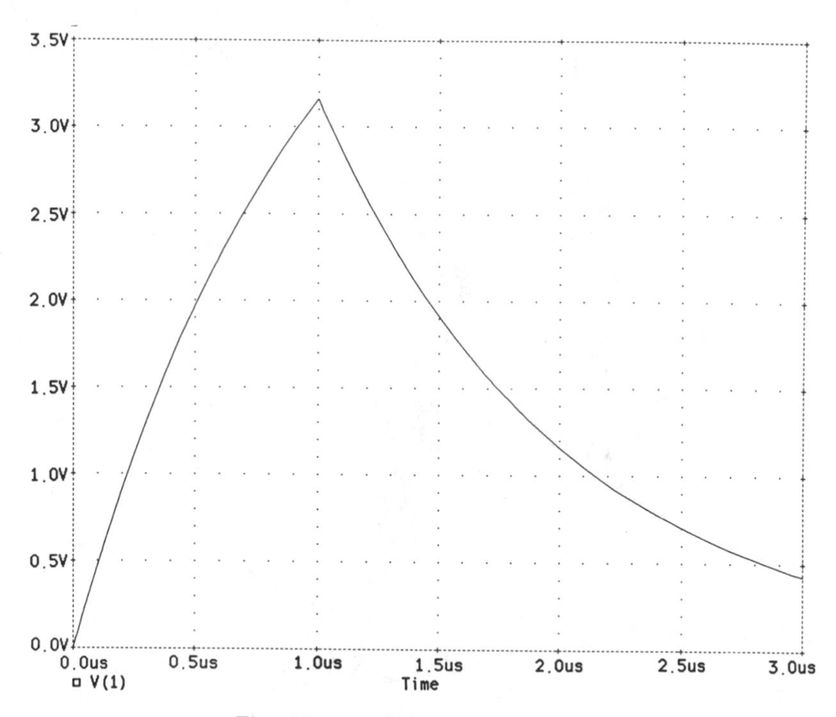

Figure 6-12. EXP(0 5) example.

6.1.4 PIECE-WISE LINEAR FUNCTION

The piece-wise linear (PWL) function is a highly powerful and versatile waveform definition. The PWL capability can be made to replicate any of the waveforms discussed in this section. Furthermore, the PWL can be made to simulate any waveform desired. To realize this capability, an ability to define time and voltage or time and current must exist. Any waveform has, at any given time, a specified voltage or current value. The PWL is structured to permit this data duality to exist. The general form is shown below:

PWL (T1 V1 [T2 V2 T3 V3 T4 V4 ...])

Each pair of values (Ti, Vi) specifies that the value of the source voltage is Vi at time Ti. The value of the source at intermediate values of time is determined by linear interpolation between subsequent pairs of input values. The generalized waveform of the PWL is shown in Figure 6-13. The figure shows only seven pairs of data. No limit, on the number of data pairs exists in SPICE. If data exceeds the length of the current input data line, the user may extend the line by the use of a "+" in the first column of the following line. This process is identical to the process used in transistor model descriptions.

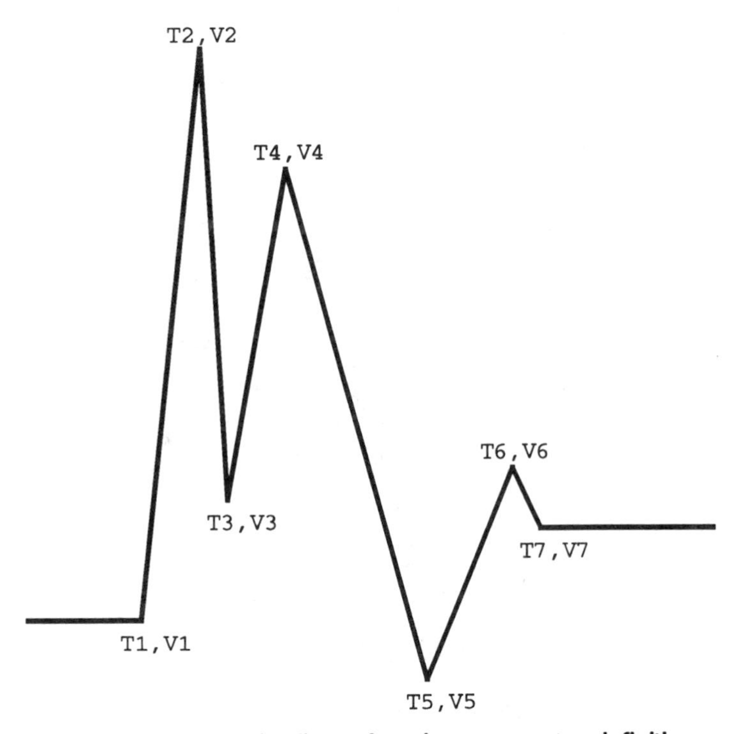

Figure 6-13. Piece-wise linear function parameter definition.

If we were to describe a PWL waveform as shown in Figure 6-14, the following information can be extracted about the waveform:

- At time T1 = 0 sec, the voltage V1 = 0 V
- At time T2 = 1 sec, the voltage V2 = 1000 V
- At time T3 = 2 sec, the voltage V3 = -500 V
- At time T4 = 3 sec, the voltage V4 = 700 V
- At time T5 = 4 sec, the voltage V5 = -100 V
- At time T6 = 5 sec, the voltage V6 = 100 V

and, implied At time T7 = infinity, the voltage V7 = 0 V

The above waveform could be described as follows and is graphically depicted in Figure 6-14.

VPWL 4 0 PWL(0S 0 1S 1000 2S -500 3S 700 4S -100 5S 100 6S 0)

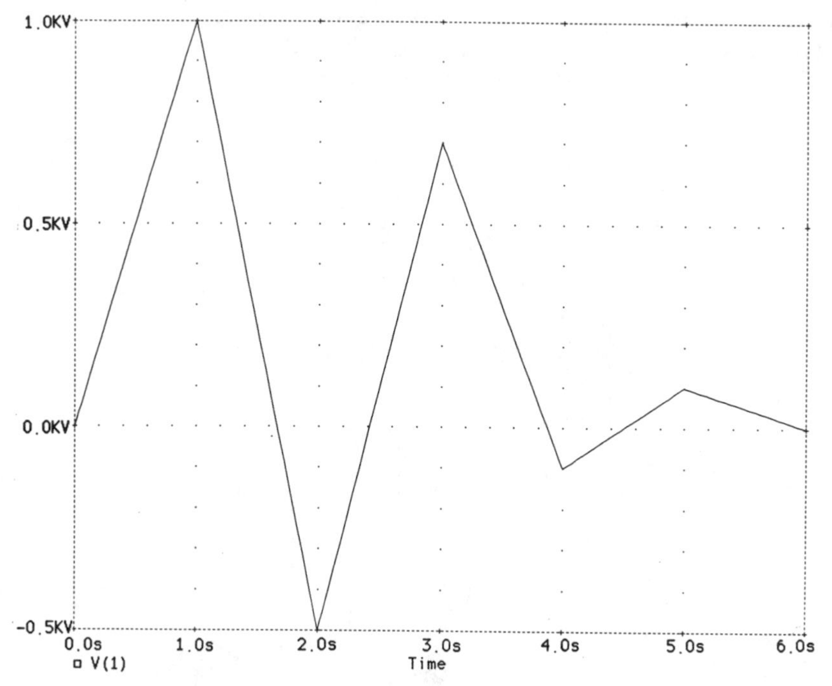

Figure 6-14. PWL(0 0 1 1K 2 -500 3 700 4 -100 5 100 6 0) example.

The following statements focus on the main attributes of a PWL waveform and should be carefully considered when working with this function.

1. The waveform is not repetitious. If the duration of the analysis exceeds the last value specified of Ti, then the last value specified of Vi will be used and a flat horizontal waveform will be produced.

2. Since linear interpretation is used between points, only transition points, where the slope of Vi is changing to a new value, need be specified.

3. There is no limit to the number of time/voltage pairs that may be specified. If the pairs exceed approximately 70 columns, the subsequent line extensions may be realized using the + extension as discussed previously.

4. The time value for each subsequent pair of data must reflect a time greater than in the previous pair. Values of time cannot be negative.

An example of a variable frequency waveform is depicted in Figure 6-15. The figure consists of a variable-frequency square wave. Such a definition might be used to verify a digital frequency discriminator circuit, a phase lock loop response, on any circuitry where a variable-frequency waveform was required. The reader is reminded that, while any waveform can be simulated using a PWL function, the waveform is not cyclic unless the necessary data points are defined to allow the necessary repetition over the desired time interval.

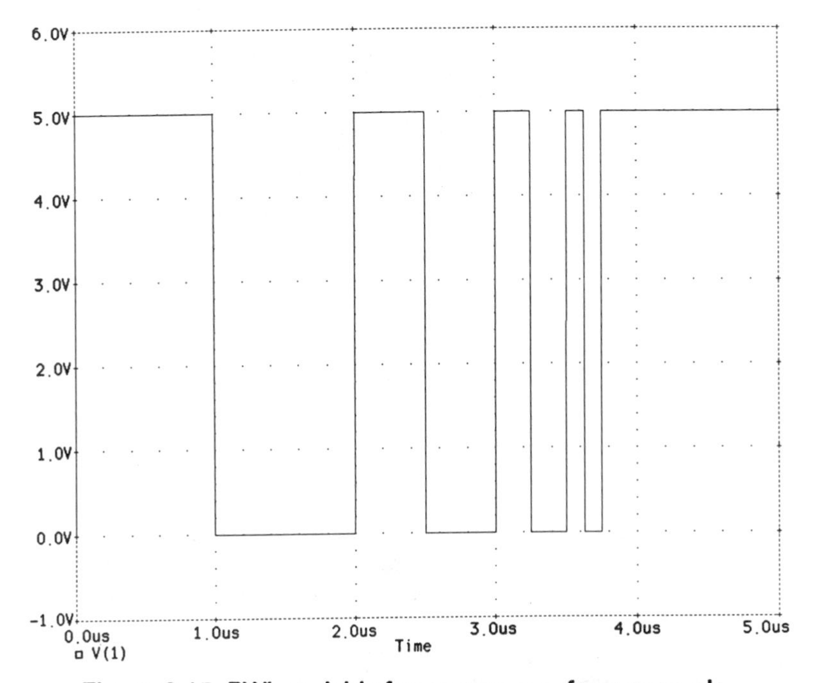

Figure 6-15. PWL variable frequency waveform example.

6.1.5 SINGLE-FREQUENCY FM

The last of the five waveform functions is the SINGLE FREQUENCY FM. An FM waveform consists of a DC offset, amplitude, carrier frequency, modulation index, and a signal frequency. These FM waveform components can be specified in SPICE. The result is a single modulating frequency FM signal suitable for verifying FM detector and filter circuitry. In general applications of SPICE, this waveform is rarely used due to it's highly specialized nature. The general format of the waveform definition is shown below with the corresponding parameters in Table 6-4.

SFFM (VO VA FC MDI FS)

Table 6-4. EXP Definition Parameters

Parameters	Default Values	Units
VO (offset)		V or A
VA (amplitude)		V or A
FC (carrier frequency)	1/TSTOP	Hz
MDI (modulation index)		
FS (signal frequency)	1/TSTOP	Hz

The shape of the waveform is described by the following equation:

$$value = VO + VA*\sin[(2*\pi*FC*time) + MDI*\sin(2*\pi*FS*time)]$$

Because of the complex appearance of the waveform, the reader is provided with a high-resolution graphic plot (Figure 6-16). The waveform begins at 0 ms with a DC amplitude of 0 V. This starting amplitude is a result of VO. The waveform rises to an amplitude of 2 V due to VA and presents a carrier frequency of 10 KHz. The carrier frequency is not obvious in the waveform because of the affects of the modulation index MDI and the carrier frequency FS. The described single-frequency FM (SFFM) waveform could be described as follows.

```
VSSFM 5 0 SFFM(0 2 10K 5 1K)
```

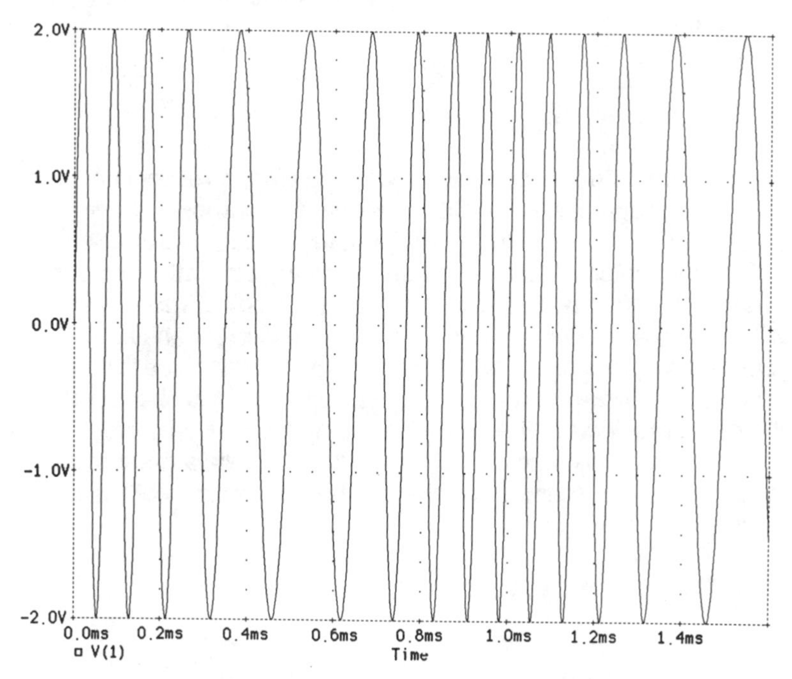

Figure 6-16. SFFM(0 2 10K 5 1K) example.

The specification of VO, VA, and FC are mandatory with the balance of the parameters being optional. If only the mandatory parameters are specified, then a sinusoidal waveform with offset VO, amplitude VA, and frequency FA will be generated. The reader is reminded that the peak-to-peak value of the sinusoidal wave is 2 * VA and is symmetrical about VO. If the above waveform example was redefined as shown below, a single-frequency sinusoidal waveform would result. This waveform is shown in Figure 6-17.

```
VSSFM 5 0 SFFM(0 2 10K)
```

6.1.6 SUMMARY

The emphasis of the previous five subsections has been on the creation of the transient waveforms in a voltage realm. Should the user desire to specify the waveform as a current, then any voltage designation will become a current designation and the units of amplitude will be in Amperes.

The waveforms discussed may be used with any voltage/current source, but only one waveform may be specified per source. If the user desires to use more than one waveform on a circuit, then separate sources must be specified. For example, if a FET switch were to be gated on and off by a digital (pulse) waveform and a sinusoidal wave injected into the switch input, then two voltage sources would have to be specified. Each source would have a separate name and node connection. If two waveforms are to be summed together, the sources should be connected to the same point using resistors tied together at a summing junction.

6.2 FOURIER ANALYSIS

SPICE can perform a Fourier analysis on any waveform present in a transient analysis. A Fourier analysis, in SPICE, consists of the program computing the amplitude and phase of any output waveform, its fundamental frequency, plus the next eight harmonics. Assume you are modeling an audio amplifier. The output is node 10 and you are injecting a 1 KHz sinusoidal wave into the input. SPICE will perform a Fourier analysis on node 10 looking at the amplitude and phase components of the output. The amount of energy that exists at 1 KHz, 2 KHz, ..., 9 KHz is analyzed and a resulting calculation of total harmonic distortion (THD) being provided to the user. The sample circuit presented in this chapter makes use of this function so the reader can gain a greater appreciation of the power of this analysis capability. The general form of the Fourier command is shown below.

.FOUR FREQ OV1 [OV2 OV3 ...]

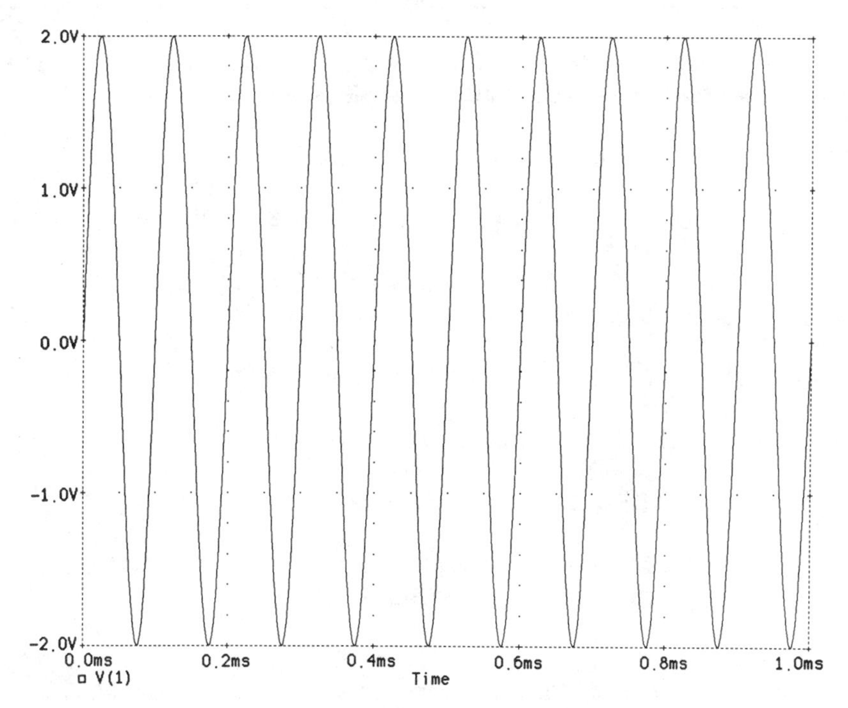

Figure 6-17. SFFM(0 2 10K) example.

This statement controls whether SPICE performs a Fourier analysis as part of the transient analysis. FREQ is the fundamental frequency, and OV1, OV2, OV3 are the output variables for which the analysis is desired. Notice that the number of outputs that can receive a Fourier analysis is unlimited. The Fourier analysis is performed over the interval TSTOP minus one full period starting at TSTOP and working backward in time. TSTOP is the final time specified for the transient analysis (see Section 6.3), and the period is one cycle of the fundamental frequency. The DC component and the first nine components are determined. For maximum accuracy, TMAX (see the .TRAN statement in Section 6.3) should be set to a maximum value of period/100.0 (or less for very high Q circuits).

If we were to request a Fourier analysis on the amplifier example above, then the following statement would be placed in the input file for SPICE.

```
.FOUR 1K V(10)
```

6.3 TRANSIENT ANALYSIS STATEMENTS

In the previous discussions associated with AC and DC analyses, a natural grouping of commands has been used. The grouping of statements is necessary to cause a specified type of SPICE analysis to occur. In a DC analysis, the following three statement were found to be necessary.

1. VIN 1 0 DC 3 (input source definition)

2. .DC VIN -2 4 .01 (source to be swept & sweep range)

3. .PLOT DC V(4) V(6) and/or .PRINT DC V(4) V(6) (node/source to be plotted or printed)

In an AC analysis, the following statements were found to be necessary.

1. VIN 1 0 AC 1 (input source definition)

2. .AC DEC 50 100 1K (frequency sweep range)

3. .PLOT AC VM(4) VDB(6) and/or .PRINT AC VM(4) VDB(6) (node/source to be plotted or printed)

The transient analysis requires, typically, the following three pieces of information which are related in concept to those of the AC and DC analyses.

1. *Input source definition*. This is optional in a transient analysis. If the circuit, being analyzed, has no input, as in an oscillator, then no source definition is appropriate.

2. *Time duration* of the analysis and specified time increments.

3. *Node/source specification* to be plotted or printed.

6.3.1 .TRAN STATEMENT

The time duration of the analysis, and the increment, are specified using a .TRAN statement. The two required pieces of information are the stop time TSTOP and the incremental step size TSTEP. Assume a 1 KHz sinusoidal wave is injected into the input of an amplifier. The output is should be a sinusoidal wave, which has an output period of 1/1 KHz = 1 ms. Therefore, TSTOP should be at least 1 ms. To compensate for phase delay and parametric circuit parameters, it is recommended that TSTOP be $\geq$ period * 2. As a rule of thumb, TSTEP should be at least TSTOP/100. This will assure that, at least, 101 points are computed and displayed.

The transient analysis represents a dynamically adjusting analysis. The analysis computes an initial operating point at the start of the analysis (time zero) and looks at the next increment of time, specified by TSTEP. If the next increment of time indicates that a rapid change has occurred at some node, such as a fast rising/falling pulse edge, then the value of TSTEP is reduced to assure that certain internal accuracy criteria are met. Internally, SPICE assures that certain $\Delta I/\Delta T$ and $\Delta V/\Delta T$ limits are not exceeded. This is to assure that the voltage at any node or the current through any element will not change more than some maximum percentage from its previous value. If little percentage change occurs, in voltage or current, then the time increment, TSTEP, is increased up to a maximum value of TSTEP * 2.

The general form of the .TRAN statement is shown below.

.TRAN TSTEP TSTOP [TSTART [TMAX]] [UIC]

where TSTEP is the printing or plotting increment for the analysis output. TSTOP is the analysis end time, and TSTART is the initial analysis data recording time. If TSTART is omitted, it is assumed to be zero. The transient analysis always begins at time zero. In the time interval before TSTART, the circuit is analyzed (to reach the analysis state of TSTART), but no output data is stored. In the time interval from TSTART to TSTOP, the circuit is analyzed and outputs are stored at TSTEP increments. TMAX is the maximum step size that SPICE will use (for default, the program chooses either TSTEP or (TSTOP - TSTART)/50.0, whichever is smaller). TMAX is useful when the user desires to guarantee a fixed computing interval other than the increment defined by TSTEP.

No significant savings of time can be realized by specifying a value of TSTART. The effect of using TSTART is that the printed or plotted output will begin at TSTART, end at TSTOP, and be printed in TSTEP increments. This may be useful if only a specific time increment is of interest. The reason no significant saving in time occurs is that, because of its iterative nature, SPICE must compute all the data up to TSTART to establish a set of parametric conditions.

UIC (use initial conditions) is an optional keyword, which indicates that the user does not want SPICE to solve for the quiescent operating point before beginning the transient analysis. If this keyword is specified, SPICE uses the values specified using IC=... on the various elements as the initial transient condition and proceeds with the analysis. If the .IC statement has been specified, then the parameters specified in the .IC statement are used to compute the initial conditions for the devices.

Possible examples of the .TRAN statement are shown below.

1. .TRAN 1NS 100NS specifies a transient analysis that goes from 0 nsec to 100 nsec in 1 nsec steps (101 points)

2. .TRAN 1NS 1000NS 500NS specifies a transient analysis which starts at 500 nsec and ends at 1 μsec (1000 nsec) in 1 nsec steps (501 points)

3. .TRAN 10NS 1US UIC specifies a transient analysis which starts at 0 nsec, runs for 1 μsec, and uses a computed interval of 10 nsec. In addition, the analysis will use the initial conditions specified for circuit components.

6.3.2 .IC STATEMENT

The .IC statement, as discussed previously, is used if an initial set of conditions are desired at a specified node. Assume you want to view the discharge characteristics of a parallel RC circuit. The RC circuit output is at node 4 and starts its discharge at 10 V. The .IC statement could be used to set node 4 to 10 V. The value is set before the start of the transient analysis and an analysis of the discharge is then performed. If the capacitor had an initial condition set (i.e., CX 1 2 10U IC=10V), the use of the UIC option in the .TRAN statement would negate the need for the use of the .IC statement. The general form of the .IC statement is shown below.

.IC V(NODNUM)=VAL V(NODNUM)=VAL ...

As stated above, this statement is for setting a set of initial conditions in a transient analysis and is only applicable to a transient analysis. The .IC statement has two different interpretations. The two interpretations of this statement are defined below.

1. When the UIC parameter is specified on the .TRAN statement, then the node voltages specified on the .IC statement are used to compute the capacitor, diode, BJT, JFET, and MOSFET initial conditions. This is equivalent to specifying the IC=... parameter on each device statement, but is much more convenient. The IC=... parameter can still be specified and will take precedence over the .IC values. Since no DC bias (initial transient) solution is computed before the transient analysis, one should take care to specify all DC source voltages on the .IC statement if they are to be used to compute device initial conditions.

2. When the UIC parameter is not specified on the .TRAN statement, the DC bias (initial transient) solution will be computed before the transient analysis. In this case, the node voltages specified on the .IC statement will be forced to the desired initial values during the bias solution. During the transient analysis, the constraint on these node voltages is removed.

Assume you want to set the starting transient analysis voltages for nodes 2, 3, and 4 to 5.0 V, -5.0 V, and 0.23 V respectively. The following .IC statement could accomplish this.

```
.IC V(2)=5 V(3)=-5 V(4)=.23
```

6.3.3 PLOT AND PRINT STATEMENTS

The .PLOT and .PRINT statements are specified using a .PLOT TRAN or .PRINT TRAN command. Assume you want to plot and print the transient analysis results at node 6 (voltage versus time) and the current flowing through VS (current versus time). The following statements could be used.

```
.PLOT TRAN V(6) I(VS)
.PRINT TRAN V(6) I(VS)
```

The format is identical to the print and plot formats of the DC analysis except for the use of TRAN instead of DC.

6.3.4 RELATIONSHIP BETWEEN DC, AC, AND TRANSIENT ANALYSES FORMATS

At this point the formats and requirements for the DC, AC, and transient analyses have been discussed. SPICE uses a common format for all three types of analyses, which the reader should recognize by now. Table 6-5 shows the available parameters for each of the analyses and focuses on the common attributes of each.

The reader is directed toward the table. SPICE uses a repeating sequence of commands in each type of analysis. The sequence consists of a stimulus source definition, if applicable, which serves as the driver for the analysis. Associated with each stimulus is a sweep definition that specifies a sweep range over which the analysis will be performed. Lastly, each analysis must specify a set of outputs to be observed during the analysis process. The required parameters for each type of analysis are shown in Table 6-6. Note that an input stimulus is only mandatory on a DC analysis. If an AC analysis lacks the necessary stimulus definition no meaningful analysis will result.

6.4 SAMPLE TRANSIENT ANALYSIS

The use of the new material, presented in this section, can best be illustrated using a bandpass active filter design. The sample analysis presents the design and design equations for a bandpass filter. The resulting design is analyzed from an AC and transient analysis standpoint, that is, the AC analysis looks at AC gain, filter bandwidth, and "Q", and the transient analysis simulates the injection of a symmetrical square wave at the bandpass frequency and the verification of a sinusoidal wave output. A Fourier analysis will be performed on the output to verify filter integrity.

Table 6-5. Relationship of Required Parameters Between Analysis Types.

Parameter	AC Analysis	DC Analysis	Transient Analysis
	Statement	Statement	Statement
Input stimulus	VX N+ N- AC VALUE PHASE or IX N+ N- AC VALUE PHASE	VX N+ N- DC VALUE or IX N+ N- DC VALUE	VXXXXXXX N+ N- PULSE V1 V2 TD TR TF PW PER SIN VO VA FREQ TD KD EXP V1 V2 TD1 T1 TD2 T2 PWL T1 V1 T2 V2 ... TN VN SFFM VO VA FREQ MDI FS or IXXXXXXX N+ N- PULSE I1 I2 TD TR TF PW PER SIN IO IA FREQ TD KD EXP I1 I2 TD1 T1 TD2 T2 PWL T1 I1 T2 I2 ... TN IN SFFM IO IA FREQ MDI FS
Sweep range	.AC DEC NP FS FE .AC OCT NP FS FE .AC LIN NP FS FE	.DC SOURCE START STOP INC	.TRAN TSTEP TSTOP
Output node/source	.PRINT AC OV1 [OV2 ... and/or .PLOT AC OV1 [OV2 ...	.PRINT DC OV1 [OV2 ... and/or .PLOT DC OV1 [OV2 ...	.PRINT TRAN OV1 [OV2 ... and/or .PLOT TRAN OV1 [OV2 ...

6.4.1 CIRCUIT CALCULATIONS

The circuit, to be analyzed, is shown in Figure 6-18. The figure shows a two pole bandpass filter with the following characteristics:

1. Gain = 2
2. Q = 10
3. C1 = C2 = C = 0.02 μF
4. Center frequency = FC = 1 KHz

The equations necessary to compute the values of R1, R2, RF, and BW are shown below. The equations are quite standard and no attempt will be made to derive them. The reader is referred to any standard filter text for further information associated with this design.

BW = bandwidth = FC/Q
BW = 1 KHz/10 = 100

$R1 = Q/(2*\pi*FC*Gain*C)$
$R1 = 39.8\ K\Omega$

$R2 = Q/(2*\pi*FC*C*(2*Q^2 - Gain))$
$R2 = 401.9\ \Omega$

$RF = 2*Q/(2*\pi*FC*C)$
$RF = 159\ K\Omega$

Table 6-6. Comparison of Analysis Types and Effects of Parameters.

Parameter	AC Analysis		DC Analysis		Transient Analysis	
	Req	Comment	Req	Comment	Req	Comment
Input stimulus	No	An absence of a source will cause the analysis to default to a 0 VAC source	Yes	If no source is specified then the analysis will not run.	No	If the circuit is an oscillator then no source is necessary.
Sweep range	Yes	Must contain type of sweep, # points, start, and stop frequency	Yes	Must specify source to be swept, the sweep range, and the sweep increment.	Yes	Must specify an analysis end time and a step size.
Output node/source	Yes	Must define node or source data to be printed or plotted and whether the data is to be magnitude, phase, DB, real or imaginary.	Yes	Must define printed or plotted data and the node/source to be analyzed.	Yes	Must define printed or plotted data and the node/source to be analyzed.

The computed values have been included in Figure 6-18. The circuit description is shown in Section 6.4.2.

6.4.2 SAMPLE CIRCUIT INPUT DESCRIPTION

The SPICE circuit model is shown in Figure 6-18 and is described following the figure.

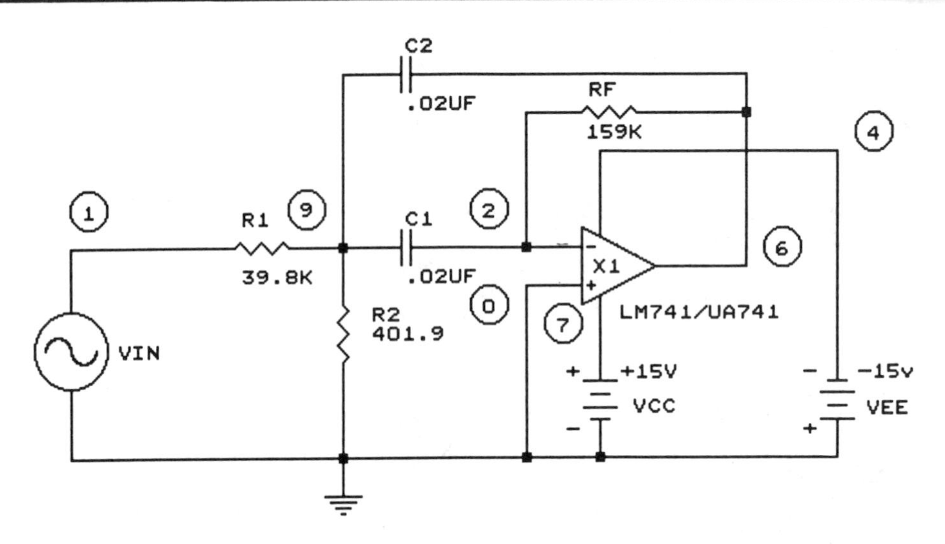

Figure 6-18. Transient analysis circuit (1 KHz bandpass filter).

```
1KHZ BANDPASS FILTER
VIN 1 0 DC 1 AC 1 PULSE(-1 1 0 .1N .1N .5M 1M)
* THIS IS WHERE THE 1KHZ PULSE IS DEFINED
* -----------------------------------------------------
VCC 7 0 15
VEE 4 0 -15
R1 1 9 39.8K
R2 9 0 401.9
RF 2 6 159K
C1 9 2 .02U
C2 9 6 .02U
* THE ABOVE ARE THE COMPUTED VALUES OF THE RESISTORS & CAPACITORS
* -----------------------------------------------------
X1 0 2 6 7 4 LM741
* THIS CALLS THE SUB-CIRCUIT LM741
* -----------------------------------------------------
.SUBCKT LM741 10 20 30 40 50
.MODEL DP D(IS=1E-14 N=2.118)
.MODEL DN D(IS=.971E-14 N=2.118)
.MODEL DV D(IS=6E-15 N=.1 XTI=.1)
.MODEL DI D(IS=1E-14 N=1)
VP 10 11 0
VN 20 21 0
DP 11 1 DP
DN 21 1 DN
FA 20 0 VN .276
I1 1 0 140.5N
```

```
C1 1 0 33.3FF IC=-.863
FP 4 0 VP 162.2
FN 0 4 VN 161.0
GC 0 4 1 0 6.6N
RT 4 0 1
CT 4 0 80N
G2 0 2 4 0 1
R2 2 0 100K
CC 2 3 30P
GO 3 0 2 0 161
RD 3 0 60
DH 3 5 DV
DL 6 3 DV
VH 40 5 .7
VL 6 50 2.2
D1 3 9 DI
D2 9 3 DI
EX 9 0 POLY(2) 3 0 3 30 0 1 -1.64
RO 3 30 40
.ENDS LM741
* ------------------------------------------------
.AC LIN 101 500 1500
.PLOT AC VM(6)
.PRINT AC VM(6) VP(6) VDB(6)
* THIS DEFINES THE SWEEP & OUTPUTS OF THE AC ANALYSIS
.TRAN .1M 4M
.PLOT TRAN V(6)
* THIS DEFINES THE TIME INTERVAL & OUTPUT OF THE TRANSIENT ANALYSIS
.FOUR 1K V(6)
* THIS DEFINES THE FOURIER ANALYSIS ON THE OUTPUT. THIS STATEMENT
* IS ONLY USEFUL WITH A TRANSIENT ANALYSIS
.END
```

The input file has been commented into areas of specific interest. Note that the VIN statement consists of DC, AC, and PULSE definitions. The PULSE is strictly for the transient analysis and the "AC 1" is strictly for the AC analysis. When SPICE encounters a .AC and .PLOT AC statement pair, it looks for the existence of an AC source (AC 1). When it encounters a .TRAN and .PLOT TRAN statement pair, it looks for a PULSE, SIN, EXP, PWL, or SFFM designation. Note: The AC analysis knows nothing of the transient analysis and the transient analysis knows nothing of the AC analysis. The common denominator, in this case, is a voltage source VIN.

6.4.3 SAMPLE TRANSIENT ANALYSIS RESULTS

The results of running the circuit described in Section 6.4.3, are shown on the following pages. The data is analyzed, in sufficient detail, at the end of this section.

```
1KHZ BANDPASS FILTER
****     CIRCUIT DESCRIPTION
******************************************************************************
VIN 1 0 DC 1 AC 1 PULSE(-1 1 0 .1N .1N .5M 1M)
VCC 7 0 15
VEE 4 0 -15
R1 1 9 39.8K
```

```
R2 9 0 401.9
RF 2 6 159K
C1 9 2 .02U
C2 9 6 .02U
X1 0 2 6 7 4 LM741
.SUBCKT LM741 10 20 30 40 50
.MODEL DP D(IS=1E-14 N=2.118)
.MODEL DN D(IS=.971E-14 N=2.118)
.MODEL DV D(IS=6E-15 N=.1 XTI=.1)
.MODEL DI D(IS=1E-14 N=1)
VP 10 11 0
VN 20 21 0
DP 11 1 DP
DN 21 1 DN
FA 20 0 VN .276
I1 1 0 140.5NA
C1 1 0 33.3FF IC=-.863
FP 4 0 VP 162.2
FN 0 4 VN 161.0
GC 0 4 1 0 6.6N
RT 4 0 1
CT 4 0 80N
G2 0 2 4 0 1
R2 2 0 100K
CC 2 3 30P
GO 3 0 2 0 161
RD 3 0 60
DH 3 5 DV
DL 6 3 DV
VH 40 5 .7
VL 6 50 2.2
D1 3 9 DI
D2 9 3 DI
EX 9 0 POLY(2) 3 0 3 30 0 1 -1.64
RO 3 30 40
.ENDS LM741
.AC LIN 101 500 1500
.TRAN .1M 4M
.PLOT TRAN V(6)
.FOUR 1K V(6)
.PLOT AC VM(6)
.PRINT AC VM(6) VP(6) VDB(6)
.END
```

```
 1KHZ BAND PASS FILTER

 ****     Diode MODEL PARAMETERS

******************************************************************************


              X1.DP             X1.DN             X1.DV             X1.DI
       IS     10.000000E-15     9.710000E-15      6.000000E-15      10.000000E-15
        N      2.118            2.118              .1
      XTI                                          .1
```

```
1KHZ BAND PASS FILTER

****     SMALL SIGNAL BIAS SOLUTION        TEMPERATURE =   27.000 DEG C

******************************************************************************


NODE   VOLTAGE     NODE   VOLTAGE     NODE   VOLTAGE     NODE   VOLTAGE

(    1)    1.0000 (    2)      .0020 (    4)  -15.0000 (    6)      .0164

(    7)   15.0000 (    9)      .0100 ( X1.1)    -.8634 ( X1.2)-1.694E-06

( X1.3)     .0164 ( X1.4)-16.94E-12 ( X1.5)   14.3000 ( X1.6)  -12.8000

( X1.9)     .0164 (X1.11)    0.0000 (X1.21)      .0020


    VOLTAGE SOURCE CURRENTS
    NAME           CURRENT

    VIN           -2.487E-05
    VCC           -1.429E-11
    VEE            1.282E-11
    X1.VP          6.997E-08
    X1.VN          7.053E-08
    X1.VH          1.429E-11
    X1.VL          1.282E-11

    TOTAL POWER DISSIPATION   2.49E-05  WATTS
```

```
1KHZ BAND PASS FILTER

****     AC ANALYSIS                          TEMPERATURE =   27.000 DEG C
**************************************************************************
 FREQ        VM(6)       VP(6)       VDB(6)

 5.000E+02   1.336E-01  -9.387E+01  -1.748E+01
 5.100E+02   1.382E-01  -9.400E+01  -1.719E+01
 5.200E+02   1.429E-01  -9.414E+01  -1.690E+01
 5.300E+02   1.478E-01  -9.428E+01  -1.661E+01
 5.400E+02   1.529E-01  -9.443E+01  -1.631E+01
 5.500E+02   1.582E-01  -9.458E+01  -1.602E+01
 5.600E+02   1.637E-01  -9.474E+01  -1.572E+01
 5.700E+02   1.694E-01  -9.491E+01  -1.542E+01
 5.800E+02   1.754E-01  -9.508E+01  -1.512E+01
 5.900E+02   1.817E-01  -9.526E+01  -1.481E+01
 6.000E+02   1.882E-01  -9.545E+01  -1.451E+01
 6.100E+02   1.951E-01  -9.565E+01  -1.419E+01
 6.200E+02   2.023E-01  -9.586E+01  -1.388E+01
 6.300E+02   2.099E-01  -9.608E+01  -1.356E+01
 6.400E+02   2.178E-01  -9.631E+01  -1.324E+01
 6.500E+02   2.262E-01  -9.655E+01  -1.291E+01
 6.600E+02   2.351E-01  -9.681E+01  -1.258E+01
 6.700E+02   2.444E-01  -9.708E+01  -1.224E+01
 6.800E+02   2.544E-01  -9.736E+01  -1.189E+01
 6.900E+02   2.649E-01  -9.767E+01  -1.154E+01
 7.000E+02   2.761E-01  -9.800E+01  -1.118E+01
 7.100E+02   2.880E-01  -9.834E+01  -1.081E+01
 7.200E+02   3.008E-01  -9.871E+01  -1.044E+01
 7.300E+02   3.144E-01  -9.911E+01  -1.005E+01
 7.400E+02   3.291E-01  -9.954E+01  -9.653E+00
 7.500E+02   3.449E-01  -1.000E+02  -9.246E+00
 7.600E+02   3.619E-01  -1.005E+02  -8.827E+00
 7.700E+02   3.804E-01  -1.010E+02  -8.395E+00
 7.800E+02   4.005E-01  -1.016E+02  -7.949E+00
 7.900E+02   4.223E-01  -1.023E+02  -7.487E+00
 8.000E+02   4.463E-01  -1.030E+02  -7.008E+00
 8.100E+02   4.726E-01  -1.038E+02  -6.510E+00
 8.200E+02   5.016E-01  -1.046E+02  -5.993E+00
 8.300E+02   5.338E-01  -1.056E+02  -5.452E+00
 8.400E+02   5.697E-01  -1.066E+02  -4.887E+00
 8.500E+02   6.099E-01  -1.079E+02  -4.295E+00
 8.600E+02   6.551E-01  -1.092E+02  -3.674E+00
 8.700E+02   7.064E-01  -1.108E+02  -3.019E+00
 8.800E+02   7.648E-01  -1.126E+02  -2.329E+00
 8.900E+02   8.317E-01  -1.147E+02  -1.600E+00
 9.000E+02   9.089E-01  -1.172E+02  -8.297E-01
 9.100E+02   9.982E-01  -1.201E+02  -1.551E-02
 9.200E+02   1.102E+00  -1.236E+02   8.419E-01
 9.300E+02   1.222E+00  -1.278E+02   1.738E+00
 9.400E+02   1.359E+00  -1.330E+02   2.662E+00
 9.500E+02   1.511E+00  -1.393E+02   3.586E+00
 9.600E+02   1.672E+00  -1.470E+02   4.465E+00
 9.700E+02   1.825E+00  -1.562E+02   5.223E+00
 9.800E+02   1.942E+00  -1.668E+02   5.764E+00
 9.900E+02   1.994E+00  -1.783E+02   5.994E+00
```

FREQ	VM(6)	VP(6)	VDB(6)
1.000E+03	1.966E+00	1.702E+02	5.872E+00
1.010E+03	1.869E+00	1.595E+02	5.434E+00
1.020E+03	1.732E+00	1.502E+02	4.773E+00
1.030E+03	1.582E+00	1.424E+02	3.987E+00
1.040E+03	1.437E+00	1.360E+02	3.152E+00
1.050E+03	1.306E+00	1.308E+02	2.316E+00
1.060E+03	1.189E+00	1.265E+02	1.506E+00
1.070E+03	1.088E+00	1.230E+02	7.346E-01
1.080E+03	1.001E+00	1.201E+02	6.791E-03
1.090E+03	9.250E-01	1.176E+02	-6.771E-01
1.100E+03	8.591E-01	1.154E+02	-1.319E+00
1.110E+03	8.016E-01	1.136E+02	-1.921E+00
1.120E+03	7.510E-01	1.121E+02	-2.487E+00
1.130E+03	7.063E-01	1.107E+02	-3.020E+00
1.140E+03	6.666E-01	1.095E+02	-3.523E+00
1.150E+03	6.312E-01	1.084E+02	-3.997E+00
1.160E+03	5.993E-01	1.074E+02	-4.447E+00
1.170E+03	5.706E-01	1.066E+02	-4.873E+00
1.180E+03	5.446E-01	1.058E+02	-5.278E+00
1.190E+03	5.209E-01	1.051E+02	-5.664E+00
1.200E+03	4.993E-01	1.044E+02	-6.033E+00
1.210E+03	4.795E-01	1.038E+02	-6.384E+00
1.220E+03	4.613E-01	1.033E+02	-6.721E+00
1.230E+03	4.444E-01	1.028E+02	-7.044E+00
1.240E+03	4.289E-01	1.023E+02	-7.354E+00
1.250E+03	4.144E-01	1.019E+02	-7.652E+00
1.260E+03	4.009E-01	1.015E+02	-7.938E+00
1.270E+03	3.884E-01	1.012E+02	-8.215E+00
1.280E+03	3.767E-01	1.008E+02	-8.481E+00
1.290E+03	3.657E-01	1.005E+02	-8.738E+00
1.300E+03	3.553E-01	1.002E+02	-8.987E+00
1.310E+03	3.456E-01	9.990E+01	-9.228E+00
1.320E+03	3.365E-01	9.964E+01	-9.461E+00
1.330E+03	3.278E-01	9.938E+01	-9.687E+00
1.340E+03	3.197E-01	9.915E+01	-9.906E+00
1.350E+03	3.119E-01	9.892E+01	-1.012E+01
1.360E+03	3.046E-01	9.871E+01	-1.033E+01
1.370E+03	2.976E-01	9.850E+01	-1.053E+01
1.380E+03	2.910E-01	9.831E+01	-1.072E+01
1.390E+03	2.847E-01	9.813E+01	-1.091E+01
1.400E+03	2.787E-01	9.795E+01	-1.110E+01
1.410E+03	2.729E-01	9.778E+01	-1.128E+01
1.420E+03	2.674E-01	9.762E+01	-1.146E+01
1.430E+03	2.622E-01	9.747E+01	-1.163E+01
1.440E+03	2.571E-01	9.732E+01	-1.180E+01
1.450E+03	2.523E-01	9.718E+01	-1.196E+01
1.460E+03	2.477E-01	9.705E+01	-1.212E+01
1.470E+03	2.433E-01	9.692E+01	-1.228E+01
1.480E+03	2.390E-01	9.680E+01	-1.243E+01
1.490E+03	2.349E-01	9.668E+01	-1.258E+01
1.500E+03	2.309E-01	9.656E+01	-1.273E+01

```
 1KHZ BAND PASS FILTER

 ****     AC ANALYSIS                        TEMPERATURE =   27.000 DEG C
******************************************************************************

  FREQ        VM(6)

(*)---------    1.0000E-01   1.0000E+00   1.0000E+01   1.0000E+02   1.0000E+03
                          - - - - - - - - - - - - - - - - - - - - - - - - - - -
  5.000E+02  1.336E-01 . *          .            .            .            .
  5.100E+02  1.382E-01 . *          .            .            .            .
  5.200E+02  1.429E-01 .  *         .            .            .            .
  5.300E+02  1.478E-01 .  *         .            .            .            .
  5.400E+02  1.529E-01 .  *         .            .            .            .
  5.500E+02  1.582E-01 .  *         .            .            .            .
  5.600E+02  1.637E-01 .  *         .            .            .            .
  5.700E+02  1.694E-01 .  *         .            .            .            .
  5.800E+02  1.754E-01 .   *        .            .            .            .
  5.900E+02  1.817E-01 .   *        .            .            .            .
  6.000E+02  1.882E-01 .   *        .            .            .            .
  6.100E+02  1.951E-01 .   *        .            .            .            .
  6.200E+02  2.023E-01 .   *        .            .            .            .
  6.300E+02  2.099E-01 .    *       .            .            .            .
  6.400E+02  2.178E-01 .    *       .            .            .            .
  6.500E+02  2.262E-01 .    *       .            .            .            .
  6.600E+02  2.351E-01 .    *       .            .            .            .
  6.700E+02  2.444E-01 .     *      .            .            .            .
  6.800E+02  2.544E-01 .     *      .            .            .            .
  6.900E+02  2.649E-01 .     *      .            .            .            .
  7.000E+02  2.761E-01 .     *      .            .            .            .
  7.100E+02  2.880E-01 .     *      .            .            .            .
  7.200E+02  3.008E-01 .      *     .            .            .            .
  7.300E+02  3.144E-01 .      *     .            .            .            .
  7.400E+02  3.291E-01 .      *     .            .            .            .
  7.500E+02  3.449E-01 .      *     .            .            .            .
  7.600E+02  3.619E-01 .       *    .            .            .            .
  7.700E+02  3.804E-01 .       *    .            .            .            .
  7.800E+02  4.005E-01 .       *    .            .            .            .
  7.900E+02  4.223E-01 .        *   .            .            .            .
  8.000E+02  4.463E-01 .        *   .            .            .            .
  8.100E+02  4.726E-01 .        *   .            .            .            .
  8.200E+02  5.016E-01 .         *  .            .            .            .
  8.300E+02  5.338E-01 .         *  .            .            .            .
  8.400E+02  5.697E-01 .         *  .            .            .            .
  8.500E+02  6.099E-01 .          * .            .            .            .
  8.600E+02  6.551E-01 .          * .            .            .            .
  8.700E+02  7.064E-01 .           *.            .            .            .
  8.800E+02  7.648E-01 .           *.            .            .            .
  8.900E+02  8.317E-01 .           *.            .            .            .
  9.000E+02  9.089E-01 .            *            .            .            .
  9.100E+02  9.982E-01 .            *            .            .            .
  9.200E+02  1.102E+00 .            .*           .            .            .
  9.300E+02  1.222E+00 .            . *          .            .            .
  9.400E+02  1.359E+00 .            . *          .            .            .
  9.500E+02  1.511E+00 .            .  *         .            .            .
  9.600E+02  1.672E+00 .            .  *         .            .            .
  9.700E+02  1.825E+00 .            .   *        .            .            .
  9.800E+02  1.942E+00 .            .   *        .            .            .
  9.900E+02  1.994E+00 .            .   *        .            .            .
  1.000E+03  1.966E+00 .            .   *        .            .            .
  1.010E+03  1.869E+00 .            .   *        .            .            .
```

```
  FREQ        VM(6)

(*)----------     1.0000E-01   1.0000E+00  1.0000E+01   1.0000E+02  1.0000E+03

                       - - - - - - - - - - - - - - - - - - - - - - - - - -
  1.020E+03  1.732E+00 .            .   *       .            .           .
  1.030E+03  1.582E+00 .            .  *        .            .           .
  1.040E+03  1.437E+00 .            .  *        .            .           .
  1.050E+03  1.306E+00 .            . *         .            .           .
  1.060E+03  1.189E+00 .            .*          .            .           .
  1.070E+03  1.088E+00 .            .*          .            .           .
  1.080E+03  1.001E+00 .            .*          .            .           .
  1.090E+03  9.250E-01 .            *           .            .           .
  1.100E+03  8.591E-01 .            *           .            .           .
  1.110E+03  8.016E-01 .           *.           .            .           .
  1.120E+03  7.510E-01 .           *.           .            .           .
  1.130E+03  7.063E-01 .           *.           .            .           .
  1.140E+03  6.666E-01 .          * .           .            .           .
  1.150E+03  6.312E-01 .          * .           .            .           .
  1.160E+03  5.993E-01 .          * .           .            .           .
  1.170E+03  5.706E-01 .         *  .           .            .           .
  1.180E+03  5.446E-01 .         *  .           .            .           .
  1.190E+03  5.209E-01 .         *  .           .            .           .
  1.200E+03  4.993E-01 .         *  .           .            .           .
  1.210E+03  4.795E-01 .        *   .           .            .           .
  1.220E+03  4.613E-01 .        *   .           .            .           .
  1.230E+03  4.444E-01 .        *   .           .            .           .
  1.240E+03  4.289E-01 .        *   .           .            .           .
  1.250E+03  4.144E-01 .        *   .           .            .           .
  1.260E+03  4.009E-01 .       *    .           .            .           .
  1.270E+03  3.884E-01 .       *    .           .            .           .
  1.280E+03  3.767E-01 .       *    .           .            .           .
  1.290E+03  3.657E-01 .       *    .           .            .           .
  1.300E+03  3.553E-01 .       *    .           .            .           .
  1.310E+03  3.456E-01 .       *    .           .            .           .
  1.320E+03  3.365E-01 .      *     .           .            .           .
  1.330E+03  3.278E-01 .      *     .           .            .           .
  1.340E+03  3.197E-01 .      *     .           .            .           .
  1.350E+03  3.119E-01 .      *     .           .            .           .
  1.360E+03  3.046E-01 .      *     .           .            .           .
  1.370E+03  2.976E-01 .      *     .           .            .           .
  1.380E+03  2.910E-01 .      *     .           .            .           .
  1.390E+03  2.847E-01 .     *      .           .            .           .
  1.400E+03  2.787E-01 .     *      .           .            .           .
  1.410E+03  2.729E-01 .     *      .           .            .           .
  1.420E+03  2.674E-01 .     *      .           .            .           .
  1.430E+03  2.622E-01 .     *      .           .            .           .
  1.440E+03  2.571E-01 .     *      .           .            .           .
  1.450E+03  2.523E-01 .     *      .           .            .           .
  1.460E+03  2.477E-01 .     *      .           .            .           .
  1.470E+03  2.433E-01 .     *      .           .            .           .
  1.480E+03  2.390E-01 .    *       .           .            .           .
  1.490E+03  2.349E-01 .    *       .           .            .           .
  1.500E+03  2.309E-01 .    *       .           .            .           .
                       . . . . . . . . . . . . . . . . . . . . . . . . . .
```

```
 1KHZ BAND PASS FILTER

 ****     INITIAL TRANSIENT SOLUTION       TEMPERATURE =   27.000 DEG C

******************************************************************************


 NODE   VOLTAGE     NODE   VOLTAGE     NODE   VOLTAGE     NODE   VOLTAGE

(    1)   -1.0000 (    2)      .0020 (    4)  -15.0000 (    6)      .0164

(    7)   15.0000 (    9)     -.0100 ( X1.1)     -.8634 ( X1.2)-1.694E-06

( X1.3)      .0164 ( X1.4)-16.94E-12 ( X1.5)   14.3000 ( X1.6)  -12.8000

( X1.9)      .0164 (X1.11)    0.0000 (X1.21)      .0020


    VOLTAGE SOURCE CURRENTS
    NAME          CURRENT

    VIN           2.487E-05
    VCC          -1.429E-11
    VEE           1.282E-11
    X1.VP         6.997E-08
    X1.VN         7.053E-08
    X1.VH         1.429E-11
    X1.VL         1.282E-11

    TOTAL POWER DISSIPATION   2.49E-05  WATTS
```

```
1KHZ BAND PASS FILTER


****     TRANSIENT ANALYSIS              TEMPERATURE =   27.000 DEG C


******************************************************************************


   TIME        V(6)
(*)----------    -2.0000E+00  -1.0000E+00   0.0000E+00   1.0000E+00   2.0000E+00

                        - - - - - - - - - - - - - - - - - - - - - - - - - - -
  0.000E+00  1.636E-02 .            .            *            .            .
  1.000E-04 -2.065E-01 .            .         *  .            .            .
  2.000E-04 -3.271E-01 .            .        *   .            .            .
  3.000E-04 -3.242E-01 .            .        *   .            .            .
  4.000E-04 -2.067E-01 .            .         *  .            .            .
  5.000E-04 -1.706E-02 .            .            *            .            .
  6.000E-04  3.959E-01 .            .            .    *       .            .
  7.000E-04  6.520E-01 .            .            .       *    .            .
  8.000E-04  6.663E-01 .            .            .        *   .            .
  9.000E-04  4.513E-01 .            .            .     *      .            .
  1.000E-03  1.037E-01 .            .            .*           .            .
  1.100E-03 -4.821E-01 .            .      *     .            .            .
  1.200E-03 -8.624E-01 .            . *          .            .            .
  1.300E-03 -8.933E-01 .            .*           .            .            .
  1.400E-03 -6.150E-01 .            .    *       .            .            .
  1.500E-03 -1.406E-01 .            .          * .            .            .
  1.600E-03  5.912E-01 .            .            .       *    .            .
  1.700E-03  1.073E+00 .            .            .            .*           .
  1.800E-03  1.168E+00 .            .            .            . *          .
  1.900E-03  8.341E-01 .            .            .          * .            .
  2.000E-03  2.521E-01 .            .            .  *         .            .
  2.100E-03 -6.070E-01 .            .    *       .            .            .
  2.200E-03 -1.177E+00 .          * .            .            .            .
  2.300E-03 -1.311E+00 .        *   .            .            .            .
  2.400E-03 -9.901E-01 .            *            .            .            .
  2.500E-03 -3.058E-01 .            .        *   .            .            .
  2.600E-03  6.652E-01 .            .            .        *   .            .
  2.700E-03  1.320E+00 .            .            .            .   *        .
  2.800E-03  1.494E+00 .            .            .            .     *      .
  2.900E-03  1.159E+00 .            .            .            . *          .
  3.000E-03  4.271E-01 .            .            .     *      .            .
  3.100E-03 -6.490E-01 .            .    *       .            .            .
  3.200E-03 -1.393E+00 .       *    .            .            .            .
  3.300E-03 -1.594E+00 .    *       .            .            .            .
  3.400E-03 -1.244E+00 .         *  .            .            .            .
  3.500E-03 -4.813E-01 .            .      *     .            .            .
  3.600E-03  6.738E-01 .            .            .        *   .            .
  3.700E-03  1.501E+00 .            .            .            .      *     .
  3.800E-03  1.745E+00 .            .            .            .         *  .
  3.900E-03  1.394E+00 .            .            .            .    *       .
  4.000E-03  5.964E-01 .            .            .       *    .            .
                        . . . . . . . . . . . . . . . . . . . . . . . . . . .
```

```
1KHZ BAND PASS FILTER

****     FOURIER ANALYSIS                    TEMPERATURE =   27.000 DEG C

******************************************************************************


FOURIER COMPONENTS OF TRANSIENT RESPONSE V(6)


DC COMPONENT =   4.677727E-02
```

HARMONIC NO	FREQUENCY (HZ)	FOURIER COMPONENT	NORMALIZED COMPONENT	PHASE (DEG)	NORMALIZED PHASE (DEG)
1	1.000E+03	1.713E+00	1.000E+00	1.652E+02	0.000E+00
2	2.000E+03	4.048E-02	2.362E-02	-1.453E+02	-3.106E+02
3	3.000E+03	3.488E-02	2.036E-02	1.369E+02	-2.831E+01
4	4.000E+03	1.537E-02	8.972E-03	-1.568E+02	-3.220E+02
5	5.000E+03	1.555E-02	9.076E-03	1.470E+02	-1.825E+01
6	6.000E+03	9.912E-03	5.785E-03	-1.589E+02	-3.241E+02
7	7.000E+03	8.971E-03	5.236E-03	1.525E+02	-1.273E+01
8	8.000E+03	7.427E-03	4.334E-03	-1.588E+02	-3.240E+02
9	9.000E+03	4.881E-03	2.848E-03	1.659E+02	6.620E-01

```
TOTAL HARMONIC DISTORTION =   3.497517E+00 PERCENT
```

6.4.4 EXTRACTION OF TRANSIENT ANALYSIS DATA

The data results focus on the AC and transient parameters of the bandpass filter. The AC results will be reviewed first. The stated intent of the AC analysis was to determine the following:

1. Center frequency of the filter
2. Gain of the filter
3. Bandwidth of the filter

A review of the AC data of Section 6.4.3 can provide this necessary information. The data produced by the .PRINT statement will be the primary focus for this portion of the interpretation. The data shows a peak amplitude occurring at 990 Hz. The high resolution plot of amplitude versus frequency (Figure 6-19) supports this number. At 990 Hz, the output amplitude is computed to be 1.994 V. Since the input was normalized to 1 V the output amplitude is gain. The desired gain was 2.0, which is close to the 1.994.

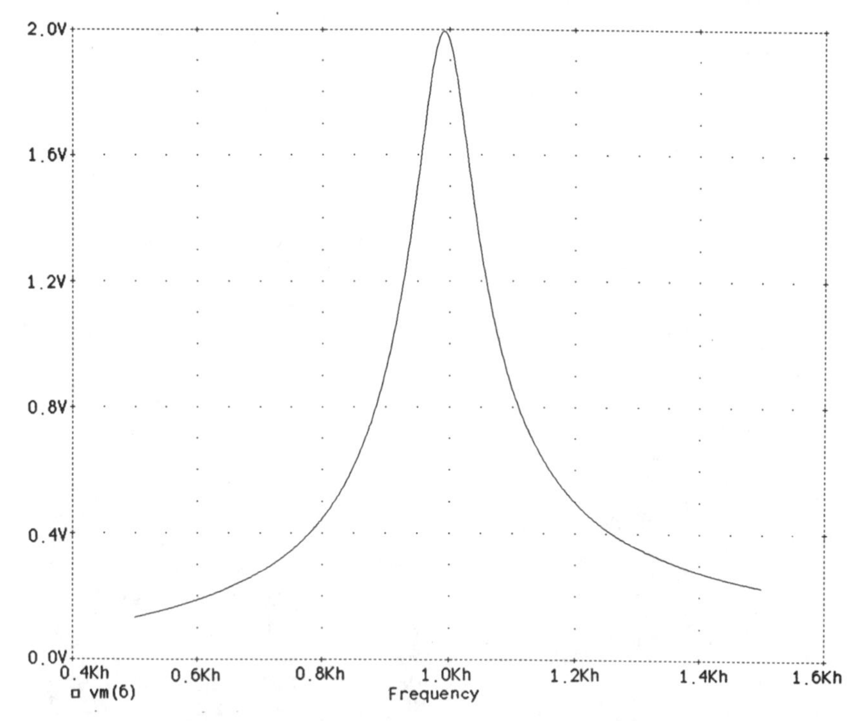

Figure 6-19. Plot of amplitude versus frequency (1 KHz bandpass filter).

The actual center frequency may not be 990 Hz, but rather, a frequency slightly above or below this number. If more accuracy were desired, a second analysis over a much tighter frequency range would be necessary. Assuming that the center frequency is as shown, the 3 dB points must next be determined. At the center frequency, the DB gain is 5.994. The two 3 dB points are at the frequencies where the gain is 5.994 - 3.0 = 2.994 dB. The bandpass filter, being typically symmetric near the peak, has two 3 dB points. This should be obvious to the reader when looking at Figure 6-20. The lower frequency point is between 940 Hz and 950 Hz. The difference in decibels, between these two points, is 3.586 - 2.662 = 0.9240. Dividing decibels by Hertz results in 0.9240 dB/10 Hz =0.0924 dB/Hz. The difference between the desired decibels and the 3.586 is 3.586 - 2.994 = 0.592 dB. Dividing 0.592 dB/0.0924 dB/Hz = 6.41 Hz. Therefore, the lower 3 dB point is 950 Hz - 6.41 Hz = 943.6 Hz. A similar calculation on the upper 3 dB point yield a 3 dB frequency of 1041.89 Hz. The bandwidth is the difference between the upper and lower 3 dB points. The bandwidth is 1041.89 - 943.6 = 98.3 Hz. To compute the predicted Q, we must divide the bandwidth into the center frequency. The computed Q is 990 Hz/98.3 Hz = 10.07 which is close to the design Q of 10.

Although not specified as required information, Figure 6-21 is a high-resolution plot of phase shift, through the filter versus frequency. The plot indicates a phase shift from -90 degrees to -270 degrees around the point of resonance. This is an expected response with a -180 degrees phase shift occurring at the resonant frequency.

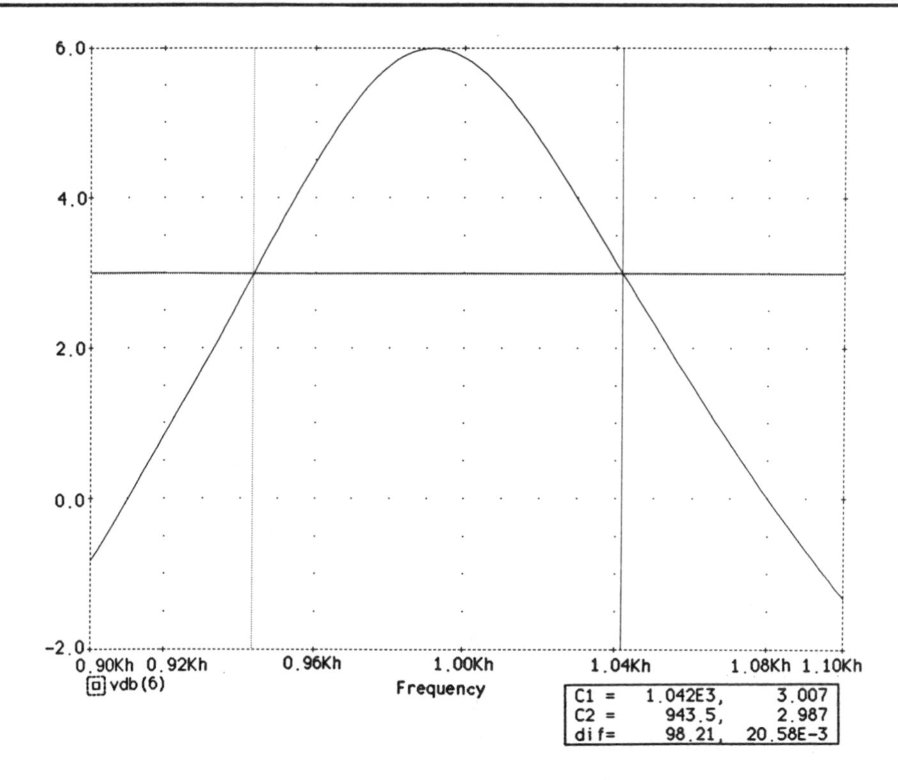

Figure 6-20. Plot of decibel gain versus frequency (1 KHz bandpass filter).

Table 6-7 is a summary of the extracted data and lists the differences between the design and measured parameters. As can be seen slight differences exist between the design parameters and the extracted analysis data. This difference exists due to the ideal nature of the operational amplifier assumed in the equations compared to a realistic operational model used in the analysis. Actual breadboarding of the circuit will favor the SPICE results more than the theoretical results.

Table 6-7. Comparison of Desired versus Actual Results.

Parameter	Design	Measured
Q	10.00	10.07
FC	1 KHz	990 Hz
GAIN	2.000	1.994

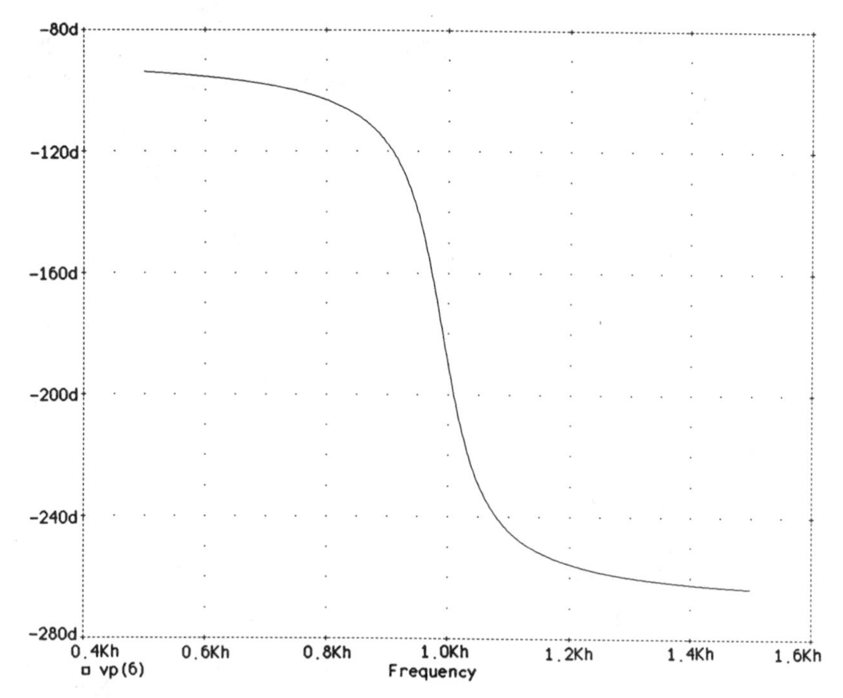

Figure 6-21. Plot of phase shift versus frequency (1 KHz bandpass filter).

The transient analysis is the next analysis presented. The plot of the output node 6 over time is shown. A high-resolution version of the same plot is shown in Figure 6-22. The plot shows a sinusoidal wave with increasing amplitude. This sinusoidal wave was created from a 1 KHz input square wave. Since a square wave is made up of a fundamental and an infinite number of harmonics, if the harmonics are removed, a sinusoidal wave should result. The amplitude of the sinusoidal wave increases with each cycle due to the narrowband characteristics of the filter. This is a normal response. The square wave input was a 2 V peak-to-peak square wave with a zero offset and a 1 msec period. This equated to a 1.273 V sine wave component at 1 KHz. The output would be expected to be approximately 5 V peak to peak because the bandpass filter has a gain of 2. The transient analysis plot shows the output sinusoidal wave approaching 5 V peak to peak as expected and will reach that level in several more cycles. Therefore we have correlated the gain of the AC analysis to that of the transient analysis.

The final output data is the Fourier analysis of the sinusoidal wave. The analysis shows magnitude (Fourier component) and phase of the fundamental frequency (1 KHz) and normalizes the magnitude (Normalized component) and phase for the reader's convenience. The magnitudes and phase of the next eight harmonics are also shown. The analysis indicates that the output of the filter has a DC component of 46.777 mV. This is the amplifier DC offset voltage due to input bias currents and input offset voltage. The end of the Fourier analysis reveals a total harmonic distortion (THD) of 3.4975 percent, which indicates that the filter is working well to eliminate harmonic components of the input square wave.

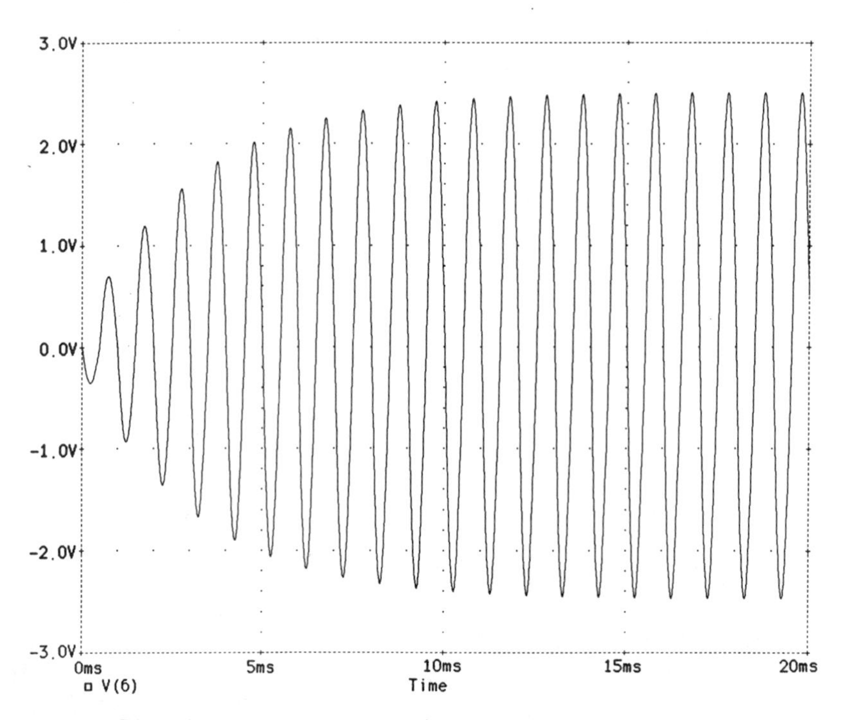

Figure 6-22. Plot of output amplitude V(6) versus time (1 KHz bandpass filter).

6.5 SUMMARY

This chapter has presented a combined AC and transient analysis of a band-pass filter. In doing this, many commands have been used and are summarized below.

1. Voltage source definition

```
VXXXXXXX N+ N- [DC] VALUE [ACMAG [ACPHASE]]
                [PULSE V1 V2 [TD [TR [TF [PW [PER]]]]]]
         or     [SIN VO VA [FREQ [TD [KD]]]]
         or     [EXP V1 V2 [TD1 [T1 [TD2 [T2]]]]]
         or     [PWL T1 V1 T2 V2 ... TN VN]
         or     [SFFM VO VA FREQ [MDI [FS]]]
```

2. Current source definition

```
IXXXXXXX N+ N- [DC] VALUE [ACMAG [ACPHASE]]
                    [PULSE I1 I2 [TD [TR [TF [PW [PER]]]]]]
          or        [SIN IO IA [FREQ [TD [KD]]]]
          or        [EXP I1 I2 [TD1 [T1 [TD2 [T2]]]]]
          or        [PWL T1 I1 T2 I2 ... TN IN]
          or        [SFFM IO IA FREQ [MDI [FS]]]
```

3. Transient analysis statement

```
.TRAN TSTEP TSTOP [TSTART [TMAX]] [UIC]
```

4. Print statement

```
.PRINT TRAN V(X) I(VX)
```

5. Plot statement

```
.PLOT TRAN V(X) I(VX)
```

6. Initial conditions statement

```
.IC V(NODNUM)=VAL V(NODNUM)=VAL ...
```

7. Fourier analysis statement

```
.FOUR FREQ OV1 [OV2 OV3 ...]
```

7

COMBINED AC,DC, AND TRANSIENT ANALYSIS

This chapter presents the concept of a combined AC, DC, and transient analysis. In addition, the discussions of remaining SPICE-oriented statements will be discussed.

7.1 TEMPERATURE COMMAND

Under default conditions, SPICE uses 27°C for the analysis temperature. As previously discussed, all semiconductor devices and resistors are temperature dependent. This dependence is built into the basic semiconductor model and can only be removed with some difficulty. SPICE permits the user to evaluate the effects of temperature on a circuit. This is accomplished by using a .TEMP statement, which overrides the default value of 27°C:

.TEMP T1 [T2] [T3] [TN]

The values of T1, T2, T3, and TN are the temperatures at which the analysis will be performed. The specification of at least one temperature is mandatory with additional temperatures being optional. Assume you are designing a circuit that will have to work over the range of -55°C to 125°C, which is the standard military temperature range. The analysis should also establish a room temperature reference point of 25°C. The following statement will satisfy this requirement.

```
.TEMP 25 -55 125
```

The reader is reminded that these represent ambient temperatures. If a semiconductor device is dissipating 500 mW and has experienced a 30° C internal temperature rise, SPICE has no way to take this into account. The actual device temperature, under these conditions, would be the ambient temperature plus 30° C. The only choice the user has is to specify the analysis temperature at 30° C hotter than the planned ambient. A problem might arise doing this. The other devices in the circuit are analyzed at a temperature 30° C hotter than actual and may yield unrealistic results. No solution exists to solve this problem.

When a temperature analysis is performed, SPICE performs the requested analysis at each temperature specified. If three temperatures are specified, then three analyses will be performed. This should caution the reader to restrict the number of temperatures.

7.2 TEMPERATURE DEPENDENCE

As stated previously, SPICE uses 27° C as the default temperature, but the analysis temperature may be specified in the program. SPICE has built-in temperature parameters associated with all of its semiconductor models. For example, if the temperature goes up, then the base-emitter junction potential (VBE) decreases and the collector-base leakage increases. The parameters associated with temperature effects are shown as follows.

Temperature appears explicitly in the exponential terms of the BJT (bipolar transistor) and diode model equations. In addition, saturation currents have a built-in temperature dependence. The temperature dependence of the saturation current in the BJT models is determined by

$$IS(T1) = IS(T0) * (T1/T0)^{XT1} * \exp(q * EG * (T1-T0)/(k * T1 * T0))$$

where k is Boltzmanns constant, q is the electronic charge, EG is the energy gap, T1 is the temperature of interest (in Kelvins), T0 is the ambient temperature (in Kelvins), and XT1 is the saturation current temperature exponent (usually equal to 3). The temperature dependence of forward and reverse beta is determined by

$$beta(T1) = beta(T0) * (T1/T0)^{XTB}$$

where XTB is the user supplied model parameter. The temperature effects on beta are carried out by appropriate adjustment to the values of BF, ISE, BR, ISC.

The temperature dependence of the saturation current in the junction diode model is determined by

$$IS(T1) = IS(T0) * (T1/T0)^{XT1/n} * \exp(q * EG * (T1-T0)/(k * n * T1 * T0))$$

where n is the emission coefficient, which is a model parameter.

The temperature appears explicitly in the value of the junction potential, PHI, for all models. The temperature dependence is determined by

$$PHI(TEMP) = k * TEMP/q * \log(Na * Nd/(Ni(TEMP)^2)$$

Where TEMP is the circuit evaluation temperature, Na and Nd are the acceptor and donor impurity densities, and Ni is the intrinsic concentration.

Temperature appears explicitly in the value of surface mobility, UO, for the MOSFET model. The temperature dependence is determined by

$$UO(TEMP) = UO(Tnom)/(TEMP/Tnom)^{1.5}$$

where Tnom is the nominal temperature (27° C).

The temperature affects on resistors is determined by

$$value(TEMP) = value(27^\circ C)*[1+TC1*(TEMP-27^\circ C)+TC2*(TEMP-27^\circ C)^2]$$

where TC1 is the first-order (linear) temperature coefficient and
TC2 is the second-order (quadratic) temperature coefficient.

7.3 OTHER SPICE COMMANDS

So far, we have discussed all major SPICE commands. Two commands and an element which have not been discussed are as follows.

1. **.NOISE** - AC noise analysis
2. **.DISTO** - AC distortion analysis
3. **TXXXXXXX** - Transmission line model

These find limited use in most analyses and are discussed here in the interest of completeness.

7.3.1 .NOISE STATEMENT

The **.NOISE** statement causes a noise analysis to be performed, on the circuit, in conjunction with an AC analysis. For the statement to be accepted, an **.AC** statement must exist as part of the circuit analysis input file. The general form of the **.NOISE** statement is shown following.

.NOISE V(N1,[N2]) INSRC SINT

V(N1,N2) is an output node and defines the summing point. INSRC is the name of the independent voltage or current source at which equivalent noise input will be calculated. INSRC is not a noise generator but only a source to compute the equivalent input noise. SINT is the summary interval. SPICE computes the equivalent output noise at the specified output as well as the equivalent input noise at the specified input. In addition, the contributions of every noise generator in the circuit are listed at every SINT frequency point. For example, if SINT is 10, then every 10th frequency in the print interval, defined by the .AC statement, produces a detailed table showing the contribution of each of the circuit noise generator sources. If SINT is zero or absent then no summary output is made.

SPICE uses resistors and semiconductor devices as noise sources. Each noise generator's contribution (resistor or semiconductor) is calculated and propagated to the output node(s) for each frequency defined by the AC analysis. The propagated noise values are summed using root mean square (RMS) as a basis and the gain from the input source to output node is also computed. From this information, an equivalent input noise and total output noise are computed.

The output noise and the equivalent input noise may be optionally printed and/or plotted. If INSRC is a current source, then the input noise units are in Amperes/Hertz$^{.5}$. If the input is a voltage source then the input noise units are in Volts/Hertz$^{.5}$. The output noise units are always in Volts/Hertz$^{.5}$. An example of a noise analysis statement is shown as follows.

```
.NOISE V(5) VIN 10
```

7.3.2 .DISTO STATEMENT

This SPICE function performs an AC distortion analysis looking at sums and differences of certain frequencies at an output. This function has been dropped in may versions of SPICE. The description is included for completeness, but the reader is encouraged to carefully review the capabilities of the SPICE package used for applicability. The general form is shown as follows:

```
.DISTO RLOAD (INTER {SWK2 [REFPWR (SPW2)]})
```

This distortion function controls whether SPICE will compute the distortion characteristics of the circuit in a small-signal mode as part of the AC small-signal sinusoidal steady-state analysis. The analysis is performed assuming that only one or two signal frequencies are imposed at an input. Assume the two frequencies to be F1 (the nominal analysis frequency) and F2 (SKW2 * F1).

RLOAD is the name of the output load resistor into which all distortion power products are to be computed, INTER is the interval at which the summary output of the contributions of all nonlinear devices to the total distortion is to be output - If a value is omitted or set to zero, no summary output will be made, REFPWR is the reference power level used in computing the distortion products - the default value is 1 mW/ dB, SKW2 is the ratio of F2 to F1 - the default value is 0.9 (i.e., F2 = 0.9 * F1), SPW2 is the amplitude of F2 - the default value is 1.0.

The distortion function computes the following distortion measures:

HD2	-	The magnitude of the frequency component 2 * F1 assuming that F2 is not present.
HD3	-	The magnitude of the frequency component 3 * F1 assuming that F2 is not present.
SIM2	-	The magnitude of the frequency component F1 + F2.
DIM2	-	The magnitude of the frequency component F1 - F2.
DIM3	-	The magnitude of the frequency component 2 * F1 - F2.

The distortion measures HD2, HD3, SIM2, DIM2, and DIM3 may also be printed and/or plotted (see the description of the .PRINT and .PLOT commands). An example of the distortion function is as follows:

```
.DISTO RX 4 0.9 0.001 0.7
```

7.3.3 TRANSMISSION LINES (LOSSLESS)

SPICE contains a lossless transmission line model. The basic format is the same as for any SPICE component. The transmission line model is a two-port (four-lead) model and simulates a transmission line at a characteristic impedance. Parameters associated with delay and line length may also be specified. The component finds use in high-frequency radio frequency applications. Because of its lossless nature, it may not suitably simulate the required environment. An acceptable alternative may be to model a transmission line using resistors, capacitors, and inductors. The correct selection of components can produce a model that more closely approximates an actual situation without a significant level of effort. In creating a transmission line using RLC components, the a simple approach may be the definition and use of subcircuits to simulate lumped elements.

The general form of the SPICE lossless transmission line is as follows:

TXXXXXXX N1 N2 N3 N4 Z0=VALUE [TD=VALUE] [F=FREQ [NL=NRMLEN]] [IC=V1,I1,V2,I2]

where N1 and N2 are nodes of port 1, N3 and N4 are nodes of port 2, Z0 is the characteristic impedance of the transmission line, TD is the transmission delay, F is the optional frequency in conjunction with NL, NL is the normalized electrical length, and IC are the initial conditions of V1, I1, V2, I2.

The length of the line may be expressed in either of two forms: The transmission delay in seconds, TD, may be specified directly (as TD = 10 nsec, for example)or a frequency F may be given, together with NL. NL is the normalized electrical length of the transmission line with respect to the wavelength in the line at the frequency F.

If a frequency is specified but NL is omitted, a value for NL of 0.25 is assumed. This default assumes a quarter-wave frequency for NL. Note that although both forms for expressing the line length are indicated as optional, one of the two must be specified. This element models only one propagating mode. If all four nodes are distinct in the actual circuit, then two nodes may be excited. To simulate such a situation, two transmission line elements are required.

The optional initial condition specification consists of the voltage and current at each of the transmission line ports. The initial conditions, if any, apply only if the UIC option is specified on the .TRAN statement.
Possible examples of a delay line definition are

```
TA 1 2 3 4 Z0=50 TD=235NS
T1 1 0 2 0 Z0=75 F=49.9MEG
TRANS 2 99 3 88 Z0=93 F=4MEG NL=0.5
```

7.4 COMBINED PARAMETER ANALYSIS

The circuit shown in Figure 7-1 will be used to demonstrate the full SPICE analysis capability. The circuit will be analyzed using the following commands:

```
.OP             .TF
.SENS           .TEMP
.DC             .PLOT DC
.PRINT DC       .AC
.PLOT AC        .PRINT AC
.TRAN           .PLOT TRAN
.PRINT TRAN     .OPTIONS
.FOUR
```

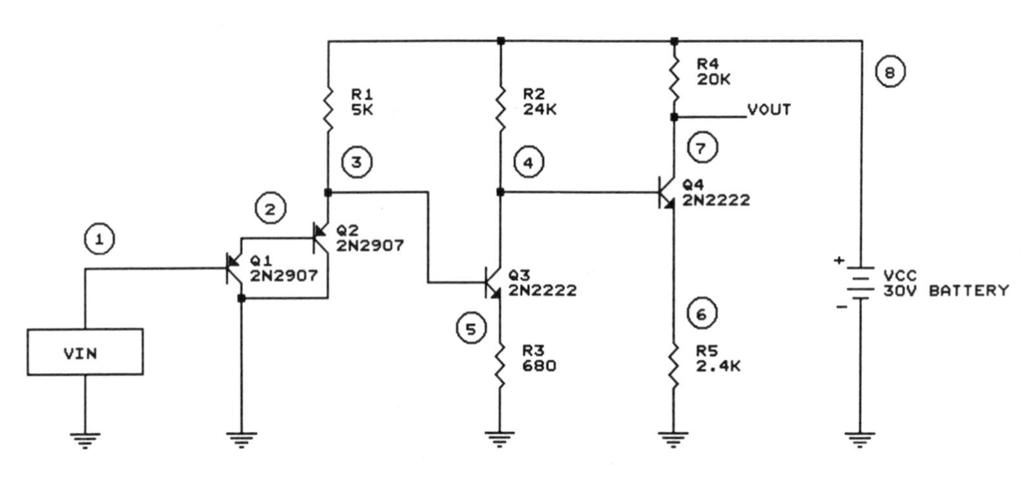

Figure 7-1. Transistor amplifier example.

The .SUBCKT command has been excluded for convenience. The circuit, of Figure 7-1, can be described as follows.

```
FOUR TRANSISTOR AMPLIFIER
VCC 8 0 30
VIN 1 0 DC 0 AC 1 SIN(20E-3 10E-3 1K 0 0)
Q1 0 1 2 Q2N2907A
Q2 0 2 3 Q2N2907A
Q3 4 3 5 Q2N2222A
Q4 7 4 6 Q2N2222A

* TEMPERATURE COEFFICIENT OF 100PPM
R1 8 3 5K TC=1E-4
R2 8 4 24K TC=1E-4
R3 5 0 680 TC=1E-4
R4 8 7 20K TC=1E-4
R5 6 0 2.4K TC=1E-4

* TRANSISTOR DEFINITIONS
.MODEL Q2N2222A NPN(IS=14.34F XTI=3 EG=1.11 VAF=74.03 BF=255.9
+ NE=1.307 ISE=14.34F IKF=.2847 XTB=1.5 BR=6.092 NC=2 ISC=0 IKR=0
+ RC=1 CJC=7.306P MJC=.3416 VJC=.75 FC=.5 CJE=22.01P MJE=.377
+ VJE=.75 TR=46.91N TF=411.1P ITF=.6 VTF=1.7 XTF=3 RB=10)
.MODEL Q2N2907A PNP(IS=650.6E-18 XTI=3 EG=1.11 VAF=115.7 BF=231.7
+ NE=1.829 ISE=54.81F IKF=1.079 XTB=1.5 BR=3.563 NC=2 ISC=0
+ IKR=0 RC=.715 CJC=14.76P MJC=.5383 VJC=.75 FC=.5 CJE=19.82P
+ MJE=.3357 VJE=.75 TR=111.3N TF=603.7P ITF=.65 VTF=5 XTF=1.7
+ RB=10)
```

```
* GENERAL COMMANDS
.TEMP 25 70 0
.OPTIONS NODE LIST LIMPTS=5001 ITL5=0

* OPERATING POINT TYPE COMMANDS
.OP
.TF V(7) VIN
.SENS V(7)

* TRANSIENT ANALYSIS COMMANDS
.TRAN 40.0U 4.0M 0 40.0U
.FOUR 1.0K V(7)
.PRINT TRAN V(7)
.PLOT TRAN V(7)

* AC ANALYSIS COMMANDS
.AC DEC 10 10 1MEG
.PRINT AC VM(7) VP(7) VDB(7) IM(VIN)
.PLOT AC VDB(7) VP(7) VM(7) IM(VIN)

* DC ANALYSIS COMMANDS
.DC VIN -.2 .2 2.5M
.PRINT DC V(7)
.PLOT DC V(7)
.END
```

The analysis is attempting to find out information related to the following:

1. Input linearity range and the DC gain of the circuit
2. Operating point information and circuit power consumption
3. Input resistance, output resistance, and DC gain at the operating point parameters
4. The component that produces the greatest effect on the output
5. The AC midband gain (1000 Hz) and the 3-dB point of the output.
6. The peak to peak output with a 20 mV peak-to-peak input sine wave applied and the associated gain
7. The total harmonic distortion at the output
8. The effect of three temperatures on these results and a resulting summary of those affects.

7.5 COMBINED ANALYSIS RESULTS AND CONCLUSIONS

The results of the analysis are shown below. Because of the extensive analysis requested, and the three temperatures requested, the reader will find the analysis results to be lengthy.

7.5.1 SPICE ANALYSIS RESULTS

```
 FOUR TRANSISTOR AMPLIFIER

 ****     CIRCUIT DESCRIPTION

******************************************************************************

VCC 8 0 30
VIN 1 0 DC 0 AC 1 SIN(20E-3 10E-3 1K 0 0)
Q1 0 1 2 Q2N2907A
Q2 0 2 3 Q2N2907A
Q3 4 3 5 Q2N2222A
Q4 7 4 6 Q2N2222A

* TEMPERATURE COEFFICIENT OF 100PPM
R1 8 3 5K TC=1E-4
R2 8 4 24K TC=1E-4
R3 5 0 680 TC=1E-4
R4 8 7 20K TC=1E-4
R5 6 0 2.4K TC=1E-4

* TRANSISTOR DEFINITIONS
.MODEL Q2N2222A NPN(IS=14.34F XTI=3 EG=1.11 VAF=74.03 BF=255.9
+ NE=1.307 ISE=14.34F IKF=.2847 XTB=1.5 BR=6.092 NC=2 ISC=0 IKR=0
+ RC=1 CJC=7.306P MJC=.3416 VJC=.75 FC=.5 CJE=22.01P MJE=.377
+ VJE=.75 TR=46.91N TF=411.1P ITF=.6 VTF=1.7 XTF=3 RB=10)
.MODEL Q2N2907A PNP(IS=650.6E-18 XTI=3 EG=1.11 VAF=115.7 BF=231.7
+ NE=1.829 ISE=54.81F IKF=1.079 XTB=1.5 BR=3.563 NC=2 ISC=0
+ IKR=0 RC=.715 CJC=14.76P MJC=.5383 VJC=.75 FC=.5 CJE=19.82P
+ MJE=.3357 VJE=.75 TR=111.3N TF=603.7P ITF=.65 VTF=5 XTF=1.7
+ RB=10)

* GENERAL COMMANDS
.TEMP 25 70 0
.OPTIONS NODE LIST LIMPTS=5001 ITL5=0

* OPERATING POINT TYPE COMMANDS
.OP
.TF V(7) VIN
.SENS V(7)

* TRANSIENT ANALYSIS COMMANDS
.TRAN 40.0U 4.0M 0 40.0U
.FOUR 1.0K V(7)
.PRINT TRAN V(7)
.PLOT TRAN V(7)

* AC ANALYSIS COMMANDS
.AC DEC 10 10 1MEG
.PRINT AC VM(7) VP(7) VDB(7) IM(VIN)
.PLOT AC VDB(7) VP(7) VM(7) IM(VIN)

* DC ANALYSIS COMMANDS
.DC VIN -.2 .2 2.5M
.PRINT DC V(7)
.PLOT DC V(7)
.END
```

```
FOUR TRANSISTOR AMPLIFIER

 ****     ELEMENT NODE TABLE

******************************************************************************

0               Q1          Q2          R3          R5          VCC
                VIN

1               Q1          VIN

2               Q1          Q2

3               Q2          Q3          R1

4               Q3          Q4          R2

5               Q3          R3

6               Q4          R5

7               Q4          R4

8               R1          R2          R4          VCC
```

```
 FOUR TRANSISTOR AMPLIFIER

 ****     BJT MODEL PARAMETERS
******************************************************************************
             Q2N2222A         Q2N2907A
             NPN              PNP
       IS     14.340000E-15   650.600000E-18
       BF    255.9            231.7
       NF      1                1
      VAF     74.03           115.7
      IKF       .2847           1.079
      ISE     14.340000E-15    54.810000E-15
       NE      1.307            1.829
       BR      6.092            3.563
       NR      1                1
       RB     10               10
      RBM     10               10
       RC      1                 .715
      CJE     22.010000E-12    19.820000E-12
      MJE       .377             .3357
      CJC      7.306000E-12    14.760000E-12
      MJC       .3416            .5383
       TF    411.100000E-12   603.700000E-12
      XTF      3                1.7
      VTF      1.7              5
      ITF       .6               .65
       TR     46.910000E-09   111.300000E-09
      XTB      1.5              1.5
```

```
FOUR TRANSISTOR AMPLIFIER
 ****     CIRCUIT ELEMENT SUMMARY
******************************************************************************
**** RESISTORS

NAME                NODES      MODEL      VALUE       TC1        TC2        TCE

R1                  8      3              5.00E+03  1.00E-04  0.00E+00  0.00E+00
R2                  8      4              2.40E+04  1.00E-04  0.00E+00  0.00E+00
R3                  5      0              6.80E+02  1.00E-04  0.00E+00  0.00E+00
R4                  8      7              2.00E+04  1.00E-04  0.00E+00  0.00E+00
R5                  6      0              2.40E+03  1.00E-04  0.00E+00  0.00E+00

**** INDEPENDENT SOURCES

NAME                NODES    DC VALUE  AC VALUE  AC PHASE
VCC                 8      0 3.00E+01  0.00E+00  0.00E+00  degrees
VIN                 1      0 0.00E+00  1.00E+00  0.00E+00  degrees

                 TRANSIENT:  SIN
                     Offset  2.00E-02
                  Amplitude  1.00E-02
                       Freq  1.00E+03
                      Delay  0.00E+00
                    Damping  0.00E+00
                      Phase  0.00E+00

**** BIPOLAR JUNCTION TRANSISTORS

NAME                C       B       E       S MODEL       AREA
Q1                  0       1       2       0 Q2N2907A    1.00E+00
Q2                  0       2       3       0 Q2N2907A    1.00E+00
Q3                  4       3       5       0 Q2N2222A    1.00E+00
Q4                  7       4       6       0 Q2N2222A    1.00E+00
```

```
FOUR TRANSISTOR AMPLIFIER
 ****     TEMPERATURE-ADJUSTED VALUES       TEMPERATURE =    25.000 DEG C
******************************************************************************
 **** RESISTORS
 NAME        VALUE
 R1         4.999E+03
 R2         2.400E+04
 R3         6.799E+02
 R4         2.000E+04
 R5         2.400E+03

 **** BJT MODEL PARAMETERS
 NAME          BF         ISE        VJE        CJE        RE         RB
               BR         ISC        VJC        CJC        RC         RBM
               IS         ISS        VJS        CJS
 Q2N2222A   2.533E+02  1.144E-14  7.535E-01  2.196E-11  0.000E+00  1.000E+01
            6.031E+00  0.000E+00  7.535E-01  7.292E-12  1.000E+00  1.000E+01
            1.054E-14  0.000E+00  7.535E-01  0.000E+00
 Q2N2907A   2.294E+02  4.678E-14  7.535E-01  1.978E-11  0.000E+00  1.000E+01
            3.527E+00  0.000E+00  7.535E-01  1.472E-11  7.150E-01  1.000E+01
            4.782E-16  0.000E+00  7.535E-01  0.000E+00
```

```
FOUR TRANSISTOR AMPLIFIER

 ****     DC TRANSFER CURVES                TEMPERATURE =   25.000 DEG C

******************************************************************************
  VIN          V(7)

  -2.000E-01   3.749E+00
  -1.975E-01   3.743E+00
  -1.950E-01   3.736E+00
  -1.925E-01   3.729E+00
  -1.900E-01   3.722E+00
  -1.875E-01   3.716E+00
  -1.850E-01   3.709E+00
  -1.825E-01   3.702E+00
  -1.800E-01   3.696E+00
  -1.775E-01   3.689E+00
  -1.750E-01   3.683E+00
  -1.725E-01   3.676E+00
  -1.700E-01   3.669E+00
  -1.675E-01   3.663E+00
  -1.650E-01   3.656E+00
  -1.625E-01   3.649E+00
  -1.600E-01   3.643E+00
  -1.575E-01   3.636E+00
  -1.550E-01   3.629E+00
  -1.525E-01   3.623E+00
  -1.500E-01   3.616E+00
  -1.475E-01   3.610E+00
  -1.450E-01   3.603E+00
  -1.425E-01   3.596E+00
  -1.400E-01   3.590E+00
  -1.375E-01   3.583E+00
  -1.350E-01   3.577E+00
  -1.325E-01   3.570E+00
  -1.300E-01   3.564E+00
  -1.275E-01   3.557E+00
  -1.250E-01   3.550E+00
  -1.225E-01   3.544E+00
  -1.200E-01   3.537E+00
  -1.175E-01   3.531E+00
  -1.150E-01   3.524E+00
  -1.125E-01   3.518E+00
  -1.100E-01   3.511E+00
  -1.075E-01   3.505E+00
  -1.050E-01   3.499E+00
  -1.025E-01   3.492E+00
  -1.000E-01   3.486E+00
  -9.750E-02   3.479E+00
  -9.500E-02   3.473E+00
  -9.250E-02   3.467E+00
  -9.000E-02   3.460E+00
  -8.750E-02   3.454E+00
  -8.500E-02   3.448E+00
  -8.250E-02   3.442E+00
  -8.000E-02   3.435E+00
  -7.750E-02   3.429E+00
  -7.500E-02   3.423E+00
  -7.250E-02   3.417E+00
  -7.000E-02   3.411E+00
  -6.750E-02   3.405E+00
  -6.500E-02   3.399E+00
```

VIN	V(7)
-6.250E-02	3.393E+00
-6.000E-02	3.387E+00
-5.750E-02	3.382E+00
-5.500E-02	3.376E+00
-5.250E-02	3.370E+00
-5.000E-02	3.365E+00
-4.750E-02	3.360E+00
-4.500E-02	3.355E+00
-4.250E-02	3.350E+00
-4.000E-02	3.345E+00
-3.750E-02	3.341E+00
-3.500E-02	3.337E+00
-3.250E-02	3.334E+00
-3.000E-02	3.331E+00
-2.750E-02	3.330E+00
-2.500E-02	3.330E+00
-2.250E-02	3.334E+00
-2.000E-02	3.350E+00
-1.750E-02	3.622E+00
-1.500E-02	4.252E+00
-1.250E-02	4.882E+00
-1.000E-02	5.514E+00
-7.500E-03	6.146E+00
-5.000E-03	6.779E+00
-2.500E-03	7.413E+00
0.000E+00	8.047E+00
2.500E-03	8.682E+00
5.000E-03	9.317E+00
7.500E-03	9.953E+00
1.000E-02	1.059E+01
1.250E-02	1.123E+01
1.500E-02	1.186E+01
1.750E-02	1.250E+01
2.000E-02	1.314E+01
2.250E-02	1.378E+01
2.500E-02	1.442E+01
2.750E-02	1.506E+01
3.000E-02	1.570E+01
3.250E-02	1.634E+01
3.500E-02	1.698E+01
3.750E-02	1.762E+01
4.000E-02	1.826E+01
4.250E-02	1.890E+01
4.500E-02	1.954E+01
4.750E-02	2.017E+01
5.000E-02	2.081E+01
5.250E-02	2.145E+01
5.500E-02	2.209E+01
5.750E-02	2.272E+01
6.000E-02	2.336E+01
6.250E-02	2.399E+01
6.500E-02	2.462E+01
6.750E-02	2.525E+01
7.000E-02	2.588E+01
7.250E-02	2.648E+01
7.500E-02	2.691E+01
7.750E-02	2.711E+01
8.000E-02	2.721E+01
8.250E-02	2.727E+01

VIN	V(7)
8.500E-02	2.731E+01
8.750E-02	2.733E+01
9.000E-02	2.735E+01
9.250E-02	2.737E+01
9.500E-02	2.738E+01
9.750E-02	2.738E+01
1.000E-01	2.738E+01
1.025E-01	2.739E+01
1.050E-01	2.738E+01
1.075E-01	2.738E+01
1.100E-01	2.738E+01
1.125E-01	2.738E+01
1.150E-01	2.737E+01
1.175E-01	2.736E+01
1.200E-01	2.736E+01
1.225E-01	2.735E+01
1.250E-01	2.734E+01
1.275E-01	2.733E+01
1.300E-01	2.732E+01
1.325E-01	2.731E+01
1.350E-01	2.731E+01
1.375E-01	2.729E+01
1.400E-01	2.728E+01
1.425E-01	2.727E+01
1.450E-01	2.726E+01
1.475E-01	2.725E+01
1.500E-01	2.724E+01
1.525E-01	2.723E+01
1.550E-01	2.722E+01
1.575E-01	2.720E+01
1.600E-01	2.719E+01
1.625E-01	2.718E+01
1.650E-01	2.717E+01
1.675E-01	2.715E+01
1.700E-01	2.714E+01
1.725E-01	2.713E+01
1.750E-01	2.711E+01
1.775E-01	2.710E+01
1.800E-01	2.709E+01
1.825E-01	2.707E+01
1.850E-01	2.706E+01
1.875E-01	2.704E+01
1.900E-01	2.703E+01
1.925E-01	2.701E+01
1.950E-01	2.700E+01
1.975E-01	2.699E+01
2.000E-01	2.697E+01

```
FOUR TRANSISTOR AMPLIFIER

 ****     DC TRANSFER CURVES                    TEMPERATURE =   25.000 DEG C

******************************************************************************

  VIN         V(7)
(*)----------    0.0000E+00    1.0000E+01    2.0000E+01    3.0000E+01    4.0000E+01
                            - - - - - - - - - - - - - - - - - - - - - - - - - - - - -
 -2.000E-01  3.749E+00 .    *        .             .             .             .
 -1.975E-01  3.743E+00 .    *        .             .             .             .
 -1.950E-01  3.736E+00 .    *        .             .             .             .
 -1.925E-01  3.729E+00 .    *        .             .             .             .
 -1.900E-01  3.722E+00 .    *        .             .             .             .
 -1.875E-01  3.716E+00 .    *        .             .             .             .
 -1.850E-01  3.709E+00 .    *        .             .             .             .
 -1.825E-01  3.702E+00 .    *        .             .             .             .
 -1.800E-01  3.696E+00 .    *        .             .             .             .
 -1.775E-01  3.689E+00 .    *        .             .             .             .
 -1.750E-01  3.683E+00 .    *        .             .             .             .
 -1.725E-01  3.676E+00 .    *        .             .             .             .
 -1.700E-01  3.669E+00 .    *        .             .             .             .
 -1.675E-01  3.663E+00 .    *        .             .             .             .
 -1.650E-01  3.656E+00 .    *        .             .             .             .
 -1.625E-01  3.649E+00 .    *        .             .             .             .
 -1.600E-01  3.643E+00 .    *        .             .             .             .
 -1.575E-01  3.636E+00 .    *        .             .             .             .
 -1.550E-01  3.629E+00 .    *        .             .             .             .
 -1.525E-01  3.623E+00 .    *        .             .             .             .
 -1.500E-01  3.616E+00 .    *        .             .             .             .
 -1.475E-01  3.610E+00 .    *        .             .             .             .
 -1.450E-01  3.603E+00 .    *        .             .             .             .
 -1.425E-01  3.596E+00 .    *        .             .             .             .
 -1.400E-01  3.590E+00 .    *        .             .             .             .
 -1.375E-01  3.583E+00 .    *        .             .             .             .
 -1.350E-01  3.577E+00 .    *        .             .             .             .
 -1.325E-01  3.570E+00 .    *        .             .             .             .
 -1.300E-01  3.564E+00 .    *        .             .             .             .
 -1.275E-01  3.557E+00 .    *        .             .             .             .
 -1.250E-01  3.550E+00 .    *        .             .             .             .
 -1.225E-01  3.544E+00 .    *        .             .             .             .
 -1.200E-01  3.537E+00 .    *        .             .             .             .
 -1.175E-01  3.531E+00 .    *        .             .             .             .
 -1.150E-01  3.524E+00 .    *        .             .             .             .
 -1.125E-01  3.518E+00 .    *        .             .             .             .
 -1.100E-01  3.511E+00 .    *        .             .             .             .
 -1.075E-01  3.505E+00 .    *        .             .             .             .
 -1.050E-01  3.499E+00 .    *        .             .             .             .
 -1.025E-01  3.492E+00 .    *        .             .             .             .
 -1.000E-01  3.486E+00 .    *        .             .             .             .
 -9.750E-02  3.479E+00 .    *        .             .             .             .
 -9.500E-02  3.473E+00 .    *        .             .             .             .
 -9.250E-02  3.467E+00 .    *        .             .             .             .
 -9.000E-02  3.460E+00 .   *         .             .             .             .
 -8.750E-02  3.454E+00 .   *         .             .             .             .
 -8.500E-02  3.448E+00 .   *         .             .             .             .
 -8.250E-02  3.442E+00 .   *         .             .             .             .
 -8.000E-02  3.435E+00 .   *         .             .             .             .
 -7.750E-02  3.429E+00 .   *         .             .             .             .
 -7.500E-02  3.423E+00 .   *         .             .             .             .
 -7.250E-02  3.417E+00 .   *         .             .             .             .
```

```
   VIN        V(7)
(*)---------    0.0000E+00 1.0000E+01 2.0000E+01 3.0000E+01 4.0000E+01
                        - - - - - - - - - - - - - - - - - - - - -
-7.000E-02  3.411E+00  .  *      .         .         .         .
-6.750E-02  3.405E+00  .  *      .         .         .         .
-6.500E-02  3.399E+00  .  *      .         .         .         .
-6.250E-02  3.393E+00  .  *      .         .         .         .
-6.000E-02  3.387E+00  .  *      .         .         .         .
-5.750E-02  3.382E+00  .  *      .         .         .         .
-5.500E-02  3.376E+00  .  *      .         .         .         .
-5.250E-02  3.370E+00  .  *      .         .         .         .
-5.000E-02  3.365E+00  .  *      .         .         .         .
-4.750E-02  3.360E+00  .  *      .         .         .         .
-4.500E-02  3.355E+00  .  *      .         .         .         .
-4.250E-02  3.350E+00  .  *      .         .         .         .
-4.000E-02  3.345E+00  .  *      .         .         .         .
-3.750E-02  3.341E+00  .  *      .         .         .         .
-3.500E-02  3.337E+00  .  *      .         .         .         .
-3.250E-02  3.334E+00  .  *      .         .         .         .
-3.000E-02  3.331E+00  .  *      .         .         .         .
-2.750E-02  3.330E+00  .  *      .         .         .         .
-2.500E-02  3.330E+00  .  *      .         .         .         .
-2.250E-02  3.334E+00  .  *      .         .         .         .
-2.000E-02  3.350E+00  .  *      .         .         .         .
-1.750E-02  3.622E+00  .   *     .         .         .         .
-1.500E-02  4.252E+00  .   *     .         .         .         .
-1.250E-02  4.882E+00  .    *    .         .         .         .
-1.000E-02  5.514E+00  .     *   .         .         .         .
-7.500E-03  6.146E+00  .     *   .         .         .         .
-5.000E-03  6.779E+00  .      *  .         .         .         .
-2.500E-03  7.413E+00  .      *  .         .         .         .
 0.000E+00  8.047E+00  .       * .         .         .         .
 2.500E-03  8.682E+00  .        *.         .         .         .
 5.000E-03  9.317E+00  .        *.         .         .         .
 7.500E-03  9.953E+00  .         *         .         .         .
 1.000E-02  1.059E+01  .         .*        .         .         .
 1.250E-02  1.123E+01  .         .*        .         .         .
 1.500E-02  1.186E+01  .         . *       .         .         .
 1.750E-02  1.250E+01  .         . *       .         .         .
 2.000E-02  1.314E+01  .         .  *      .         .         .
 2.250E-02  1.378E+01  .         .   *     .         .         .
 2.500E-02  1.442E+01  .         .   *     .         .         .
 2.750E-02  1.506E+01  .         .    *    .         .         .
 3.000E-02  1.570E+01  .         .     *   .         .         .
 3.250E-02  1.634E+01  .         .     *   .         .         .
 3.500E-02  1.698E+01  .         .      *  .         .         .
 3.750E-02  1.762E+01  .         .       * .         .         .
 4.000E-02  1.826E+01  .         .       * .         .         .
 4.250E-02  1.890E+01  .         .        *.         .         .
 4.500E-02  1.954E+01  .         .        *.         .         .
 4.750E-02  2.017E+01  .         .         *         .         .
 5.000E-02  2.081E+01  .         .         .*        .         .
 5.250E-02  2.145E+01  .         .         .*        .         .
 5.500E-02  2.209E+01  .         .         . *       .         .
 5.750E-02  2.272E+01  .         .         .  *      .         .
 6.000E-02  2.336E+01  .         .         .  *      .         .
 6.250E-02  2.399E+01  .         .         .   *     .         .
 6.500E-02  2.462E+01  .         .         .    *    .         .
 6.750E-02  2.525E+01  .         .         .    *    .         .
 7.000E-02  2.588E+01  .         .         .     *   .         .
 7.250E-02  2.648E+01  .         .         .     *   .         .
```

```
  VIN          V(7)
(*)----------    0.0000E+00   1.0000E+01   2.0000E+01   3.0000E+01   4.0000E+01

                       - - - - - - - - - - - - - - - - - - - - - - - - - - -
  7.500E-02  2.691E+01 .            .            .        *   .            .
  7.750E-02  2.711E+01 .            .            .        *   .            .
  8.000E-02  2.721E+01 .            .            .        *   .            .
  8.250E-02  2.727E+01 .            .            .        *   .            .
  8.500E-02  2.731E+01 .            .            .         *  .            .
  8.750E-02  2.733E+01 .            .            .         *  .            .
  9.000E-02  2.735E+01 .            .            .         *  .            .
  9.250E-02  2.737E+01 .            .            .         *  .            .
  9.500E-02  2.738E+01 .            .            .         *  .            .
  9.750E-02  2.738E+01 .            .            .         *  .            .
  1.000E-01  2.738E+01 .            .            .         *  .            .
  1.025E-01  2.739E+01 .            .            .         *  .            .
  1.050E-01  2.738E+01 .            .            .         *  .            .
  1.075E-01  2.738E+01 .            .            .         *  .            .
  1.100E-01  2.738E+01 .            .            .         *  .            .
  1.125E-01  2.738E+01 .            .            .         *  .            .
  1.150E-01  2.737E+01 .            .            .         *  .            .
  1.175E-01  2.736E+01 .            .            .         *  .            .
  1.200E-01  2.736E+01 .            .            .         *  .            .
  1.225E-01  2.735E+01 .            .            .         *  .            .
  1.250E-01  2.734E+01 .            .            .         *  .            .
  1.275E-01  2.733E+01 .            .            .         *  .            .
  1.300E-01  2.732E+01 .            .            .         *  .            .
  1.325E-01  2.731E+01 .            .            .         *  .            .
  1.350E-01  2.731E+01 .            .            .        *   .            .
  1.375E-01  2.729E+01 .            .            .        *   .            .
  1.400E-01  2.728E+01 .            .            .        *   .            .
  1.425E-01  2.727E+01 .            .            .        *   .            .
  1.450E-01  2.726E+01 .            .            .        *   .            .
  1.475E-01  2.725E+01 .            .            .        *   .            .
  1.500E-01  2.724E+01 .            .            .        *   .            .
  1.525E-01  2.723E+01 .            .            .        *   .            .
  1.550E-01  2.722E+01 .            .            .        *   .            .
  1.575E-01  2.720E+01 .            .            .        *   .            .
  1.600E-01  2.719E+01 .            .            .        *   .            .
  1.625E-01  2.718E+01 .            .            .        *   .            .
  1.650E-01  2.717E+01 .            .            .        *   .            .
  1.675E-01  2.715E+01 .            .            .        *   .            .
  1.700E-01  2.714E+01 .            .            .        *   .            .
  1.725E-01  2.713E+01 .            .            .        *   .            .
  1.750E-01  2.711E+01 .            .            .        *   .            .
  1.775E-01  2.710E+01 .            .            .        *   .            .
  1.800E-01  2.709E+01 .            .            .        *   .            .
  1.825E-01  2.707E+01 .            .            .        *   .            .
  1.850E-01  2.706E+01 .            .            .        *   .            .
  1.875E-01  2.704E+01 .            .            .        *   .            .
  1.900E-01  2.703E+01 .            .            .        *   .            .
  1.925E-01  2.701E+01 .            .            .        *   .            .
  1.950E-01  2.700E+01 .            .            .        *   .            .
  1.975E-01  2.699E+01 .            .            .        *   .            .
  2.000E-01  2.697E+01 .            .            .        *   .            .
                       - - - - - - - - - - - - - - - - - - - - - - - - - - -
```

```
FOUR TRANSISTOR AMPLIFIER

 ****     SMALL SIGNAL BIAS SOLUTION        TEMPERATURE =   25.000 DEG C

******************************************************************************

 NODE   VOLTAGE     NODE   VOLTAGE     NODE   VOLTAGE     NODE   VOLTAGE

(    1)    0.0000  (    2)     .6344  (    3)    1.4081  (    4)    3.3016

(    5)     .7566  (    6)    2.6512  (    7)    8.0469  (    8)   30.0000

    VOLTAGE SOURCE CURRENTS
    NAME         CURRENT

    VCC         -7.930E-03
    VIN          1.444E-07

    TOTAL POWER DISSIPATION   2.38E-01  WATTS
```

```
FOUR TRANSISTOR AMPLIFIER

 ****     OPERATING POINT INFORMATION       TEMPERATURE =   25.000 DEG C

******************************************************************************

**** BIPOLAR JUNCTION TRANSISTORS

NAME         Q1           Q2           Q3           Q4
MODEL        Q2N2907A     Q2N2907A     Q2N2222A     Q2N2222A
IB          -1.44E-07    -2.54E-05     7.32E-06     7.03E-06
IC          -2.53E-05    -5.69E-03     1.11E-03     1.10E-03
VBE         -6.34E-01    -7.74E-01     6.51E-01     6.50E-01
VBC          0.00E+00     6.34E-01    -1.89E+00    -4.75E+00
VCE         -6.34E-01    -1.41E+00     2.55E+00     5.40E+00
BETADC       1.75E+02     2.23E+02     1.51E+02     1.56E+02
GM           9.85E-04     2.20E-01     4.29E-02     4.26E-02
RPI          1.99E+05     1.02E+03     3.89E+03     4.05E+03
RX           1.00E+01     1.00E+01     1.00E+01     1.00E+01
RO           4.57E+06     2.05E+04     6.87E+04     7.18E+04
CBE          3.13E-11     1.67E-10     5.40E-11     5.38E-11
CBC          1.47E-11     1.06E-11     4.75E-12     3.70E-12
CBX          0.00E+00     0.00E+00     0.00E+00     0.00E+00
CJS          0.00E+00     0.00E+00     0.00E+00     0.00E+00
BETAAC       1.96E+02     2.25E+02     1.67E+02     1.73E+02
FT           3.41E+06     1.98E+08     1.16E+08     1.18E+08

 ****     SMALL-SIGNAL CHARACTERISTICS

      V(7)/VIN =  2.538E+02

      INPUT RESISTANCE AT VIN =  1.442E+08

      OUTPUT RESISTANCE AT V(7) =  1.963E+04
```

```
FOUR TRANSISTOR AMPLIFIER

 ****     DC SENSITIVITY ANALYSIS          TEMPERATURE =   25.000 DEG C

********************************************************************************
DC SENSITIVITIES OF OUTPUT V(7)

          ELEMENT         ELEMENT        ELEMENT        NORMALIZED
          NAME            VALUE          SENSITIVITY    SENSITIVITY
                                         (VOLTS/UNIT) (VOLTS/PERCENT)
          R1              4.999E+03      -2.626E-03     -1.313E-01
          R2              2.400E+04       8.350E-03      2.004E+00
          R3              6.799E+02      -2.833E-01     -1.926E+00
          R4              2.000E+04      -1.078E-03     -2.155E-01
          R5              2.400E+03       8.425E-03      2.022E-01
          VCC             3.000E+01      -6.064E+00     -1.819E+00
          VIN             0.000E+00       2.538E+02      0.000E+00
Q1
          RB              1.000E+01       3.666E-05      3.666E-06
          RC              7.150E-01       1.419E-06      1.015E-08
          RE              0.000E+00       0.000E+00      0.000E+00
          BF              2.294E+02       1.221E-04      2.801E-04
          ISE             4.678E-14      -1.853E+11     -8.669E-05
          BR              3.527E+00      -6.322E-17     -2.230E-18
          ISC             0.000E+00       0.000E+00      0.000E+00
          IS              4.782E-16      -1.363E+16     -6.518E-02
          NE              1.829E+00       6.399E-02      1.170E-03
          NC              2.000E+00       0.000E+00      0.000E+00
          IKF             1.079E+00      -1.410E-04     -1.522E-06
          IKR             0.000E+00       0.000E+00      0.000E+00
          VAF             1.157E+02      -8.070E-09     -9.337E-09
          VAR             0.000E+00       0.000E+00      0.000E+00
Q2
          RB              1.000E+01       6.460E-03      6.460E-04
          RC              7.150E-01       6.380E-04      4.561E-06
          RE              0.000E+00       0.000E+00      0.000E+00
          BF              2.294E+02      -2.774E-02     -6.364E-02
          ISE             4.678E-14       3.617E+12      1.692E-03
          BR              3.527E+00       9.956E-12      3.512E-13
          ISC             0.000E+00       0.000E+00      0.000E+00
          IS              4.782E-16      -1.384E+16     -6.618E-02
          NE              1.829E+00      -1.523E+00     -2.786E-02
          NC              2.000E+00       0.000E+00      0.000E+00
          IKF             1.079E+00      -6.274E-02     -6.770E-04
          IKR             0.000E+00       0.000E+00      0.000E+00
          VAF             1.157E+02       6.114E-04      7.074E-04
          VAR             0.000E+00       0.000E+00      0.000E+00
Q3
          RB              1.000E+01      -1.862E-03     -1.862E-04
          RC              1.000E+00      -1.128E-04     -1.128E-06
          RE              0.000E+00       0.000E+00      0.000E+00
          BF              2.533E+02       3.000E-03      7.599E-03
          ISE             1.144E-14      -4.734E+13     -5.417E-03
          BR              6.031E+00      -5.358E-11     -3.231E-12
          ISC             0.000E+00       0.000E+00      0.000E+00
          IS              1.054E-14       6.599E+14      6.955E-02
          NE              1.307E+00       8.041E+00      1.051E-01
          NC              2.000E+00       0.000E+00      0.000E+00
          IKF             2.847E-01       1.022E-01      2.910E-04
          IKR             0.000E+00       0.000E+00      0.000E+00
          VAF             7.403E+01      -2.607E-03     -1.930E-03
          VAR             0.000E+00       0.000E+00      0.000E+00
```

DC SENSITIVITIES OF OUTPUT V(7)

ELEMENT NAME	ELEMENT VALUE	ELEMENT SENSITIVITY (VOLTS/UNIT)	NORMALIZED SENSITIVITY (VOLTS/PERCENT)
Q4			
RB	1.000E+01	5.348E-05	5.348E-06
RC	1.000E+00	2.034E-05	2.034E-07
RE	0.000E+00	0.000E+00	0.000E+00
BF	2.533E+02	-3.202E-03	-8.111E-03
ISE	1.144E-14	5.105E+13	5.842E-03
BR	6.031E+00	5.788E-11	3.491E-12
ISC	0.000E+00	0.000E+00	0.000E+00
IS	1.054E-14	-6.100E+13	-6.429E-03
NE	1.307E+00	-8.656E+00	-1.131E-01
NC	2.000E+00	0.000E+00	0.000E+00
IKF	2.847E-01	-1.844E-02	-5.250E-05
IKR	0.000E+00	0.000E+00	0.000E+00
VAF	7.403E+01	1.187E-03	8.788E-04
VAR	0.000E+00	0.000E+00	0.000E+00

```
FOUR TRANSISTOR AMPLIFIER

 ****     AC ANALYSIS                             TEMPERATURE =   25.000 DEG C

******************************************************************************
 FREQ         VM(7)        VP(7)        VDB(7)       IM(VIN)

  1.000E+01    2.538E+02   -3.485E-03    4.809E+01    6.995E-09
  1.259E+01    2.538E+02   -4.388E-03    4.809E+01    7.031E-09
  1.585E+01    2.538E+02   -5.524E-03    4.809E+01    7.088E-09
  1.995E+01    2.538E+02   -6.954E-03    4.809E+01    7.177E-09
  2.512E+01    2.538E+02   -8.755E-03    4.809E+01    7.317E-09
  3.162E+01    2.538E+02   -1.102E-02    4.809E+01    7.532E-09
  3.981E+01    2.538E+02   -1.388E-02    4.809E+01    7.861E-09
  5.012E+01    2.538E+02   -1.747E-02    4.809E+01    8.357E-09
  6.310E+01    2.538E+02   -2.199E-02    4.809E+01    9.087E-09
  7.943E+01    2.538E+02   -2.769E-02    4.809E+01    1.014E-08
  1.000E+02    2.538E+02   -3.485E-02    4.809E+01    1.161E-08
  1.259E+02    2.538E+02   -4.388E-02    4.809E+01    1.362E-08
  1.585E+02    2.538E+02   -5.524E-02    4.809E+01    1.630E-08
  1.995E+02    2.538E+02   -6.954E-02    4.809E+01    1.983E-08
  2.512E+02    2.538E+02   -8.755E-02    4.809E+01    2.439E-08
  3.162E+02    2.538E+02   -1.102E-01    4.809E+01    3.025E-08
  3.981E+02    2.538E+02   -1.388E-01    4.809E+01    3.771E-08
  5.012E+02    2.538E+02   -1.747E-01    4.809E+01    4.717E-08
  6.310E+02    2.538E+02   -2.199E-01    4.809E+01    5.915E-08
  7.943E+02    2.538E+02   -2.769E-01    4.809E+01    7.428E-08
  1.000E+03    2.538E+02   -3.485E-01    4.809E+01    9.336E-08
  1.259E+03    2.538E+02   -4.388E-01    4.809E+01    1.174E-07
  1.585E+03    2.538E+02   -5.524E-01    4.809E+01    1.477E-07
  1.995E+03    2.538E+02   -6.954E-01    4.809E+01    1.859E-07
  2.512E+03    2.538E+02   -8.754E-01    4.809E+01    2.340E-07
  3.162E+03    2.538E+02   -1.102E+00    4.809E+01    2.945E-07
  3.981E+03    2.538E+02   -1.387E+00    4.809E+01    3.707E-07
  5.012E+03    2.537E+02   -1.746E+00    4.809E+01    4.667E-07
  6.310E+03    2.537E+02   -2.198E+00    4.809E+01    5.875E-07
  7.943E+03    2.536E+02   -2.767E+00    4.808E+01    7.396E-07
  1.000E+04    2.534E+02   -3.482E+00    4.808E+01    9.310E-07
  1.259E+04    2.531E+02   -4.380E+00    4.807E+01    1.172E-06
  1.585E+04    2.527E+02   -5.509E+00    4.805E+01    1.476E-06
  1.995E+04    2.521E+02   -6.924E+00    4.803E+01    1.858E-06
  2.512E+04    2.511E+02   -8.695E+00    4.800E+01    2.339E-06
  3.162E+04    2.496E+02   -1.090E+01    4.794E+01    2.944E-06
  3.981E+04    2.472E+02   -1.364E+01    4.786E+01    3.707E-06
  5.012E+04    2.435E+02   -1.701E+01    4.773E+01    4.667E-06
  6.310E+04    2.380E+02   -2.110E+01    4.753E+01    5.876E-06
  7.943E+04    2.301E+02   -2.597E+01    4.724E+01    7.399E-06
  1.000E+05    2.189E+02   -3.164E+01    4.681E+01    9.317E-06
  1.259E+05    2.042E+02   -3.801E+01    4.620E+01    1.173E-05
  1.585E+05    1.859E+02   -4.489E+01    4.538E+01    1.478E-05
  1.995E+05    1.648E+02   -5.198E+01    4.434E+01    1.862E-05
  2.512E+05    1.424E+02   -5.897E+01    4.307E+01    2.347E-05
  3.162E+05    1.203E+02   -6.561E+01    4.160E+01    2.961E-05
  3.981E+05    9.970E+01   -7.178E+01    3.997E+01    3.738E-05
  5.012E+05    8.146E+01   -7.745E+01    3.822E+01    4.726E-05
  6.310E+05    6.584E+01   -8.271E+01    3.637E+01    5.989E-05
  7.943E+05    5.280E+01   -8.771E+01    3.445E+01    7.616E-05
  1.000E+06    4.208E+01   -9.262E+01    3.248E+01    9.732E-05
```

```
FOUR TRANSISTOR AMPLIFIER

 ****     AC ANALYSIS                          TEMPERATURE =   25.000 DEG C
******************************************************************************
 LEGEND:         *: VDB(7)        +: VP(7)          =: VM(7)          $: IM(VIN)

  FREQ        VDB(7)
(*)----------     1.0000E+01   2.0000E+01   3.0000E+01   4.0000E+01   5.0000E+01
(+)----------    -2.0000E+02  -1.5000E+02  -1.0000E+02  -5.0000E+01   0.0000E+00
(=)----------     1.0000E+00   1.0000E+01   1.0000E+02   1.0000E+03   1.0000E+04
($)----------     1.0000E-09   1.0000E-07   1.0000E-05   1.0000E-03   1.0000E-01

                      - - - - - - - - - - - - - - - - - - - - - - - - - - - - -
  1.000E+01 4.809E+01 .      $      .            .     =      .          * +
  1.259E+01 4.809E+01 .      $      .            .     =      .          * +
  1.585E+01 4.809E+01 .      $      .            .     =      .          * +
  1.995E+01 4.809E+01 .      $      .            .     =      .          * +
  2.512E+01 4.809E+01 .      $      .            .     =      .          * +
  3.162E+01 4.809E+01 .      $      .            .     =      .          * +
  3.981E+01 4.809E+01 .      $      .            .     =      .          * +
  5.012E+01 4.809E+01 .      $      .            .     =      .          * +
  6.310E+01 4.809E+01 .       $     .            .     =      .          * +
  7.943E+01 4.809E+01 .       $     .            .     =      .          * +
  1.000E+02 4.809E+01 .       $     .            .     =      .          * +
  1.259E+02 4.809E+01 .        $    .            .     =      .          * +
  1.585E+02 4.809E+01 .        $    .            .     =      .          * +
  1.995E+02 4.809E+01 .         $   .            .     =      .          * +
  2.512E+02 4.809E+01 .          $  .            .     =      .          * +
  3.162E+02 4.809E+01 .          $  .            .     =      .          * +
  3.981E+02 4.809E+01 .           $ .            .     =      .          * +
  5.012E+02 4.809E+01 .           $ .            .     =      .          * +
  6.310E+02 4.809E+01 .            $.            .     =      .          * +
  7.943E+02 4.809E+01 .             $            .     =      .          * +
  1.000E+03 4.809E+01 .             $            .     =      .          * +
  1.259E+03 4.809E+01 .             .$           .     =      .          * +
  1.585E+03 4.809E+01 .             . $          .     =      .          * +
  1.995E+03 4.809E+01 .             . $          .     =      .          * +
  2.512E+03 4.809E+01 .             .  $         .     =      .          * +
  3.162E+03 4.809E+01 .             .   $        .     =      .          * +
  3.981E+03 4.809E+01 .             .   $        .     =      .          * +
  5.012E+03 4.809E+01 .             .    $       .     =      .          * +
  6.310E+03 4.809E+01 .             .    $       .     =      .          *+.
  7.943E+03 4.808E+01 .             .     $      .     =      .          *+.
  1.000E+04 4.808E+01 .             .      $     .     =      .         * +.
  1.259E+04 4.807E+01 .             .      $     .     =      .         * +.
  1.585E+04 4.805E+01 .             .       $    .     =      .         * +.
  1.995E+04 4.803E+01 .             .        $   .     =      .         *+ .
  2.512E+04 4.800E+01 .             .        $   .     =      .         *+ .
  3.162E+04 4.794E+01 .             .         $  .     =      .         X  .
  3.981E+04 4.786E+01 .             .          $ .     =      .        +*  .
  5.012E+04 4.773E+01 .             .          $ .     =      .        +*  .
  6.310E+04 4.753E+01 .             .           $.    =       .       + *  .
  7.943E+04 4.724E+01 .             .            $    =       .     +  *   .
  1.000E+05 4.681E+01 .             .            $    =       .    +   *   .
  1.259E+05 4.620E+01 .             .            .$   =       .  +    *    .
  1.585E+05 4.538E+01 .             .            . $ =        .+     *     .
  1.995E+05 4.434E+01 .             .            . $=        +.     *      .
  2.512E+05 4.307E+01 .             .            . =$       + .   *        .
  3.162E+05 4.160E+01 .             .            . = $    +   . *          .
  3.981E+05 3.997E+01 .             .            =   $  +     *            .
  5.012E+05 3.822E+01 .             .           =.    $+    * .            .
  6.310E+05 3.637E+01 .             .          = .   + $ *    .            .
  7.943E+05 3.445E+01 .             .         =  .  +  X      .            .
  1.000E+06 3.248E+01 .             .        =   . +*  $      .            .
```

```
FOUR TRANSISTOR AMPLIFIER

 ****     INITIAL TRANSIENT SOLUTION       TEMPERATURE =   25.000 DEG C

******************************************************************************

 NODE   VOLTAGE     NODE   VOLTAGE     NODE   VOLTAGE     NODE   VOLTAGE

(    1)     .0200  (    2)     .6544  (    3)    1.4281  (    4)    2.6773

(    5)     .7757  (    6)    2.0356  (    7)   13.1410  (    8)   30.0000

    VOLTAGE SOURCE CURRENTS
    NAME          CURRENT

    VCC          -7.697E-03
    VIN           1.443E-07

    TOTAL POWER DISSIPATION   2.31E-01  WATTS
```

```
FOUR TRANSISTOR AMPLIFIER

 ****     TRANSIENT ANALYSIS                  TEMPERATURE =   25.000 DEG C

**********************************************************************************

  TIME        V(7)

   0.000E+00   1.314E+01
   4.000E-05   1.376E+01
   8.000E-05   1.435E+01
   1.200E-04   1.487E+01
   1.600E-04   1.528E+01
   2.000E-04   1.555E+01
   2.400E-04   1.568E+01
   2.800E-04   1.564E+01
   3.200E-04   1.544E+01
   3.600E-04   1.511E+01
   4.000E-04   1.464E+01
   4.400E-04   1.409E+01
   4.800E-04   1.347E+01
   5.200E-04   1.284E+01
   5.600E-04   1.222E+01
   6.000E-04   1.166E+01
   6.400E-04   1.120E+01
   6.800E-04   1.085E+01
   7.200E-04   1.065E+01
   7.600E-04   1.061E+01
   8.000E-04   1.073E+01
   8.400E-04   1.099E+01
   8.800E-04   1.140E+01
   9.200E-04   1.191E+01
   9.600E-04   1.250E+01
   1.000E-03   1.313E+01
   1.040E-03   1.376E+01
   1.080E-03   1.435E+01
   1.120E-03   1.487E+01
   1.160E-03   1.528E+01
   1.200E-03   1.555E+01
   1.240E-03   1.568E+01
   1.280E-03   1.564E+01
   1.320E-03   1.544E+01
   1.360E-03   1.511E+01
   1.400E-03   1.464E+01
   1.440E-03   1.409E+01
   1.480E-03   1.347E+01
   1.520E-03   1.284E+01
   1.560E-03   1.222E+01
   1.600E-03   1.166E+01
   1.640E-03   1.120E+01
   1.680E-03   1.085E+01
   1.720E-03   1.065E+01
   1.760E-03   1.061E+01
   1.800E-03   1.073E+01
   1.840E-03   1.099E+01
   1.880E-03   1.140E+01
   1.920E-03   1.191E+01
   1.960E-03   1.250E+01
```

```
TIME        V(7)

 2.000E-03   1.313E+01
 2.040E-03   1.376E+01
 2.080E-03   1.435E+01
 2.120E-03   1.487E+01
 2.160E-03   1.528E+01
 2.200E-03   1.555E+01
 2.240E-03   1.568E+01
 2.280E-03   1.564E+01
 2.320E-03   1.544E+01
 2.360E-03   1.511E+01
 2.400E-03   1.464E+01
 2.440E-03   1.409E+01
 2.480E-03   1.347E+01
 2.520E-03   1.284E+01
 2.560E-03   1.222E+01
 2.600E-03   1.166E+01
 2.640E-03   1.120E+01
 2.680E-03   1.085E+01
 2.720E-03   1.065E+01
 2.760E-03   1.061E+01
 2.800E-03   1.073E+01
 2.840E-03   1.099E+01
 2.880E-03   1.140E+01
 2.920E-03   1.191E+01
 2.960E-03   1.250E+01
 3.000E-03   1.313E+01
 3.040E-03   1.376E+01
 3.080E-03   1.435E+01
 3.120E-03   1.487E+01
 3.160E-03   1.528E+01
 3.200E-03   1.555E+01
 3.240E-03   1.568E+01
 3.280E-03   1.564E+01
 3.320E-03   1.544E+01
 3.360E-03   1.511E+01
 3.400E-03   1.464E+01
 3.440E-03   1.409E+01
 3.480E-03   1.347E+01
 3.520E-03   1.284E+01
 3.560E-03   1.222E+01
 3.600E-03   1.166E+01
 3.640E-03   1.120E+01
 3.680E-03   1.085E+01
 3.720E-03   1.065E+01
 3.760E-03   1.061E+01
 3.800E-03   1.073E+01
 3.840E-03   1.099E+01
 3.880E-03   1.140E+01
 3.920E-03   1.191E+01
 3.960E-03   1.250E+01
 4.000E-03   1.313E+01
```

```
FOUR TRANSISTOR AMPLIFIER

 ****     TRANSIENT ANALYSIS                    TEMPERATURE =   25.000 DEG C
******************************************************************************
  TIME        V(7)
(*)----------     1.0000E+01    1.2000E+01    1.4000E+01    1.6000E+01    1.8000E+01
                  - - - - - - - - - - - - - - - - - - - - - - - - - - - - - - - - - - -
  0.000E+00  1.314E+01 .             .      *      .             .             .
  4.000E-05  1.376E+01 .             .          *  .             .             .
  8.000E-05  1.435E+01 .             .             . *           .             .
  1.200E-04  1.487E+01 .             .             .     *       .             .
  1.600E-04  1.528E+01 .             .             .       *     .             .
  2.000E-04  1.555E+01 .             .             .         *   .             .
  2.400E-04  1.568E+01 .             .             .          *  .             .
  2.800E-04  1.564E+01 .             .             .          *  .             .
  3.200E-04  1.544E+01 .             .             .        *    .             .
  3.600E-04  1.511E+01 .             .             .      *      .             .
  4.000E-04  1.464E+01 .             .             .   *         .             .
  4.400E-04  1.409E+01 .             .             .*            .             .
  4.800E-04  1.347E+01 .             .          *  .             .             .
  5.200E-04  1.284E+01 .             .     *       .             .             .
  5.600E-04  1.222E+01 .             .*            .             .             .
  6.000E-04  1.166E+01 .           * .             .             .             .
  6.400E-04  1.120E+01 .        *    .             .             .             .
  6.800E-04  1.085E+01 .      *      .             .             .             .
  7.200E-04  1.065E+01 .   *         .             .             .             .
  7.600E-04  1.061E+01 .   *         .             .             .             .
  8.000E-04  1.073E+01 .    *        .             .             .             .
  8.400E-04  1.099E+01 .      *      .             .             .             .
  8.800E-04  1.140E+01 .         *   .             .             .             .
  9.200E-04  1.191E+01 .            *.             .             .             .
  9.600E-04  1.250E+01 .             .  *          .             .             .
  1.000E-03  1.313E+01 .             .      *      .             .             .
  1.040E-03  1.376E+01 .             .          *  .             .             .
  1.080E-03  1.435E+01 .             .             . *           .             .
  1.120E-03  1.487E+01 .             .             .     *       .             .
  1.160E-03  1.528E+01 .             .             .       *     .             .
  1.200E-03  1.555E+01 .             .             .         *   .             .
  1.240E-03  1.568E+01 .             .             .          *  .             .
  1.280E-03  1.564E+01 .             .             .          *  .             .
  1.320E-03  1.544E+01 .             .             .        *    .             .
  1.360E-03  1.511E+01 .             .             .      *      .             .
  1.400E-03  1.464E+01 .             .             .   *         .             .
  1.440E-03  1.409E+01 .             .             .*            .             .
  1.480E-03  1.347E+01 .             .          *  .             .             .
  1.520E-03  1.284E+01 .             .     *       .             .             .
  1.560E-03  1.222E+01 .             .*            .             .             .
  1.600E-03  1.166E+01 .           * .             .             .             .
  1.640E-03  1.120E+01 .        *    .             .             .             .
  1.680E-03  1.085E+01 .      *      .             .             .             .
  1.720E-03  1.065E+01 .   *         .             .             .             .
  1.760E-03  1.061E+01 .   *         .             .             .             .
  1.800E-03  1.073E+01 .    *        .             .             .             .
  1.840E-03  1.099E+01 .      *      .             .             .             .
  1.880E-03  1.140E+01 .         *   .             .             .             .
  1.920E-03  1.191E+01 .            *.             .             .             .
  1.960E-03  1.250E+01 .             .  *          .             .             .
  2.000E-03  1.313E+01 .             .      *      .             .             .
  2.040E-03  1.376E+01 .             .          *  .             .             .
  2.080E-03  1.435E+01 .             .             . *           .             .
  2.120E-03  1.487E+01 .             .             .     *       .             .
  2.160E-03  1.528E+01 .             .             .       *     .             .
  2.200E-03  1.555E+01 .             .             .         *   .             .
  2.240E-03  1.568E+01 .             .             .          *  .             .
```

```
 TIME         V(7)
(*)----------    1.0000E+01   1.2000E+01   1.4000E+01   1.6000E+01   1.8000E+01

                     - - - - - - - - - - - - - - - - - - - - - - - - - - - - - -
 2.280E-03  1.564E+01 .         .         .       * .         .
 2.320E-03  1.544E+01 .         .         .      *  .         .
 2.360E-03  1.511E+01 .         .         .    *    .         .
 2.400E-03  1.464E+01 .         .         .  *      .         .
 2.440E-03  1.409E+01 .         .         .*        .         .
 2.480E-03  1.347E+01 .         .       * .         .         .
 2.520E-03  1.284E+01 .         .   *     .         .         .
 2.560E-03  1.222E+01 .         .*        .         .         .
 2.600E-03  1.166E+01 .       * .         .         .         .
 2.640E-03  1.120E+01 .     *   .         .         .         .
 2.680E-03  1.085E+01 .   *     .         .         .         .
 2.720E-03  1.065E+01 .  *      .         .         .         .
 2.760E-03  1.061E+01 .  *      .         .         .         .
 2.800E-03  1.073E+01 .   *     .         .         .         .
 2.840E-03  1.099E+01 .    *    .         .         .         .
 2.880E-03  1.140E+01 .      *  .         .         .         .
 2.920E-03  1.191E+01 .        *.         .         .         .
 2.960E-03  1.250E+01 .         . *       .         .         .
 3.000E-03  1.313E+01 .         .    *    .         .         .
 3.040E-03  1.376E+01 .         .       * .         .         .
 3.080E-03  1.435E+01 .         .         . *       .         .
 3.120E-03  1.487E+01 .         .         .   *     .         .
 3.160E-03  1.528E+01 .         .         .     *   .         .
 3.200E-03  1.555E+01 .         .         .       * .         .
 3.240E-03  1.568E+01 .         .         .       * .         .
 3.280E-03  1.564E+01 .         .         .       * .         .
 3.320E-03  1.544E+01 .         .         .      *  .         .
 3.360E-03  1.511E+01 .         .         .    *    .         .
 3.400E-03  1.464E+01 .         .         .  *      .         .
 3.440E-03  1.409E+01 .         .         .*        .         .
 3.480E-03  1.347E+01 .         .       * .         .         .
 3.520E-03  1.284E+01 .         .   *     .         .         .
 3.560E-03  1.222E+01 .         .*        .         .         .
 3.600E-03  1.166E+01 .       * .         .         .         .
 3.640E-03  1.120E+01 .     *   .         .         .         .
 3.680E-03  1.085E+01 .   *     .         .         .         .
 3.720E-03  1.065E+01 .  *      .         .         .         .
 3.760E-03  1.061E+01 .  *      .         .         .         .
 3.800E-03  1.073E+01 .   *     .         .         .         .
 3.840E-03  1.099E+01 .    *    .         .         .         .
 3.880E-03  1.140E+01 .      *  .         .         .         .
 3.920E-03  1.191E+01 .        *.         .         .         .
 3.960E-03  1.250E+01 .         . *       .         .         .
 4.000E-03  1.313E+01 .         .    *    .         .         .
                      . . . . . . . . . . . . . . . . . . . . . . . . . . . . . .
```

```
FOUR TRANSISTOR AMPLIFIER

 ****     FOURIER ANALYSIS                    TEMPERATURE =   25.000 DEG C

******************************************************************************

FOURIER COMPONENTS OF TRANSIENT RESPONSE V(7)

 DC COMPONENT =    1.314226E+01

 HARMONIC    FREQUENCY     FOURIER     NORMALIZED      PHASE        NORMALIZED
    NO         (HZ)       COMPONENT    COMPONENT       (DEG)        PHASE (DEG)

     1       1.000E+03    2.541E+00    1.000E+00    -2.922E-01     0.000E+00
     2       2.000E+03    1.349E-03    5.310E-04    -7.299E+01    -7.270E+01
     3       3.000E+03    1.614E-04    6.353E-05     4.369E+01     4.398E+01
     4       4.000E+03    7.795E-05    3.068E-05     7.576E+01     7.605E+01
     5       5.000E+03    2.522E-05    9.928E-06     4.944E+01     4.973E+01
     6       6.000E+03    2.233E-05    8.790E-06     7.433E+01     7.462E+01
     7       7.000E+03    2.199E-05    8.656E-06     1.299E+02     1.302E+02
     8       8.000E+03    4.296E-05    1.691E-05     3.836E+01     3.865E+01
     9       9.000E+03    4.615E-05    1.816E-05     1.156E+02     1.159E+02

     TOTAL HARMONIC DISTORTION =   5.365102E-02 PERCENT
```

```
FOUR TRANSISTOR AMPLIFIER

 ****      TEMPERATURE-ADJUSTED VALUES      TEMPERATURE =   70.000 DEG C

******************************************************************************

 **** RESISTORS
 NAME          VALUE
 R1          5.022E+03
 R2          2.410E+04
 R3          6.829E+02
 R4          2.009E+04
 R5          2.410E+03


 **** BJT MODEL PARAMETERS
 NAME            BF          ISE         VJE         CJE          RE          RB
                 BR          ISC         VJC         CJC          RC          RBM
                 IS          ISS         VJS         CJS
 Q2N2222A    3.128E+02   9.767E-13   6.738E-01   2.300E-11   0.000E+00   1.000E+01
             7.447E+00   0.000E+00   6.738E-01   7.603E-12   1.000E+00   1.000E+01
             4.640E-12   0.000E+00   6.738E-01   0.000E+00
 Q2N2907A    2.832E+02   1.057E-12   6.738E-01   2.061E-11   0.000E+00   1.000E+01
             4.355E+00   0.000E+00   6.738E-01   1.570E-11   7.150E-01   1.000E+01
             2.105E-13   0.000E+00   6.738E-01   0.000E+00
```

```
FOUR TRANSISTOR AMPLIFIER

 ****     DC TRANSFER CURVES                 TEMPERATURE =   70.000 DEG C

******************************************************************************

  VIN          V(7)

 -2.000E-01   3.941E+00
 -1.975E-01   3.934E+00
 -1.950E-01   3.928E+00
 -1.925E-01   3.921E+00
 -1.900E-01   3.914E+00
 -1.875E-01   3.908E+00
 -1.850E-01   3.901E+00
 -1.825E-01   3.894E+00
 -1.800E-01   3.888E+00
 -1.775E-01   3.881E+00
 -1.750E-01   3.875E+00
 -1.725E-01   3.868E+00
 -1.700E-01   3.861E+00
 -1.675E-01   3.855E+00
 -1.650E-01   3.848E+00
 -1.625E-01   3.841E+00
 -1.600E-01   3.835E+00
 -1.575E-01   3.828E+00
 -1.550E-01   3.821E+00
 -1.525E-01   3.815E+00
 -1.500E-01   3.808E+00
 -1.475E-01   3.801E+00
 -1.450E-01   3.795E+00
 -1.425E-01   3.788E+00
 -1.400E-01   3.781E+00
 -1.375E-01   3.775E+00
 -1.350E-01   3.768E+00
 -1.325E-01   3.761E+00
 -1.300E-01   3.755E+00
 -1.275E-01   3.748E+00
 -1.250E-01   3.742E+00
 -1.225E-01   3.735E+00
 -1.200E-01   3.728E+00
 -1.175E-01   3.722E+00
 -1.150E-01   3.715E+00
 -1.125E-01   3.708E+00
 -1.100E-01   3.702E+00
 -1.075E-01   3.695E+00
 -1.050E-01   3.689E+00
 -1.025E-01   3.682E+00
 -1.000E-01   3.675E+00
 -9.750E-02   3.669E+00
 -9.500E-02   3.662E+00
 -9.250E-02   3.656E+00
 -9.000E-02   3.649E+00
 -8.750E-02   3.642E+00
 -8.500E-02   3.636E+00
 -8.250E-02   3.629E+00
 -8.000E-02   3.623E+00
 -7.750E-02   3.616E+00
 -7.500E-02   3.610E+00
 -7.250E-02   3.603E+00
 -7.000E-02   3.596E+00
```

```
VIN          V(7)

-6.750E-02   3.590E+00
-6.500E-02   3.583E+00
-6.250E-02   3.577E+00
-6.000E-02   3.570E+00
-5.750E-02   3.564E+00
-5.500E-02   3.557E+00
-5.250E-02   3.551E+00
-5.000E-02   3.544E+00
-4.750E-02   3.538E+00
-4.500E-02   3.532E+00
-4.250E-02   3.525E+00
-4.000E-02   3.519E+00
-3.750E-02   3.512E+00
-3.500E-02   3.506E+00
-3.250E-02   3.499E+00
-3.000E-02   3.493E+00
-2.750E-02   3.487E+00
-2.500E-02   3.480E+00
-2.250E-02   3.474E+00
-2.000E-02   3.468E+00
-1.750E-02   3.462E+00
-1.500E-02   3.455E+00
-1.250E-02   3.449E+00
-1.000E-02   3.443E+00
-7.500E-03   3.437E+00
-5.000E-03   3.431E+00
-2.500E-03   3.425E+00
 0.000E+00   3.419E+00
 2.500E-03   3.413E+00
 5.000E-03   3.407E+00
 7.500E-03   3.401E+00
 1.000E-02   3.395E+00
 1.250E-02   3.389E+00
 1.500E-02   3.384E+00
 1.750E-02   3.378E+00
 2.000E-02   3.373E+00
 2.250E-02   3.367E+00
 2.500E-02   3.362E+00
 2.750E-02   3.357E+00
 3.000E-02   3.352E+00
 3.250E-02   3.348E+00
 3.500E-02   3.344E+00
 3.750E-02   3.340E+00
 4.000E-02   3.336E+00
 4.250E-02   3.334E+00
 4.500E-02   3.332E+00
 4.750E-02   3.331E+00
 5.000E-02   3.333E+00
 5.250E-02   3.340E+00
 5.500E-02   3.363E+00
 5.750E-02   3.708E+00
 6.000E-02   4.344E+00
 6.250E-02   4.980E+00
 6.500E-02   5.617E+00
 6.750E-02   6.254E+00
 7.000E-02   6.892E+00
 7.250E-02   7.531E+00
 7.500E-02   8.170E+00
 7.750E-02   8.810E+00
 8.000E-02   9.450E+00
```

VIN	V(7)
8.250E-02	1.009E+01
8.500E-02	1.073E+01
8.750E-02	1.137E+01
9.000E-02	1.201E+01
9.250E-02	1.266E+01
9.500E-02	1.330E+01
9.750E-02	1.394E+01
1.000E-01	1.458E+01
1.025E-01	1.523E+01
1.050E-01	1.587E+01
1.075E-01	1.651E+01
1.100E-01	1.715E+01
1.125E-01	1.780E+01
1.150E-01	1.844E+01
1.175E-01	1.908E+01
1.200E-01	1.972E+01
1.225E-01	2.036E+01
1.250E-01	2.100E+01
1.275E-01	2.164E+01
1.300E-01	2.228E+01
1.325E-01	2.292E+01
1.350E-01	2.355E+01
1.375E-01	2.419E+01
1.400E-01	2.482E+01
1.425E-01	2.543E+01
1.450E-01	2.595E+01
1.475E-01	2.625E+01
1.500E-01	2.640E+01
1.525E-01	2.648E+01
1.550E-01	2.654E+01
1.575E-01	2.658E+01
1.600E-01	2.661E+01
1.625E-01	2.663E+01
1.650E-01	2.665E+01
1.675E-01	2.666E+01
1.700E-01	2.667E+01
1.725E-01	2.667E+01
1.750E-01	2.667E+01
1.775E-01	2.668E+01
1.800E-01	2.668E+01
1.825E-01	2.667E+01
1.850E-01	2.667E+01
1.875E-01	2.667E+01
1.900E-01	2.666E+01
1.925E-01	2.666E+01
1.950E-01	2.665E+01
1.975E-01	2.665E+01
2.000E-01	2.664E+01

```
FOUR TRANSISTOR AMPLIFIER

 ****     DC TRANSFER CURVES                 TEMPERATURE =   70.000 DEG C

******************************************************************************

  VIN        V(7)

(*)----------    0.0000E+00   5.0000E+00   1.0000E+01   1.5000E+01   2.0000E+01
                            - - - - - - - - - - - - - - - - - - - - - - - - - -
 -2.000E-01  3.941E+00 .    *
 -1.975E-01  3.934E+00 .    *        .            .            .            .
 -1.950E-01  3.928E+00 .    *        .            .            .            .
 -1.925E-01  3.921E+00 .    *        .            .            .            .
 -1.900E-01  3.914E+00 .    *        .            .            .            .
 -1.875E-01  3.908E+00 .    *        .            .            .            .
 -1.850E-01  3.901E+00 .    *        .            .            .            .
 -1.825E-01  3.894E+00 .    *        .            .            .            .
 -1.800E-01  3.888E+00 .    *        .            .            .            .
 -1.775E-01  3.881E+00 .    *        .            .            .            .
 -1.750E-01  3.875E+00 .    *        .            .            .            .
 -1.725E-01  3.868E+00 .    *        .            .            .            .
 -1.700E-01  3.861E+00 .    *        .            .            .            .
 -1.675E-01  3.855E+00 .    *        .            .            .            .
 -1.650E-01  3.848E+00 .    *        .            .            .            .
 -1.625E-01  3.841E+00 .    *        .            .            .            .
 -1.600E-01  3.835E+00 .    *        .            .            .            .
 -1.575E-01  3.828E+00 .    *        .            .            .            .
 -1.550E-01  3.821E+00 .    *        .            .            .            .
 -1.525E-01  3.815E+00 .    *        .            .            .            .
 -1.500E-01  3.808E+00 .    *        .            .            .            .
 -1.475E-01  3.801E+00 .    *        .            .            .            .
 -1.450E-01  3.795E+00 .    *        .            .            .            .
 -1.425E-01  3.788E+00 .    *        .            .            .            .
 -1.400E-01  3.781E+00 .    *        .            .            .            .
 -1.375E-01  3.775E+00 .    *        .            .            .            .
 -1.350E-01  3.768E+00 .    *        .            .            .            .
 -1.325E-01  3.761E+00 .    *        .            .            .            .
 -1.300E-01  3.755E+00 .    *        .            .            .            .
 -1.275E-01  3.748E+00 .    *        .            .            .            .
 -1.250E-01  3.742E+00 .    *        .            .            .            .
 -1.225E-01  3.735E+00 .    *        .            .            .            .
 -1.200E-01  3.728E+00 .    *        .            .            .            .
 -1.175E-01  3.722E+00 .    *        .            .            .            .
 -1.150E-01  3.715E+00 .    *        .            .            .            .
 -1.125E-01  3.708E+00 .    *        .            .            .            .
 -1.100E-01  3.702E+00 .    *        .            .            .            .
 -1.075E-01  3.695E+00 .    *        .            .            .            .
 -1.050E-01  3.689E+00 .    *        .            .            .            .
 -1.025E-01  3.682E+00 .    *        .            .            .            .
 -1.000E-01  3.675E+00 .    *        .            .            .            .
 -9.750E-02  3.669E+00 .    *        .            .            .            .
 -9.500E-02  3.662E+00 .    *        .            .            .            .
 -9.250E-02  3.656E+00 .    *        .            .            .            .
 -9.000E-02  3.649E+00 .    *        .            .            .            .
 -8.750E-02  3.642E+00 .    *        .            .            .            .
 -8.500E-02  3.636E+00 .    *        .            .            .            .
 -8.250E-02  3.629E+00 .    *        .            .            .            .
 -8.000E-02  3.623E+00 .    *        .            .            .            .
 -7.750E-02  3.616E+00 .    *        .            .            .            .
 -7.500E-02  3.610E+00 .    *        .            .            .            .
 -7.250E-02  3.603E+00 .    *        .            .            .            .
 -7.000E-02  3.596E+00 .    *        .            .            .            .
```

```
   VIN          V(7)

(*)----------    0.0000E+00   5.0000E+00   1.0000E+01   1.5000E+01   2.0000E+01

                          - - - - - - - - - - - - - - - - - - - - - - - - - - -
 -6.750E-02  3.590E+00 .    *       .            .            .             .
 -6.500E-02  3.583E+00 .    *       .            .            .             .
 -6.250E-02  3.577E+00 .    *       .            .            .             .
 -6.000E-02  3.570E+00 .    *       .            .            .             .
 -5.750E-02  3.564E+00 .    *       .            .            .             .
 -5.500E-02  3.557E+00 .    *       .            .            .             .
 -5.250E-02  3.551E+00 .    *       .            .            .             .
 -5.000E-02  3.544E+00 .    *       .            .            .             .
 -4.750E-02  3.538E+00 .    *       .            .            .             .
 -4.500E-02  3.532E+00 .    *       .            .            .             .
 -4.250E-02  3.525E+00 .    *       .            .            .             .
 -4.000E-02  3.519E+00 .    *       .            .            .             .
 -3.750E-02  3.512E+00 .    *       .            .            .             .
 -3.500E-02  3.506E+00 .    *       .            .            .             .
 -3.250E-02  3.499E+00 .    *       .            .            .             .
 -3.000E-02  3.493E+00 .    *       .            .            .             .
 -2.750E-02  3.487E+00 .    *       .            .            .             .
 -2.500E-02  3.480E+00 .    *       .            .            .             .
 -2.250E-02  3.474E+00 .    *       .            .            .             .
 -2.000E-02  3.468E+00 .    *       .            .            .             .
 -1.750E-02  3.462E+00 .    *       .            .            .             .
 -1.500E-02  3.455E+00 .   *        .            .            .             .
 -1.250E-02  3.449E+00 .   *        .            .            .             .
 -1.000E-02  3.443E+00 .   *        .            .            .             .
 -7.500E-03  3.437E+00 .   *        .            .            .             .
 -5.000E-03  3.431E+00 .   *        .            .            .             .
 -2.500E-03  3.425E+00 .   *        .            .            .             .
  0.000E+00  3.419E+00 .   *        .            .            .             .
  2.500E-03  3.413E+00 .   *        .            .            .             .
  5.000E-03  3.407E+00 .   *        .            .            .             .
  7.500E-03  3.401E+00 .   *        .            .            .             .
  1.000E-02  3.395E+00 .   *        .            .            .             .
  1.250E-02  3.389E+00 .   *        .            .            .             .
  1.500E-02  3.384E+00 .   *        .            .            .             .
  1.750E-02  3.378E+00 .   *        .            .            .             .
  2.000E-02  3.373E+00 .   *        .            .            .             .
  2.250E-02  3.367E+00 .   *        .            .            .             .
  2.500E-02  3.362E+00 .   *        .            .            .             .
  2.750E-02  3.357E+00 .   *        .            .            .             .
  3.000E-02  3.352E+00 .   *        .            .            .             .
  3.250E-02  3.348E+00 .   *        .            .            .             .
  3.500E-02  3.344E+00 .   *        .            .            .             .
  3.750E-02  3.340E+00 .   *        .            .            .             .
  4.000E-02  3.336E+00 .   *        .            .            .             .
  4.250E-02  3.334E+00 .   *        .            .            .             .
  4.500E-02  3.332E+00 .   *        .            .            .             .
  4.750E-02  3.331E+00 .   *        .            .            .             .
  5.000E-02  3.333E+00 .   *        .            .            .             .
  5.250E-02  3.340E+00 .   *        .            .            .             .
  5.500E-02  3.363E+00 .   *        .            .            .             .
  5.750E-02  3.708E+00 .    *       .            .            .             .
  6.000E-02  4.344E+00 .     *      .            .            .             .
  6.250E-02  4.980E+00 .     *      .            .            .             .
  6.500E-02  5.617E+00 .      *     .            .            .             .
  6.750E-02  6.254E+00 .       *    .            .            .             .
  7.000E-02  6.892E+00 .        *   .            .            .             .
```

```
   VIN        V(7)

(*)----------     0.0000E+00   5.0000E+00   1.0000E+01   1.5000E+01   2.0000E+01
                     - - - - - - - - - - - - - - - - - - - - - - - - - -
 7.250E-02  7.531E+00 .         *  .           .            .           .
 7.500E-02  8.170E+00 .          * .           .            .           .
 7.750E-02  8.810E+00 .          * .           .            .           .
 8.000E-02  9.450E+00 .           *.           .            .           .
 8.250E-02  1.009E+01 .            *           .            .           .
 8.500E-02  1.073E+01 .            .*          .            .           .
 8.750E-02  1.137E+01 .            . *         .            .           .
 9.000E-02  1.201E+01 .            .  *        .            .           .
 9.250E-02  1.266E+01 .            .  *        .            .           .
 9.500E-02  1.330E+01 .            .   *       .            .           .
 9.750E-02  1.394E+01 .            .    *      .            .           .
 1.000E-01  1.458E+01 .            .     *     .            .           .
 1.025E-01  1.523E+01 .            .      *    .            .           .
 1.050E-01  1.587E+01 .            .       *   .            .           .
 1.075E-01  1.651E+01 .            .       *   .            .           .
 1.100E-01  1.715E+01 .            .        *  .            .           .
 1.125E-01  1.780E+01 .            .         * .            .           .
 1.150E-01  1.844E+01 .            .         * .            .           .
 1.175E-01  1.908E+01 .            .          *.            .           .
 1.200E-01  1.972E+01 .            .           *            .           .
 1.225E-01  2.036E+01 .            .           *            .           .
 1.250E-01  2.100E+01 .            .           .*           .           .
 1.275E-01  2.164E+01 .            .           . *          .           .
 1.300E-01  2.228E+01 .            .           .  *         .           .
 1.325E-01  2.292E+01 .            .           .   *        .           .
 1.350E-01  2.355E+01 .            .           .    *       .           .
 1.375E-01  2.419E+01 .            .           .    *       .           .
 1.400E-01  2.482E+01 .            .           .     *      .           .
 1.425E-01  2.543E+01 .            .           .      *     .           .
 1.450E-01  2.595E+01 .            .           .       *    .           .
 1.475E-01  2.625E+01 .            .           .       *    .           .
 1.500E-01  2.640E+01 .            .           .       *    .           .
 1.525E-01  2.648E+01 .            .           .       *    .           .
 1.550E-01  2.654E+01 .            .           .        *   .           .
 1.575E-01  2.658E+01 .            .           .        *   .           .
 1.600E-01  2.661E+01 .            .           .        *   .           .
 1.625E-01  2.663E+01 .            .           .        *   .           .
 1.650E-01  2.665E+01 .            .           .        *   .           .
 1.675E-01  2.666E+01 .            .           .        *   .           .
 1.700E-01  2.667E+01 .            .           .        *   .           .
 1.725E-01  2.667E+01 .            .           .        *   .           .
 1.750E-01  2.667E+01 .            .           .        *   .           .
 1.775E-01  2.668E+01 .            .           .        *   .           .
 1.800E-01  2.668E+01 .            .           .        *   .           .
 1.825E-01  2.667E+01 .            .           .        *   .           .
 1.850E-01  2.667E+01 .            .           .        *   .           .
 1.875E-01  2.667E+01 .            .           .        *   .           .
 1.900E-01  2.666E+01 .            .           .        *   .           .
 1.925E-01  2.666E+01 .            .           .        *   .           .
 1.950E-01  2.665E+01 .            .           .        *   .           .
 1.975E-01  2.665E+01 .            .           .        *   .           .
 2.000E-01  2.664E+01 .            .           .        *   .           .
                      . . . . . . . . . . . . . . . . . . . . . . . . . .
```

```
FOUR TRANSISTOR AMPLIFIER

 ****     SMALL SIGNAL BIAS SOLUTION      TEMPERATURE =   70.000 DEG C

******************************************************************************

 NODE   VOLTAGE     NODE   VOLTAGE     NODE   VOLTAGE     NODE   VOLTAGE

(    1)    0.0000  (    2)     .5440  (    3)    1.2545  (    4)    3.9654

(    5)     .6880  (    6)    3.3780  (    7)    3.4186  (    8)   30.0000

    VOLTAGE SOURCE CURRENTS
    NAME          CURRENT

    VCC          -8.128E-03
    VIN           9.730E-08

    TOTAL POWER DISSIPATION   2.44E-01  WATTS
```

```
FOUR TRANSISTOR AMPLIFIER

 ****     OPERATING POINT INFORMATION      TEMPERATURE =   70.000 DEG C

******************************************************************************
**** BIPOLAR JUNCTION TRANSISTORS

NAME          Q1           Q2           Q3           Q4
MODEL         Q2N2907A     Q2N2907A     Q2N2222A     Q2N2222A
IB           -9.73E-08    -2.07E-05     5.37E-06     7.81E-05
IC           -2.06E-05    -5.70E-03     1.00E-03     1.32E-03
VBE          -5.44E-01    -7.11E-01     5.67E-01     5.87E-01
VBC           0.00E+00     5.44E-01    -2.71E+00     5.47E-01
VCE          -5.44E-01    -1.25E+00     3.28E+00     4.07E-02
BETADC        2.11E+02     2.76E+02     1.87E+02     1.69E+01
GM            6.95E-04     1.92E-01     3.38E-02     4.67E-02
RPI           3.43E+05     1.45E+03     6.11E+03     3.26E+03
RX            1.00E+01     1.00E+01     1.00E+01     1.00E+01
RO            5.62E+06     2.04E+04     7.66E+04     5.90E+01
CBE           3.18E-11     1.51E-10     5.14E-11     6.43E-11
CBC           1.57E-11     1.14E-11     4.38E-12     8.17E-10
CBX           0.00E+00     0.00E+00     0.00E+00     0.00E+00
CJS           0.00E+00     0.00E+00     0.00E+00     0.00E+00
BETAAC        2.39E+02     2.78E+02     2.06E+02     1.53E+02
FT            2.33E+06     1.87E+08     9.63E+07     8.44E+06

 ****     SMALL-SIGNAL CHARACTERISTICS

      V(7)/VIN = -2.398E+00

      INPUT RESISTANCE AT VIN =  2.197E+08

      OUTPUT RESISTANCE AT V(7) =  2.010E+03
```

```
FOUR TRANSISTOR AMPLIFIER

 ****     DC SENSITIVITY ANALYSIS           TEMPERATURE =   70.000 DEG C

******************************************************************************

DC SENSITIVITIES OF OUTPUT V(7)

         ELEMENT         ELEMENT          ELEMENT        NORMALIZED
          NAME            VALUE         SENSITIVITY      SENSITIVITY
                                       (VOLTS/UNIT)   (VOLTS/PERCENT)

          R1            5.022E+03        2.837E-05       1.425E-03
          R2            2.410E+04       -7.719E-05      -1.861E-02
          R3            6.829E+02        2.424E-03       1.655E-02
          R4            2.009E+04       -1.324E-04      -2.659E-02
          R5            2.410E+03        1.160E-03       2.795E-02
          VCC           3.000E+01        1.666E-01       4.997E-02
          VIN           0.000E+00       -2.398E+00       0.000E+00
Q1
          RB            1.000E+01       -2.333E-07      -2.333E-08
          RC            7.150E-01       -1.256E-08      -8.979E-11
          RE            0.000E+00        0.000E+00       0.000E+00
          BF            2.832E+02       -8.744E-07      -2.477E-06
          ISE           1.057E-12        7.966E+07       8.418E-07
          BR            4.355E+00        1.236E-16       5.385E-18
          ISC           0.000E+00        0.000E+00       0.000E+00
          IS            2.105E-13        3.367E+11       7.088E-04
          NE            1.829E+00       -4.630E-04      -8.468E-06
          NC            2.000E+00        0.000E+00       0.000E+00
          IKF           1.079E+00        1.248E-06       1.346E-08
          IKR           0.000E+00        0.000E+00       0.000E+00
          VAF           1.157E+02        7.247E-11       8.384E-11
          VAR           0.000E+00        0.000E+00       0.000E+00
Q2
          RB            1.000E+01       -4.957E-05      -4.957E-06
          RC            7.150E-01       -6.953E-06      -4.971E-08
          RE            0.000E+00        0.000E+00       0.000E+00
          BF            2.832E+02        2.440E-04       6.912E-04
          ISE           1.057E-12       -1.737E+09      -1.835E-05
          BR            4.355E+00       -3.838E-11      -1.672E-12
          ISC           0.000E+00        0.000E+00       0.000E+00
          IS            2.105E-13        3.418E+11       7.197E-04
          NE            1.829E+00        1.318E-02       2.411E-04
          NC            2.000E+00        0.000E+00       0.000E+00
          IKF           1.079E+00        6.838E-04       7.378E-06
          IKR           0.000E+00        0.000E+00       0.000E+00
          VAF           1.157E+02       -5.696E-06      -6.590E-06
          VAR           0.000E+00        0.000E+00       0.000E+00
Q3
          RB            1.000E+01        1.291E-05       1.291E-06
          RC            1.000E+00        1.039E-06       1.039E-08
          RE            0.000E+00        0.000E+00       0.000E+00
          BF            3.128E+02       -1.677E-05      -5.246E-05
          ISE           9.767E-13        3.929E+09       3.837E-05
          BR            7.447E+00        1.482E-10       1.104E-11
          ISC           0.000E+00        0.000E+00       0.000E+00
          IS            4.640E-12       -1.596E+10      -7.408E-04
          NE            1.307E+00       -4.304E-02      -5.625E-04
          NC            2.000E+00        0.000E+00       0.000E+00
          IKF           2.847E-01       -9.428E-04      -2.684E-06
          IKR           0.000E+00        0.000E+00       0.000E+00
          VAF           7.403E+01        3.797E-05       2.811E-05
          VAR           0.000E+00        0.000E+00       0.000E+00
```

DC SENSITIVITIES OF OUTPUT V(7)

	ELEMENT NAME	ELEMENT VALUE	ELEMENT SENSITIVITY (VOLTS/UNIT)	NORMALIZED SENSITIVITY (VOLTS/PERCENT)
Q4				
	RB	1.000E+01	-5.661E-06	-5.661E-07
	RC	1.000E+00	1.191E-03	1.191E-05
	RE	0.000E+00	0.000E+00	0.000E+00
	BF	3.128E+02	-5.317E-06	-1.663E-05
	ISE	9.767E-13	1.061E+09	1.036E-05
	BR	7.447E+00	-2.612E-03	-1.945E-04
	ISC	0.000E+00	0.000E+00	0.000E+00
	IS	4.640E-12	2.894E+08	1.343E-05
	NE	1.307E+00	-1.204E-02	-1.573E-04
	NC	2.000E+00	0.000E+00	0.000E+00
	IKF	2.847E-01	-4.617E-04	-1.315E-06
	IKR	0.000E+00	0.000E+00	0.000E+00
	VAF	7.403E+01	-2.002E-06	-1.482E-06
	VAR	0.000E+00	0.000E+00	0.000E+00

FOUR TRANSISTOR AMPLIFIER

**** AC ANALYSIS TEMPERATURE = 70.000 DEG C

****

FREQ	VM(7)	VP(7)	VDB(7)	IM(VIN)
1.000E+01	2.398E+00	-1.800E+02	7.597E+00	4.659E-09
1.259E+01	2.398E+00	-1.800E+02	7.597E+00	4.721E-09
1.585E+01	2.398E+00	-1.800E+02	7.597E+00	4.817E-09
1.995E+01	2.398E+00	-1.800E+02	7.597E+00	4.965E-09
2.512E+01	2.398E+00	-1.800E+02	7.597E+00	5.191E-09
3.162E+01	2.398E+00	-1.800E+02	7.597E+00	5.530E-09
3.981E+01	2.398E+00	-1.800E+02	7.597E+00	6.029E-09
5.012E+01	2.398E+00	-1.800E+02	7.597E+00	6.744E-09
6.310E+01	2.398E+00	-1.800E+02	7.597E+00	7.744E-09
7.943E+01	2.398E+00	-1.800E+02	7.597E+00	9.106E-09
1.000E+02	2.398E+00	-1.800E+02	7.597E+00	1.092E-08
1.259E+02	2.398E+00	-1.800E+02	7.597E+00	1.330E-08
1.585E+02	2.398E+00	-1.800E+02	7.597E+00	1.638E-08
1.995E+02	2.398E+00	-1.800E+02	7.597E+00	2.033E-08
2.512E+02	2.398E+00	-1.800E+02	7.597E+00	2.535E-08
3.162E+02	2.398E+00	-1.800E+02	7.597E+00	3.173E-08
3.981E+02	2.398E+00	-1.800E+02	7.597E+00	3.979E-08
5.012E+02	2.398E+00	-1.800E+02	7.597E+00	4.997E-08
6.310E+02	2.398E+00	-1.800E+02	7.597E+00	6.281E-08
7.943E+02	2.398E+00	-1.800E+02	7.597E+00	7.900E-08
1.000E+03	2.398E+00	-1.800E+02	7.597E+00	9.939E-08
1.259E+03	2.398E+00	-1.800E+02	7.597E+00	1.251E-07
1.585E+03	2.398E+00	-1.800E+02	7.597E+00	1.574E-07
1.995E+03	2.398E+00	-1.800E+02	7.597E+00	1.982E-07
2.512E+03	2.398E+00	-1.800E+02	7.597E+00	2.494E-07
3.162E+03	2.398E+00	-1.800E+02	7.598E+00	3.140E-07
3.981E+03	2.398E+00	-1.800E+02	7.598E+00	3.953E-07
5.012E+03	2.398E+00	-1.800E+02	7.598E+00	4.976E-07
6.310E+03	2.398E+00	-1.800E+02	7.598E+00	6.265E-07
7.943E+03	2.398E+00	-1.799E+02	7.598E+00	7.887E-07
1.000E+04	2.398E+00	-1.799E+02	7.598E+00	9.929E-07
1.259E+04	2.398E+00	-1.799E+02	7.598E+00	1.250E-06
1.585E+04	2.399E+00	-1.799E+02	7.599E+00	1.574E-06
1.995E+04	2.399E+00	-1.798E+02	7.600E+00	1.981E-06
2.512E+04	2.399E+00	-1.798E+02	7.601E+00	2.494E-06
3.162E+04	2.400E+00	-1.798E+02	7.604E+00	3.140E-06
3.981E+04	2.401E+00	-1.797E+02	7.607E+00	3.953E-06
5.012E+04	2.402E+00	-1.796E+02	7.613E+00	4.977E-06
6.310E+04	2.405E+00	-1.795E+02	7.622E+00	6.267E-06
7.943E+04	2.409E+00	-1.794E+02	7.636E+00	7.892E-06
1.000E+05	2.415E+00	-1.793E+02	7.658E+00	9.939E-06
1.259E+05	2.424E+00	-1.792E+02	7.690E+00	1.252E-05
1.585E+05	2.437E+00	-1.791E+02	7.737E+00	1.578E-05
1.995E+05	2.456E+00	-1.791E+02	7.803E+00	1.989E-05
2.512E+05	2.480E+00	-1.793E+02	7.890E+00	2.510E-05
3.162E+05	2.511E+00	-1.796E+02	7.997E+00	3.171E-05
3.981E+05	2.545E+00	1.796E+02	8.113E+00	4.015E-05
5.012E+05	2.577E+00	1.785E+02	8.224E+00	5.098E-05
6.310E+05	2.603E+00	1.769E+02	8.310E+00	6.501E-05
7.943E+05	2.616E+00	1.749E+02	8.354E+00	8.338E-05
1.000E+06	2.613E+00	1.725E+02	8.342E+00	1.077E-04

```
FOUR TRANSISTOR AMPLIFIER

 ****     AC ANALYSIS                                    TEMPERATURE =   70.000 DEG C
******************************************************************************
LEGEND: *: VDB(7)        +: VP(7)          =: VM(7)          $: IM(VIN)

  FREQ        VDB(7)
(*)----------    5.0000E+00   6.0000E+00   7.0000E+00   8.0000E+00   9.0000E+00
(+)----------   -2.0000E+02  -1.0000E+02   0.0000E+00   1.0000E+02   2.0000E+02
(=)----------    1.0000E+00   1.0000E+01   1.0000E+02   1.0000E+03   1.0000E+04
($)----------    1.0000E-09   1.0000E-07   1.0000E-05   1.0000E-03   1.0000E-01

                       - - - - - - - - - - - - - - - - - - - - - - - - - - -
  1.000E+01  7.597E+00 .  + X       .            .        *    .            .
  1.259E+01  7.597E+00 .  + X       .            .        *    .            .
  1.585E+01  7.597E+00 .  + X       .            .        *    .            .
  1.995E+01  7.597E+00 .  + X       .            .        *    .            .
  2.512E+01  7.597E+00 .  + X       .            .        *    .            .
  3.162E+01  7.597E+00 .  + X       .            .        *    .            .
  3.981E+01  7.597E+00 .  + =$      .            .        *    .            .
  5.012E+01  7.597E+00 .  + =$      .            .        *    .            .
  6.310E+01  7.597E+00 .  + =$      .            .        *    .            .
  7.943E+01  7.597E+00 .  + = $     .            .        *    .            .
  1.000E+02  7.597E+00 .  + = $     .            .        *    .            .
  1.259E+02  7.597E+00 .  + =  $    .            .        *    .            .
  1.585E+02  7.597E+00 .  + =  $    .            .        *    .            .
  1.995E+02  7.597E+00 .  + =   $   .            .        *    .            .
  2.512E+02  7.597E+00 .  + =    $  .            .        *    .            .
  3.162E+02  7.597E+00 .  + =    $  .            .        *    .            .
  3.981E+02  7.597E+00 .  + =     $ .            .        *    .            .
  5.012E+02  7.597E+00 .  + =      $.            .        *    .            .
  6.310E+02  7.597E+00 .  + =      $.            .        *    .            .
  7.943E+02  7.597E+00 .  + =       $            .        *    .            .
  1.000E+03  7.597E+00 .  + =       $            .        *    .            .
  1.259E+03  7.597E+00 .  + =       .$           .        *    .            .
  1.585E+03  7.597E+00 .  + =       . $          .        *    .            .
  1.995E+03  7.597E+00 .  + =       . $          .        *    .            .
  2.512E+03  7.597E+00 .  + =       .  $         .        *    .            .
  3.162E+03  7.598E+00 .  + =       .   $        .        *    .            .
  3.981E+03  7.598E+00 .  + =       .   $        .        *    .            .
  5.012E+03  7.598E+00 .  + =       .    $       .        *    .            .
  6.310E+03  7.598E+00 .  + =       .     $      .        *    .            .
  7.943E+03  7.598E+00 .  + =       .     $      .        *    .            .
  1.000E+04  7.598E+00 .  + =       .      $     .        *    .            .
  1.259E+04  7.598E+00 .  + =       .       $    .        *    .            .
  1.585E+04  7.599E+00 .  + =       .       $    .        *    .            .
  1.995E+04  7.600E+00 .  + =       .        $   .        *    .            .
  2.512E+04  7.601E+00 .  + =       .         $  .        *    .            .
  3.162E+04  7.604E+00 .  + =       .         $  .        *    .            .
  3.981E+04  7.607E+00 .  + =       .          $ .        *    .            .
  5.012E+04  7.613E+00 .  + =       .           $.        *    .            .
  6.310E+04  7.622E+00 .  + =       .           $.        *    .            .
  7.943E+04  7.636E+00 .  + =       .            $        *    .            .
  1.000E+05  7.658E+00 .  + =       .            $         *   .            .
  1.259E+05  7.690E+00 .  + =       .            .$        *   .            .
  1.585E+05  7.737E+00 .  +  =      .            . $        *  .            .
  1.995E+05  7.803E+00 .  +  =      .            . $       **p1603X.        .
  2.512E+05  7.890E+00 .  +  =      .            .  $         *.            .
  3.162E+05  7.997E+00 .  +  =      .            .   $         *            .
  3.981E+05  8.113E+00 .     =      .            .   $         .*        +  .
  5.012E+05  8.224E+00 .     =      .            .    $        .  *      +  .
  6.310E+05  8.310E+00 .     =      .            .     $       .   *     +  .
  7.943E+05  8.354E+00 .     =      .            .     $       .    *    +  .
  1.000E+06  8.342E+00 .     =      .            .      $      .   *    +   .
```

```
FOUR TRANSISTOR AMPLIFIER

 ****     INITIAL TRANSIENT SOLUTION       TEMPERATURE =   70.000 DEG C

******************************************************************************

 NODE   VOLTAGE     NODE   VOLTAGE     NODE   VOLTAGE     NODE   VOLTAGE

(    1)     .0200  (    2)     .5640  (    3)    1.2745  (    4)    3.9062

(    5)     .7071  (    6)    3.3223  (    7)    3.3728  (    8)   30.0000

    VOLTAGE SOURCE CURRENTS

    NAME         CURRENT

    VCC         -8.129E-03
    VIN          9.721E-08

    TOTAL POWER DISSIPATION   2.44E-01  WATTS
```

```
FOUR TRANSISTOR AMPLIFIER

 ****     TRANSIENT ANALYSIS                    TEMPERATURE =   70.000 DEG C
******************************************************************************
  TIME         V(7)

   0.000E+00   3.373E+00
   4.000E-05   3.367E+00
   8.000E-05   3.363E+00
   1.200E-04   3.359E+00
   1.600E-04   3.356E+00
   2.000E-04   3.354E+00
   2.400E-04   3.353E+00
   2.800E-04   3.353E+00
   3.200E-04   3.354E+00
   3.600E-04   3.357E+00
   4.000E-04   3.361E+00
   4.400E-04   3.365E+00
   4.800E-04   3.370E+00
   5.200E-04   3.375E+00
   5.600E-04   3.381E+00
   6.000E-04   3.386E+00
   6.400E-04   3.390E+00
   6.800E-04   3.393E+00
   7.200E-04   3.395E+00
   7.600E-04   3.395E+00
   8.000E-04   3.394E+00
   8.400E-04   3.391E+00
   8.800E-04   3.388E+00
   9.200E-04   3.383E+00
   9.600E-04   3.378E+00
   1.000E-03   3.373E+00
   1.040E-03   3.368E+00
   1.080E-03   3.363E+00
   1.120E-03   3.359E+00
   1.160E-03   3.356E+00
   1.200E-03   3.354E+00
   1.240E-03   3.353E+00
   1.280E-03   3.353E+00
   1.320E-03   3.354E+00
   1.360E-03   3.357E+00
   1.400E-03   3.361E+00
   1.440E-03   3.365E+00
   1.480E-03   3.370E+00
   1.520E-03   3.375E+00
   1.560E-03   3.381E+00
   1.600E-03   3.386E+00
   1.640E-03   3.390E+00
   1.680E-03   3.393E+00
   1.720E-03   3.395E+00
   1.760E-03   3.395E+00
   1.800E-03   3.394E+00
   1.840E-03   3.391E+00
   1.880E-03   3.388E+00
   1.920E-03   3.383E+00
   1.960E-03   3.378E+00
   2.000E-03   3.373E+00
   2.040E-03   3.368E+00
   2.080E-03   3.363E+00
   2.120E-03   3.359E+00
   2.160E-03   3.356E+00
```

```
 TIME         V(7)

  2.200E-03   3.354E+00
  2.240E-03   3.353E+00
  2.280E-03   3.353E+00
  2.320E-03   3.354E+00
  2.360E-03   3.357E+00
  2.400E-03   3.361E+00
  2.440E-03   3.365E+00
  2.480E-03   3.370E+00
  2.520E-03   3.375E+00
  2.560E-03   3.381E+00
  2.600E-03   3.386E+00
  2.640E-03   3.390E+00
  2.680E-03   3.393E+00
  2.720E-03   3.395E+00
  2.760E-03   3.395E+00
  2.800E-03   3.394E+00
  2.840E-03   3.391E+00
  2.880E-03   3.388E+00
  2.920E-03   3.383E+00
  2.960E-03   3.378E+00
  3.000E-03   3.373E+00
  3.040E-03   3.368E+00
  3.080E-03   3.363E+00
  3.120E-03   3.359E+00
  3.160E-03   3.356E+00
  3.200E-03   3.354E+00
  3.240E-03   3.353E+00
  3.280E-03   3.353E+00
  3.320E-03   3.354E+00
  3.360E-03   3.357E+00
  3.400E-03   3.361E+00
  3.440E-03   3.365E+00
  3.480E-03   3.370E+00
  3.520E-03   3.375E+00
  3.560E-03   3.381E+00
  3.600E-03   3.386E+00
  3.640E-03   3.390E+00
  3.680E-03   3.393E+00
  3.720E-03   3.395E+00
  3.760E-03   3.395E+00
  3.800E-03   3.394E+00
  3.840E-03   3.391E+00
  3.880E-03   3.388E+00
  3.920E-03   3.383E+00
  3.960E-03   3.378E+00
  4.000E-03   3.373E+00
```

```
FOUR TRANSISTOR AMPLIFIER

 ****     TRANSIENT ANALYSIS              TEMPERATURE =   70.000 DEG C
******************************************************************************

  TIME         V(7)

(*)----------    3.3400E+00   3.3600E+00   3.3800E+00   3.4000E+00   3.4200E+00

                       - - - - - - - - - - - - - - - - - - - - - - - - - - -
  0.000E+00  3.373E+00 .            .       *    .            .            .
  4.000E-05  3.367E+00 .            .    *       .            .            .
  8.000E-05  3.363E+00 .            . *          .            .            .
  1.200E-04  3.359E+00 .           *.            .            .            .
  1.600E-04  3.356E+00 .         *  .            .            .            .
  2.000E-04  3.354E+00 .        *   .            .            .            .
  2.400E-04  3.353E+00 .       *    .            .            .            .
  2.800E-04  3.353E+00 .       *    .            .            .            .
  3.200E-04  3.354E+00 .        *   .            .            .            .
  3.600E-04  3.357E+00 .          * .            .            .            .
  4.000E-04  3.361E+00 .            *            .            .            .
  4.400E-04  3.365E+00 .            .  *         .            .            .
  4.800E-04  3.370E+00 .            .      *     .            .            .
  5.200E-04  3.375E+00 .            .         *  .            .            .
  5.600E-04  3.381E+00 .            .            *            .            .
  6.000E-04  3.386E+00 .            .            .   *        .            .
  6.400E-04  3.390E+00 .            .            .     *      .            .
  6.800E-04  3.393E+00 .            .            .       *    .            .
  7.200E-04  3.395E+00 .            .            .        *   .            .
  7.600E-04  3.395E+00 .            .            .         *  .            .
  8.000E-04  3.394E+00 .            .            .        *   .            .
  8.400E-04  3.391E+00 .            .            .      *     .            .
  8.800E-04  3.388E+00 .            .            .    *       .            .
  9.200E-04  3.383E+00 .            .            . *          .            .
  9.600E-04  3.378E+00 .            .           *.            .            .
  1.000E-03  3.373E+00 .            .       *    .            .            .
  1.040E-03  3.368E+00 .            .    *       .            .            .
  1.080E-03  3.363E+00 .            . *          .            .            .
  1.120E-03  3.359E+00 .           *.            .            .            .
  1.160E-03  3.356E+00 .         *  .            .            .            .
  1.200E-03  3.354E+00 .        *   .            .            .            .
  1.240E-03  3.353E+00 .       *    .            .            .            .
  1.280E-03  3.353E+00 .       *    .            .            .            .
  1.320E-03  3.354E+00 .        *   .            .            .            .
  1.360E-03  3.357E+00 .          * .            .            .            .
  1.400E-03  3.361E+00 .            *            .            .            .
  1.440E-03  3.365E+00 .            .  *         .            .            .
  1.480E-03  3.370E+00 .            .      *     .            .            .
  1.520E-03  3.375E+00 .            .         *  .            .            .
  1.560E-03  3.381E+00 .            .            *            .            .
  1.600E-03  3.386E+00 .            .            .   *        .            .
  1.640E-03  3.390E+00 .            .            .     *      .            .
  1.680E-03  3.393E+00 .            .            .       *    .            .
  1.720E-03  3.395E+00 .            .            .        *   .            .
  1.760E-03  3.395E+00 .            .            .         *  .            .
  1.800E-03  3.394E+00 .            .            .        *   .            .
  1.840E-03  3.391E+00 .            .            .      *     .            .
  1.880E-03  3.388E+00 .            .            .    *       .            .
  1.920E-03  3.383E+00 .            .            . *          .            .
  1.960E-03  3.378E+00 .            .           *.            .            .
  2.000E-03  3.373E+00 .            .       *    .            .            .
  2.040E-03  3.368E+00 .            .    *       .            .            .
  2.080E-03  3.363E+00 .            . *          .            .            .
```

```
 TIME        V(7)

(*)----------     3.3400E+00  3.3600E+00  3.3800E+00  3.4000E+00  3.4200E+00
                  - - - - - - - - - - - - - - - - - - - - - - - - - - - - - - -
 2.120E-03  3.359E+00 .          *.           .           .           .
 2.160E-03  3.356E+00 .        *  .           .           .           .
 2.200E-03  3.354E+00 .       *   .           .           .           .
 2.240E-03  3.353E+00 .      *    .           .           .           .
 2.280E-03  3.353E+00 .      *    .           .           .           .
 2.320E-03  3.354E+00 .       *   .           .           .           .
 2.360E-03  3.357E+00 .         * .           .           .           .
 2.400E-03  3.361E+00 .           *           .           .           .
 2.440E-03  3.365E+00 .           .  *        .           .           .
 2.480E-03  3.370E+00 .           .      *    .           .           .
 2.520E-03  3.375E+00 .           .         * .           .           .
 2.560E-03  3.381E+00 .           .           *           .           .
 2.600E-03  3.386E+00 .           .           .   *       .           .
 2.640E-03  3.390E+00 .           .           .     *     .           .
 2.680E-03  3.393E+00 .           .           .       *   .           .
 2.720E-03  3.395E+00 .           .           .        *  .           .
 2.760E-03  3.395E+00 .           .           .         * .           .
 2.800E-03  3.394E+00 .           .           .        *  .           .
 2.840E-03  3.391E+00 .           .           .      *    .           .
 2.880E-03  3.388E+00 .           .           .    *      .           .
 2.920E-03  3.383E+00 .           .           . *         .           .
 2.960E-03  3.378E+00 .           .          *.           .           .
 3.000E-03  3.373E+00 .           .       *   .           .           .
 3.040E-03  3.368E+00 .           .    *      .           .           .
 3.080E-03  3.363E+00 .           . *         .           .           .
 3.120E-03  3.359E+00 .          *.           .           .           .
 3.160E-03  3.356E+00 .        *  .           .           .           .
 3.200E-03  3.354E+00 .       *   .           .           .           .
 3.240E-03  3.353E+00 .      *    .           .           .           .
 3.280E-03  3.353E+00 .      *    .           .           .           .
 3.320E-03  3.354E+00 .       *   .           .           .           .
 3.360E-03  3.357E+00 .         * .           .           .           .
 3.400E-03  3.361E+00 .           *           .           .           .
 3.440E-03  3.365E+00 .           .  *        .           .           .
 3.480E-03  3.370E+00 .           .      *    .           .           .
 3.520E-03  3.375E+00 .           .         * .           .           .
 3.560E-03  3.381E+00 .           .           *           .           .
 3.600E-03  3.386E+00 .           .           .   *       .           .
 3.640E-03  3.390E+00 .           .           .     *     .           .
 3.680E-03  3.393E+00 .           .           .       *   .           .
 3.720E-03  3.395E+00 .           .           .        *  .           .
 3.760E-03  3.395E+00 .           .           .         * .           .
 3.800E-03  3.394E+00 .           .           .        *  .           .
 3.840E-03  3.391E+00 .           .           .      *    .           .
 3.880E-03  3.388E+00 .           .           .    *      .           .
 3.920E-03  3.383E+00 .           .           . *         .           .
 3.960E-03  3.378E+00 .           .          *.           .           .
 4.000E-03  3.373E+00 .           .       *   .           .           .
                  - - - - - - - - - - - - - - - - - - - - - - - - - - - - - - -
```

FOUR TRANSISTOR AMPLIFIER

**** FOURIER ANALYSIS TEMPERATURE = 70.000 DEG C

****

FOURIER COMPONENTS OF TRANSIENT RESPONSE V(7)

DC COMPONENT = 3.373256E+00

HARMONIC NO	FREQUENCY (HZ)	FOURIER COMPONENT	NORMALIZED COMPONENT	PHASE (DEG)	NORMALIZED PHASE (DEG)
1	1.000E+03	2.125E-02	1.000E+00	-1.800E+02	0.000E+00
2	2.000E+03	5.027E-04	2.365E-02	-9.028E+01	8.969E+01
3	3.000E+03	4.806E-05	2.261E-03	-1.784E+02	1.592E+00
4	4.000E+03	3.321E-06	1.562E-04	9.271E+01	2.727E+02
5	5.000E+03	1.141E-06	5.366E-05	-9.234E+01	8.763E+01
6	6.000E+03	1.173E-06	5.521E-05	-8.692E+01	9.305E+01
7	7.000E+03	1.187E-06	5.584E-05	-8.393E+01	9.604E+01
8	8.000E+03	1.373E-06	6.460E-05	-9.871E+01	8.126E+01
9	9.000E+03	1.228E-06	5.778E-05	-7.659E+01	1.034E+02

TOTAL HARMONIC DISTORTION = 2.376150E+00 PERCENT

FOUR TRANSISTOR AMPLIFIER

**** TEMPERATURE-ADJUSTED VALUES TEMPERATURE = 0.000 DEG C

****

**** RESISTORS

NAME	VALUE
R1	4.987E+03
R2	2.394E+04
R3	6.782E+02
R4	1.595E+04
R5	2.394E+03

**** BJT MODEL PARAMETERS

NAME	BF / BR / IS	ISE / ISC / ISS	VJE / VJC / VJS	CJE / CJC / CJS	RE / RC	RB / RBM
	BF	ISE	VJE	CJE	RE	RB
	BR	ISC	VJC	CJC	RC	RBM
	IS	ISS	VJS	CJS		
Q2N2222A	2.222E+02	5.181E-16	7.965E-01	2.141E-11	0.000E+00	1.000E+01
	5.289E+00	0.000E+00	7.965E-01	7.124E-12	1.000E+00	1.000E+01
	1.554E-16	0.000E+00	7.965E-01	0.000E+00		
Q2N2907A	2.012E+02	5.319E-15	7.965E-01	1.934E-11	0.000E+00	1.000E+01
	3.093E+00	0.000E+00	7.965E-01	1.418E-11	7.150E-01	1.000E+01
	7.049E-18	0.000E+00	7.965E-01	0.000E+00		

```
FOUR TRANSISTOR AMPLIFIER

 ****     DC TRANSFER CURVES                    TEMPERATURE =    0.000 DEG C

******************************************************************************

  VIN          V(7)

  -2.000E-01   3.645E+00
  -1.975E-01   3.639E+00
  -1.950E-01   3.632E+00
  -1.925E-01   3.625E+00
  -1.900E-01   3.619E+00
  -1.875E-01   3.612E+00
  -1.850E-01   3.606E+00
  -1.825E-01   3.599E+00
  -1.800E-01   3.592E+00
  -1.775E-01   3.586E+00
  -1.750E-01   3.579E+00
  -1.725E-01   3.572E+00
  -1.700E-01   3.566E+00
  -1.675E-01   3.559E+00
  -1.650E-01   3.553E+00
  -1.625E-01   3.546E+00
  -1.600E-01   3.540E+00
  -1.575E-01   3.533E+00
  -1.550E-01   3.527E+00
  -1.525E-01   3.520E+00
  -1.500E-01   3.514E+00
  -1.475E-01   3.507E+00
  -1.450E-01   3.501E+00
  -1.425E-01   3.494E+00
  -1.400E-01   3.488E+00
  -1.375E-01   3.481E+00
  -1.350E-01   3.475E+00
  -1.325E-01   3.468E+00
  -1.300E-01   3.462E+00
  -1.275E-01   3.456E+00
  -1.250E-01   3.449E+00
  -1.225E-01   3.443E+00
  -1.200E-01   3.437E+00
  -1.175E-01   3.431E+00
  -1.150E-01   3.424E+00
  -1.125E-01   3.418E+00
  -1.100E-01   3.412E+00
  -1.075E-01   3.406E+00
  -1.050E-01   3.400E+00
  -1.025E-01   3.394E+00
  -1.000E-01   3.388E+00
  -9.750E-02   3.383E+00
  -9.500E-02   3.377E+00
  -9.250E-02   3.371E+00
  -9.000E-02   3.366E+00
  -8.750E-02   3.360E+00
  -8.500E-02   3.355E+00
  -8.250E-02   3.350E+00
  -8.000E-02   3.345E+00
  -7.750E-02   3.341E+00
  -7.500E-02   3.337E+00
  -7.250E-02   3.333E+00
  -7.000E-02   3.330E+00
```

VIN	V(7)
-6.750E-02	3.329E+00
-6.500E-02	3.329E+00
-6.250E-02	3.333E+00
-6.000E-02	3.352E+00
-5.750E-02	3.724E+00
-5.500E-02	4.349E+00
-5.250E-02	4.974E+00
-5.000E-02	5.601E+00
-4.750E-02	6.229E+00
-4.500E-02	6.857E+00
-4.250E-02	7.486E+00
-4.000E-02	8.116E+00
-3.750E-02	8.747E+00
-3.500E-02	9.378E+00
-3.250E-02	1.001E+01
-3.000E-02	1.064E+01
-2.750E-02	1.128E+01
-2.500E-02	1.191E+01
-2.250E-02	1.254E+01
-2.000E-02	1.318E+01
-1.750E-02	1.381E+01
-1.500E-02	1.445E+01
-1.250E-02	1.509E+01
-1.000E-02	1.572E+01
-7.500E-03	1.636E+01
-5.000E-03	1.699E+01
-2.500E-03	1.763E+01
0.000E+00	1.827E+01
2.500E-03	1.890E+01
5.000E-03	1.954E+01
7.500E-03	2.018E+01
1.000E-02	2.081E+01
1.250E-02	2.145E+01
1.500E-02	2.208E+01
1.750E-02	2.272E+01
2.000E-02	2.335E+01
2.250E-02	2.398E+01
2.500E-02	2.461E+01
2.750E-02	2.524E+01
3.000E-02	2.587E+01
3.250E-02	2.648E+01
3.500E-02	2.706E+01
3.750E-02	2.742E+01
4.000E-02	2.757E+01
4.250E-02	2.765E+01
4.500E-02	2.769E+01
4.750E-02	2.772E+01
5.000E-02	2.774E+01
5.250E-02	2.776E+01
5.500E-02	2.777E+01
5.750E-02	2.777E+01
6.000E-02	2.778E+01
6.250E-02	2.778E+01
6.500E-02	2.778E+01
6.750E-02	2.777E+01
7.000E-02	2.777E+01
7.250E-02	2.776E+01
7.500E-02	2.776E+01
7.750E-02	2.775E+01
8.000E-02	2.774E+01

VIN	V(7)
8.250E-02	2.774E+01
8.500E-02	2.773E+01
8.750E-02	2.772E+01
9.000E-02	2.771E+01
9.250E-02	2.770E+01
9.500E-02	2.769E+01
9.750E-02	2.768E+01
1.000E-01	2.767E+01
1.025E-01	2.766E+01
1.050E-01	2.764E+01
1.075E-01	2.763E+01
1.100E-01	2.762E+01
1.125E-01	2.761E+01
1.150E-01	2.760E+01
1.175E-01	2.758E+01
1.200E-01	2.757E+01
1.225E-01	2.756E+01
1.250E-01	2.754E+01
1.275E-01	2.753E+01
1.300E-01	2.752E+01
1.325E-01	2.750E+01
1.350E-01	2.749E+01
1.375E-01	2.748E+01
1.400E-01	2.746E+01
1.425E-01	2.745E+01
1.450E-01	2.743E+01
1.475E-01	2.742E+01
1.500E-01	2.740E+01
1.525E-01	2.739E+01
1.550E-01	2.738E+01
1.575E-01	2.736E+01
1.600E-01	2.735E+01
1.625E-01	2.733E+01
1.650E-01	2.732E+01
1.675E-01	2.730E+01
1.700E-01	2.729E+01
1.725E-01	2.727E+01
1.750E-01	2.726E+01
1.775E-01	2.724E+01
1.800E-01	2.722E+01
1.825E-01	2.721E+01
1.850E-01	2.719E+01
1.875E-01	2.718E+01
1.900E-01	2.716E+01
1.925E-01	2.715E+01
1.950E-01	2.713E+01
1.975E-01	2.711E+01
2.000E-01	2.710E+01

```
FOUR TRANSISTOR AMPLIFIER

 ****     DC TRANSFER CURVES                    TEMPERATURE =    0.000 DEG C

******************************************************************************

  VIN          V(7)

(*)----------     1.5000E+01    2.0000E+01    2.5000E+01    3.0000E+01    3.5000E+01
                    - - - - - - - - - - - - - - - - - - - - - - - - - - - - - - - -
 -2.000E-01  3.645E+00 .     *       .             .             .             .
 -1.975E-01  3.639E+00 .     *       .             .             .             .
 -1.950E-01  3.632E+00 .     *       .             .             .             .
 -1.925E-01  3.625E+00 .     *       .             .             .             .
 -1.900E-01  3.619E+00 .     *       .             .             .             .
 -1.875E-01  3.612E+00 .     *       .             .             .             .
 -1.850E-01  3.606E+00 .     *       .             .             .             .
 -1.825E-01  3.599E+00 .     *       .             .             .             .
 -1.800E-01  3.592E+00 .     *       .             .             .             .
 -1.775E-01  3.586E+00 .     *       .             .             .             .
 -1.750E-01  3.579E+00 .     *       .             .             .             .
 -1.725E-01  3.572E+00 .     *       .             .             .             .
 -1.700E-01  3.566E+00 .     *       .             .             .             .
 -1.675E-01  3.559E+00 .     *       .             .             .             .
 -1.650E-01  3.553E+00 .     *       .             .             .             .
 -1.625E-01  3.546E+00 .     *       .             .             .             .
 -1.600E-01  3.540E+00 .     *       .             .             .             .
 -1.575E-01  3.533E+00 .     *       .             .             .             .
 -1.550E-01  3.527E+00 .     *       .             .             .             .
 -1.525E-01  3.520E+00 .     *       .             .             .             .
 -1.500E-01  3.514E+00 .     *       .             .             .             .
 -1.475E-01  3.507E+00 .     *       .             .             .             .
 -1.450E-01  3.501E+00 .     *       .             .             .             .
 -1.425E-01  3.494E+00 .     *       .             .             .             .
 -1.400E-01  3.488E+00 .     *       .             .             .             .
 -1.375E-01  3.481E+00 .     *       .             .             .             .
 -1.350E-01  3.475E+00 .     *       .             .             .             .
 -1.325E-01  3.468E+00 .     *       .             .             .             .
 -1.300E-01  3.462E+00 .     *       .             .             .             .
 -1.275E-01  3.456E+00 .    *        .             .             .             .
 -1.250E-01  3.449E+00 .    *        .             .             .             .
 -1.225E-01  3.443E+00 .    *        .             .             .             .
 -1.200E-01  3.437E+00 .    *        .             .             .             .
 -1.175E-01  3.431E+00 .    *        .             .             .             .
 -1.150E-01  3.424E+00 .    *        .             .             .             .
 -1.125E-01  3.418E+00 .    *        .             .             .             .
 -1.100E-01  3.412E+00 .    *        .             .             .             .
 -1.075E-01  3.406E+00 .    *        .             .             .             .
 -1.050E-01  3.400E+00 .    *        .             .             .             .
 -1.025E-01  3.394E+00 .    *        .             .             .             .
 -1.000E-01  3.388E+00 .    *        .             .             .             .
 -9.750E-02  3.383E+00 .    *        .             .             .             .
 -9.500E-02  3.377E+00 .    *        .             .             .             .
 -9.250E-02  3.371E+00 .    *        .             .             .             .
 -9.000E-02  3.366E+00 .    *        .             .             .             .
 -8.750E-02  3.360E+00 .    *        .             .             .             .
 -8.500E-02  3.355E+00 .    *        .             .             .             .
 -8.250E-02  3.350E+00 .    *        .             .             .             .
 -8.000E-02  3.345E+00 .    *        .             .             .             .
 -7.750E-02  3.341E+00 .    *        .             .             .             .
 -7.500E-02  3.337E+00 .    *        .             .             .             .
 -7.250E-02  3.333E+00 .    *        .             .             .             .
 -7.000E-02  3.330E+00 .    *        .             .             .             .
```

```
  VIN         V(7)

(*)----------      1.5000E+01   2.0000E+01   2.5000E+01   3.0000E+01   3.5000E+01
                     - - - - - - - - - - - - - - - - - - - - - - - - - - -
-6.750E-02  3.329E+00 .   *        .            .            .            .
-6.500E-02  3.329E+00 .   *        .            .            .            .
-6.250E-02  3.333E+00 .   *        .            .            .            .
-6.000E-02  3.352E+00 .   *        .            .            .            .
-5.750E-02  3.724E+00 .    *       .            .            .            .
-5.500E-02  4.349E+00 .     *      .            .            .            .
-5.250E-02  4.974E+00 .     *      .            .            .            .
-5.000E-02  5.601E+00 .      *     .            .            .            .
-4.750E-02  6.229E+00 .       *    .            .            .            .
-4.500E-02  6.857E+00 .        *   .            .            .            .
-4.250E-02  7.486E+00 .         *  .            .            .            .
-4.000E-02  8.116E+00 .          * .            .            .            .
-3.750E-02  8.747E+00 .          * .            .            .            .
-3.500E-02  9.378E+00 .           *.            .            .            .
-3.250E-02  1.001E+01 .            *            .            .            .
-3.000E-02  1.064E+01 .            .*           .            .            .
-2.750E-02  1.128E+01 .            . *          .            .            .
-2.500E-02  1.191E+01 .            . *          .            .            .
-2.250E-02  1.254E+01 .            .  *         .            .            .
-2.000E-02  1.318E+01 .            .   *        .            .            .
-1.750E-02  1.381E+01 .            .    *       .            .            .
-1.500E-02  1.445E+01 .            .     *      .            .            .
-1.250E-02  1.509E+01 .            .      *     .            .            .
-1.000E-02  1.572E+01 .            .      *     .            .            .
-7.500E-03  1.636E+01 .            .       *    .            .            .
-5.000E-03  1.699E+01 .            .        *   .            .            .
-2.500E-03  1.763E+01 .            .         *  .            .            .
 0.000E+00  1.827E+01 .            .          * .            .            .
 2.500E-03  1.890E+01 .            .           *.            .            .
 5.000E-03  1.954E+01 .            .           *.            .            .
 7.500E-03  2.018E+01 .            .            *            .            .
 1.000E-02  2.081E+01 .            .            .*           .            .
 1.250E-02  2.145E+01 .            .            . *          .            .
 1.500E-02  2.208E+01 .            .            .  *         .            .
 1.750E-02  2.272E+01 .            .            .   *        .            .
 2.000E-02  2.335E+01 .            .            .   *        .            .
 2.250E-02  2.398E+01 .            .            .    *       .            .
 2.500E-02  2.461E+01 .            .            .     *      .            .
 2.750E-02  2.524E+01 .            .            .      *     .            .
 3.000E-02  2.587E+01 .            .            .       *    .            .
 3.250E-02  2.648E+01 .            .            .       *    .            .
 3.500E-02  2.706E+01 .            .            .        *   .            .
 3.750E-02  2.742E+01 .            .            .         *  .            .
 4.000E-02  2.757E+01 .            .            .         *  .            .
 4.250E-02  2.765E+01 .            .            .         *  .            .
 4.500E-02  2.769E+01 .            .            .         *  .            .
 4.750E-02  2.772E+01 .            .            .         *  .            .
 5.000E-02  2.774E+01 .            .            .         *  .            .
 5.250E-02  2.776E+01 .            .            .         *  .            .
 5.500E-02  2.777E+01 .            .            .         *  .            .
 5.750E-02  2.777E+01 .            .            .         *  .            .
 6.000E-02  2.778E+01 .            .            .         *  .            .
 6.250E-02  2.778E+01 .            .            .         *  .            .
 6.500E-02  2.778E+01 .            .            .         *  .            .
 6.750E-02  2.777E+01 .            .            .         *  .            .
 7.000E-02  2.777E+01 .            .            .         *  .            .
```

```
   VIN          V(7)

(*)----------    1.5000E+01   2.0000E+01   2.5000E+01   3.0000E+01   3.5000E+01
                          - - - - - - - - - - - - - - - - - - - - - -
   7.250E-02  2.776E+01 .         .         .       * .         .
   7.500E-02  2.776E+01 .         .         .       * .         .
   7.750E-02  2.775E+01 .         .         .       * .         .
   8.000E-02  2.774E+01 .         .         .       * .         .
   8.250E-02  2.774E+01 .         .         .       * .         .
   8.500E-02  2.773E+01 .         .         .       * .         .
   8.750E-02  2.772E+01 .         .         .       * .         .
   9.000E-02  2.771E+01 .         .         .       * .         .
   9.250E-02  2.770E+01 .         .         .       * .         .
   9.500E-02  2.769E+01 .         .         .       * .         .
   9.750E-02  2.768E+01 .         .         .       * .         .
   1.000E-01  2.767E+01 .         .         .       * .         .
   1.025E-01  2.766E+01 .         .         .       * .         .
   1.050E-01  2.764E+01 .         .         .       * .         .
   1.075E-01  2.763E+01 .         .         .       * .         .
   1.100E-01  2.762E+01 .         .         .       * .         .
   1.125E-01  2.761E+01 .         .         .       * .         .
   1.150E-01  2.760E+01 .         .         .       * .         .
   1.175E-01  2.758E+01 .         .         .       * .         .
   1.200E-01  2.757E+01 .         .         .       * .         .
   1.225E-01  2.756E+01 .         .         .       * .         .
   1.250E-01  2.754E+01 .         .         .       * .         .
   1.275E-01  2.753E+01 .         .         .       * .         .
   1.300E-01  2.752E+01 .         .         .       * .         .
   1.325E-01  2.750E+01 .         .         .       * .         .
   1.350E-01  2.749E+01 .         .         .       * .         .
   1.375E-01  2.748E+01 .         .         .       * .         .
   1.400E-01  2.746E+01 .         .         .       * .         .
   1.425E-01  2.745E+01 .         .         .       * .         .
   1.450E-01  2.743E+01 .         .         .       * .         .
   1.475E-01  2.742E+01 .         .         .       * .         .
   1.500E-01  2.740E+01 .         .         .       * .         .
   1.525E-01  2.739E+01 .         .         .       * .         .
   1.550E-01  2.738E+01 .         .         .       * .         .
   1.575E-01  2.736E+01 .         .         .       * .         .
   1.600E-01  2.735E+01 .         .         .       * .         .
   1.625E-01  2.733E+01 .         .         .       * .         .
   1.650E-01  2.732E+01 .         .         .       * .         .
   1.675E-01  2.730E+01 .         .         .      *  .         .
   1.700E-01  2.729E+01 .         .         .      *  .         .
   1.725E-01  2.727E+01 .         .         .      *  .         .
   1.750E-01  2.726E+01 .         .         .      *  .         .
   1.775E-01  2.724E+01 .         .         .      *  .         .
   1.800E-01  2.722E+01 .         .         .      *  .         .
   1.825E-01  2.721E+01 .         .         .      *  .         .
   1.850E-01  2.719E+01 .         .         .      *  .         .
   1.875E-01  2.718E+01 .         .         .      *  .         .
   1.900E-01  2.716E+01 .         .         .      *  .         .
   1.925E-01  2.715E+01 .         .         .      *  .         .
   1.950E-01  2.713E+01 .         .         .      *  .         .
   1.975E-01  2.711E+01 .         .         .      *  .         .
   2.000E-01  2.710E+01 .         .         .      *  .         .
                        . . . . . . . . . . . . . . . . . . . . .
```

```
FOUR TRANSISTOR AMPLIFIER

 ****     SMALL SIGNAL BIAS SOLUTION       TEMPERATURE =    0.000 DEG C

******************************************************************************

 NODE   VOLTAGE     NODE   VOLTAGE     NODE   VOLTAGE     NODE   VOLTAGE


(    1)    0.0000  (    2)     .6835  (    3)    1.4916  (    4)    2.0947

(    5)     .7939  (    6)    1.4176  (    7)   18.2680  (    8)   30.0000

    VOLTAGE SOURCE CURRENTS
    NAME          CURRENT

    VCC          -7.471E-03
    VIN           1.849E-07

    TOTAL POWER DISSIPATION   2.24E-01  WATTS
```

```
FOUR TRANSISTOR AMPLIFIER

 ****     OPERATING POINT INFORMATION      TEMPERATURE =    0.000 DEG C

******************************************************************************

**** BIPOLAR JUNCTION TRANSISTORS

NAME          Q1          Q2          Q3          Q4
MODEL         Q2N2907A    Q2N2907A    Q2N2222A    Q2N2222A
IB           -1.85E-07   -2.90E-05    8.86E-06    4.05E-06
IC           -2.88E-05   -5.68E-03    1.16E-03    5.88E-04
VBE          -6.83E-01   -8.08E-01    6.98E-01    6.77E-01
VBC           0.00E+00    6.83E-01   -6.03E-01   -1.62E+01
VCE          -6.83E-01   -1.49E+00    1.30E+00    1.69E+01
BETADC        1.56E+02    1.96E+02    1.31E+02    1.45E+02
GM            1.22E-03    2.40E-01    4.91E-02    2.49E-02
RPI           1.42E+05    8.22E+02    2.94E+03    6.52E+03
RX            1.00E+01    1.00E+01    1.00E+01    1.00E+01
RO            4.02E+06    2.05E+04    6.42E+04    1.53E+05
CBE           3.10E-11    1.78E-10    5.59E-11    4.54E-11
CBC           1.42E-11    1.02E-11    5.88E-12    2.51E-12
CBX           0.00E+00    0.00E+00    0.00E+00    0.00E+00
CJS           0.00E+00    0.00E+00    0.00E+00    0.00E+00
BETAAC        1.73E+02    1.97E+02    1.45E+02    1.63E+02
FT            4.31E+06    2.03E+08    1.27E+08    8.29E+07

 ****     SMALL-SIGNAL CHARACTERISTICS


      V(7)/VIN =  2.547E+02

      INPUT RESISTANCE AT VIN =  1.112E+08

      OUTPUT RESISTANCE AT V(7) =  1.975E+04
```

```
FOUR TRANSISTOR AMPLIFIER

 ****     DC SENSITIVITY ANALYSIS                 TEMPERATURE =    0.000 DEG C
******************************************************************************

DC SENSITIVITIES OF OUTPUT V(7)

          ELEMENT         ELEMENT           ELEMENT        NORMALIZED
           NAME            VALUE          SENSITIVITY      SENSITIVITY
                                         (VOLTS/UNIT) (VOLTS/PERCENT)

           R1            4.987E+03        -2.423E-03      -1.208E-01
           R2            2.394E+04         8.749E-03       2.094E+00
           R3            6.782E+02        -2.990E-01      -2.028E+00
           R4            1.995E+04        -5.824E-04      -1.162E-01
           R5            2.394E+03         4.510E-03       1.079E-01
           VCC           3.000E+01        -6.091E+00      -1.827E+00
           VIN           0.000E+00         2.547E+02       0.000E+00
Q1
           RB            1.000E+01         4.709E-05       4.709E-06
           RC            7.150E-01         1.483E-06       1.061E-08
           RE            0.000E+00         0.000E+00       0.000E+00
           BF            2.012E+02         1.456E-04       2.928E-04
           ISE           5.319E-15        -1.607E+12      -8.547E-05
           BR            3.093E+00        -1.494E-18      -4.621E-20
           ISC           0.000E+00         0.000E+00       0.000E+00
           IS            7.049E-18        -8.500E+17      -5.992E-02
           NE            1.829E+00         7.419E-02       1.357E-03
           NC            2.000E+00         0.000E+00       0.000E+00
           IKF           1.079E+00        -1.474E-04      -1.590E-06
           IKR           0.000E+00         0.000E+00       0.000E+00
           VAF           1.157E+02        -8.343E-09      -9.653E-09
           VAR           0.000E+00         0.000E+00       0.000E+00
Q2
           RB            1.000E+01         7.380E-03       7.380E-04
           RC            7.150E-01         5.858E-04       4.188E-06
           RE            0.000E+00         0.000E+00       0.000E+00
           BF            2.012E+02        -2.912E-02      -5.858E-02
           ISE           5.319E-15         2.931E+13       1.559E-03
           BR            3.093E+00         1.545E-13       4.780E-15
           ISC           0.000E+00         0.000E+00       0.000E+00
           IS            7.049E-18        -8.631E+17      -6.084E-02
           NE            1.829E+00        -1.599E+00      -2.925E-02
           NC            2.000E+00         0.000E+00       0.000E+00
           IKF           1.079E+00        -5.761E-02      -6.216E-04
           IKR           0.000E+00         0.000E+00       0.000E+00
           VAF           1.157E+02         6.060E-04       7.011E-04
           VAR           0.000E+00         0.000E+00       0.000E+00
Q3
           RB            1.000E+01        -2.263E-03      -2.263E-04
           RC            1.000E+00        -1.162E-04      -1.162E-06
           RE            0.000E+00         0.000E+00       0.000E+00
           BF            2.222E+02         4.170E-03       9.264E-03
           ISE           5.181E-16        -1.255E+15      -6.501E-03
           BR            5.289E+00        -1.024E-12      -5.414E-14
           ISC           0.000E+00         0.000E+00       0.000E+00
           IS            1.554E-16         4.189E+16       6.509E-02
           NE            1.307E+00         1.128E+01       1.474E-01
           NC            2.000E+00         0.000E+00       0.000E+00
           IKF           2.847E-01         1.053E-01       2.998E-04
           IKR           0.000E+00         0.000E+00       0.000E+00
           VAF           7.403E+01        -8.134E-04      -6.021E-04
           VAR           0.000E+00         0.000E+00       0.000E+00
```

DC SENSITIVITIES OF OUTPUT V(7)

	ELEMENT NAME	ELEMENT VALUE	ELEMENT SENSITIVITY (VOLTS/UNIT)	NORMALIZED SENSITIVITY (VOLTS/PERCENT)
Q4				
	RB	1.000E+01	3.081E-05	3.081E-06
	RC	1.000E+00	5.839E-06	5.839E-08
	RE	0.000E+00	0.000E+00	0.000E+00
	BF	2.222E+02	-1.939E-03	-4.308E-03
	ISE	5.181E-16	7.163E+14	3.711E-03
	BR	5.289E+00	1.108E-12	5.860E-14
	ISC	0.000E+00	0.000E+00	0.000E+00
	IS	1.554E-16	-2.981E+15	-4.631E-03
	NE	1.307E+00	-6.249E+00	-8.167E-02
	NC	2.000E+00	0.000E+00	0.000E+00
	IKF	2.847E-01	-5.315E-03	-1.513E-05
	IKR	0.000E+00	0.000E+00	0.000E+00
	VAF	7.403E+01	2.169E-03	1.606E-03
	VAR	0.000E+00	0.000E+00	0.000E+00

FOUR TRANSISTOR AMPLIFIER

**** AC ANALYSIS TEMPERATURE = 0.000 DEG C

****

FREQ	VM(7)	VP(7)	VDB(7)	IM(VIN)
1.000E+01	2.547E+02	-2.625E-03	4.812E+01	9.041E-09
1.259E+01	2.547E+02	-3.304E-03	4.812E+01	9.067E-09
1.585E+01	2.547E+02	-4.160E-03	4.812E+01	9.108E-09
1.995E+01	2.547E+02	-5.237E-03	4.812E+01	9.173E-09
2.512E+01	2.547E+02	-6.593E-03	4.812E+01	9.275E-09
3.162E+01	2.547E+02	-8.300E-03	4.812E+01	9.434E-09
3.981E+01	2.547E+02	-1.045E-02	4.812E+01	9.680E-09
5.012E+01	2.547E+02	-1.315E-02	4.812E+01	1.006E-08
6.310E+01	2.547E+02	-1.656E-02	4.812E+01	1.063E-08
7.943E+01	2.547E+02	-2.085E-02	4.812E+01	1.148E-08
1.000E+02	2.547E+02	-2.625E-02	4.812E+01	1.271E-08
1.259E+02	2.547E+02	-3.304E-02	4.812E+01	1.444E-08
1.585E+02	2.547E+02	-4.160E-02	4.812E+01	1.683E-08
1.995E+02	2.547E+02	-5.237E-02	4.812E+01	2.004E-08
2.512E+02	2.547E+02	-6.593E-02	4.812E+01	2.428E-08
3.162E+02	2.547E+02	-8.300E-02	4.812E+01	2.978E-08
3.981E+02	2.547E+02	-1.045E-01	4.812E+01	3.685E-08
5.012E+02	2.547E+02	-1.315E-01	4.812E+01	4.588E-08
6.310E+02	2.547E+02	-1.656E-01	4.812E+01	5.734E-08
7.943E+02	2.547E+02	-2.085E-01	4.812E+01	7.186E-08
1.000E+03	2.547E+02	-2.625E-01	4.812E+01	9.021E-08
1.259E+03	2.547E+02	-3.304E-01	4.812E+01	1.134E-07
1.585E+03	2.547E+02	-4.160E-01	4.812E+01	1.425E-07
1.995E+03	2.547E+02	-5.237E-01	4.812E+01	1.793E-07
2.512E+03	2.547E+02	-6.593E-01	4.812E+01	2.256E-07
3.162E+03	2.547E+02	-8.299E-01	4.812E+01	2.840E-07
3.981E+03	2.547E+02	-1.045E+00	4.812E+01	3.574E-07
5.012E+03	2.547E+02	-1.315E+00	4.812E+01	4.500E-07
6.310E+03	2.546E+02	-1.656E+00	4.812E+01	5.664E-07
7.943E+03	2.546E+02	-2.084E+00	4.812E+01	7.130E-07
1.000E+04	2.545E+02	-2.623E+00	4.811E+01	8.976E-07
1.259E+04	2.543E+02	-3.301E+00	4.811E+01	1.130E-06
1.585E+04	2.541E+02	-4.153E+00	4.810E+01	1.423E-06
1.995E+04	2.537E+02	-5.224E+00	4.809E+01	1.791E-06
2.512E+04	2.532E+02	-6.568E+00	4.807E+01	2.255E-06
3.162E+04	2.523E+02	-8.250E+00	4.804E+01	2.839E-06
3.981E+04	2.509E+02	-1.035E+01	4.799E+01	3.574E-06
5.012E+04	2.488E+02	-1.296E+01	4.792E+01	4.499E-06
6.310E+04	2.455E+02	-1.617E+01	4.780E+01	5.665E-06
7.943E+04	2.405E+02	-2.010E+01	4.762E+01	7.132E-06
1.000E+05	2.333E+02	-2.480E+01	4.736E+01	8.981E-06
1.259E+05	2.230E+02	-3.032E+01	4.697E+01	1.131E-05
1.585E+05	2.092E+02	-3.659E+01	4.641E+01	1.424E-05
1.995E+05	1.917E+02	-4.344E+01	4.565E+01	1.794E-05
2.512E+05	1.712E+02	-5.060E+01	4.467E+01	2.261E-05
3.162E+05	1.489E+02	-5.778E+01	4.346E+01	2.850E-05
3.981E+05	1.265E+02	-6.470E+01	4.204E+01	3.595E-05
5.012E+05	1.053E+02	-7.121E+01	4.045E+01	4.539E-05
6.310E+05	8.630E+01	-7.727E+01	3.872E+01	5.739E-05
7.943E+05	6.991E+01	-8.296E+01	3.689E+01	7.273E-05
1.000E+06	5.614E+01	-8.843E+01	3.498E+01	9.249E-05

```
FOUR TRANSISTOR AMPLIFIER

 ****     AC ANALYSIS                             TEMPERATURE =    0.000 DEG C
******************************************************************************
 LEGEND:          *: VDB(7)       +: VP(7)         =: VM(7)         $: IM(VIN)

  FREQ        VDB(7)

(*)----------    1.0000E+01   2.0000E+01   3.0000E+01   4.0000E+01   5.0000E+01
(+)----------   -2.0000E+02  -1.5000E+02  -1.0000E+02  -5.0000E+01   0.0000E+00
(=)----------    1.0000E+00   1.0000E+01   1.0000E+02   1.0000E+03   1.0000E+04
($)----------    1.0000E-09   1.0000E-07   1.0000E-05   1.0000E-03   1.0000E-01
                         - - - - - - - - - - - - - - - - - - - - - - - - - - -
   1.000E+01   4.812E+01 .      $     .            .     =      .          * +
   1.259E+01   4.812E+01 .      $     .            .     =      .          * +
   1.585E+01   4.812E+01 .      $     .            .     =      .          * +
   1.995E+01   4.812E+01 .      $     .            .     =      .          * +
   2.512E+01   4.812E+01 .      $     .            .     =      .          * +
   3.162E+01   4.812E+01 .      $     .            .     =      .          * +
   3.981E+01   4.812E+01 .      $     .            .     =      .          * +
   5.012E+01   4.812E+01 .      $     .            .     =      .          * +
   6.310E+01   4.812E+01 .      $     .            .     =      .          * +
   7.943E+01   4.812E+01 .      $     .            .     =      .          * +
   1.000E+02   4.812E+01 .       $    .            .     =      .          * +
   1.259E+02   4.812E+01 .       $    .            .     =      .          * +
   1.585E+02   4.812E+01 .       $    .            .     =      .          * +
   1.995E+02   4.812E+01 .        $   .            .     =      .          * +
   2.512E+02   4.812E+01 .         $  .            .     =      .          * +
   3.162E+02   4.812E+01 .         $  .            .     =      .          * +
   3.981E+02   4.812E+01 .          $ .            .     =      .          * +
   5.012E+02   4.812E+01 .          $ .            .     =      .          * +
   6.310E+02   4.812E+01 .           $.            .     =      .          * +
   7.943E+02   4.812E+01 .            $            .     =      .          * +
   1.000E+03   4.812E+01 .            $            .     =      .          * +
   1.259E+03   4.812E+01 .            .$           .     =      .          * +
   1.585E+03   4.812E+01 .            . $          .     =      .          * +
   1.995E+03   4.812E+01 .            . $          .     =      .          * +
   2.512E+03   4.812E+01 .            .  $         .     =      .          * +
   3.162E+03   4.812E+01 .            .  $         .     =      .          * +
   3.981E+03   4.812E+01 .            .   $        .     =      .          * +
   5.012E+03   4.812E+01 .            .    $       .     =      .          * +
   6.310E+03   4.812E+01 .            .    $       .     =      .          * +
   7.943E+03   4.812E+01 .            .     $      .     =      .          *+.
   1.000E+04   4.811E+01 .            .      $     .     =      .          *+.
   1.259E+04   4.811E+01 .            .      $     .     =      .          *+.
   1.585E+04   4.810E+01 .            .       $    .     =      .          *+.
   1.995E+04   4.809E+01 .            .        $   .     =      .          *+.
   2.512E+04   4.807E+01 .            .        $   .     =      .         *+ .
   3.162E+04   4.804E+01 .            .         $  .     =      .         *+ .
   3.981E+04   4.799E+01 .            .          $ .     =      .         X  .
   5.012E+04   4.792E+01 .            .          $ .     =      .         X  .
   6.310E+04   4.780E+01 .            .           $.     =      .        +*  .
   7.943E+04   4.762E+01 .            .            $    =       .       + *  .
   1.000E+05   4.736E+01 .            .            $    =       .      +  *  .
   1.259E+05   4.697E+01 .            .            .$   =       .    +   *   .
   1.585E+05   4.641E+01 .            .            .$   =       .  +    *    .
   1.995E+05   4.565E+01 .            .            . $ =        . +    *     .
   2.512E+05   4.467E+01 .            .            .  $=        +     *      .
   3.162E+05   4.346E+01 .            .            .  X       + .   *        .
   3.981E+05   4.204E+01 .            .            . = $    +   .  *         .
   5.012E+05   4.045E+01 .            .            .=   $ +     .*           .
   6.310E+05   3.872E+01 .            .            =    $+    * .            .
   7.943E+05   3.689E+01 .            .          = .   + $  *   .            .
   1.000E+06   3.498E+01 .            .         =  .  +  *$     .            .
```

```
FOUR TRANSISTOR AMPLIFIER

****     INITIAL TRANSIENT SOLUTION        TEMPERATURE =    0.000 DEG C

*************************************************************************************

NODE   VOLTAGE      NODE   VOLTAGE      NODE   VOLTAGE      NODE   VOLTAGE

(    1)      .0200  (    2)      .7035  (    3)    1.5116  (    4)    1.4656

(    5)      .8131  (    6)      .8033  (    7)   23.3520  (    8)   30.0000

    VOLTAGE SOURCE CURRENTS
    NAME          CURRENT

    VCC          -7.239E-03
    VIN           1.847E-07

    TOTAL POWER DISSIPATION   2.17E-01  WATTS
```

```
FOUR TRANSISTOR AMPLIFIER

 ****     TRANSIENT ANALYSIS                  TEMPERATURE =    0.000 DEG C

******************************************************************************

  TIME        V(7)

  0.000E+00   2.335E+01
  4.000E-05   2.397E+01
  8.000E-05   2.455E+01
  1.200E-04   2.506E+01
  1.600E-04   2.546E+01
  2.000E-04   2.572E+01
  2.400E-04   2.584E+01
  2.800E-04   2.581E+01
  3.200E-04   2.562E+01
  3.600E-04   2.529E+01
  4.000E-04   2.483E+01
  4.400E-04   2.429E+01
  4.800E-04   2.368E+01
  5.200E-04   2.305E+01
  5.600E-04   2.243E+01
  6.000E-04   2.188E+01
  6.400E-04   2.142E+01
  6.800E-04   2.108E+01
  7.200E-04   2.088E+01
  7.600E-04   2.083E+01
  8.000E-04   2.095E+01
  8.400E-04   2.122E+01
  8.800E-04   2.162E+01
  9.200E-04   2.213E+01
  9.600E-04   2.272E+01
  1.000E-03   2.334E+01
  1.040E-03   2.397E+01
  1.080E-03   2.455E+01
  1.120E-03   2.506E+01
  1.160E-03   2.546E+01
  1.200E-03   2.572E+01
  1.240E-03   2.584E+01
  1.280E-03   2.581E+01
  1.320E-03   2.562E+01
  1.360E-03   2.529E+01
  1.400E-03   2.483E+01
  1.440E-03   2.429E+01
  1.480E-03   2.368E+01
  1.520E-03   2.305E+01
  1.560E-03   2.244E+01
  1.600E-03   2.188E+01
  1.640E-03   2.142E+01
  1.680E-03   2.108E+01
  1.720E-03   2.088E+01
  1.760E-03   2.083E+01
  1.800E-03   2.095E+01
  1.840E-03   2.122E+01
  1.880E-03   2.162E+01
  1.920E-03   2.213E+01
  1.960E-03   2.272E+01
  2.000E-03   2.334E+01
  2.040E-03   2.397E+01
  2.080E-03   2.455E+01
  2.120E-03   2.506E+01
  2.160E-03   2.546E+01
  2.200E-03   2.572E+01
```

```
TIME        V(7)

 2.240E-03   2.584E+01
 2.280E-03   2.581E+01
 2.320E-03   2.562E+01
 2.360E-03   2.529E+01
 2.400E-03   2.483E+01
 2.440E-03   2.429E+01
 2.480E-03   2.368E+01
 2.520E-03   2.305E+01
 2.560E-03   2.243E+01
 2.600E-03   2.188E+01
 2.640E-03   2.142E+01
 2.680E-03   2.108E+01
 2.720E-03   2.088E+01
 2.760E-03   2.083E+01
 2.800E-03   2.095E+01
 2.840E-03   2.122E+01
 2.880E-03   2.162E+01
 2.920E-03   2.213E+01
 2.960E-03   2.272E+01
 3.000E-03   2.334E+01
 3.040E-03   2.397E+01
 3.080E-03   2.455E+01
 3.120E-03   2.506E+01
 3.160E-03   2.546E+01
 3.200E-03   2.572E+01
 3.240E-03   2.584E+01
 3.280E-03   2.581E+01
 3.320E-03   2.562E+01
 3.360E-03   2.529E+01
 3.400E-03   2.483E+01
 3.440E-03   2.429E+01
 3.480E-03   2.368E+01
 3.520E-03   2.305E+01
 3.560E-03   2.243E+01
 3.600E-03   2.188E+01
 3.640E-03   2.142E+01
 3.680E-03   2.108E+01
 3.720E-03   2.088E+01
 3.760E-03   2.083E+01
 3.800E-03   2.095E+01
 3.840E-03   2.122E+01
 3.880E-03   2.162E+01
 3.920E-03   2.213E+01
 3.960E-03   2.272E+01
 4.000E-03   2.334E+01
```

```
FOUR TRANSISTOR AMPLIFIER

 ****     TRANSIENT ANALYSIS                 TEMPERATURE =    0.000 DEG C

******************************************************************************

  TIME        V(7)

(*)----------    2.0000E+01   2.2000E+01   2.4000E+01   2.6000E+01   2.8000E+01

 0.000E+00  2.335E+01 .            .        *   .            .            .
 4.000E-05  2.397E+01 .            .            *            .            .
 8.000E-05  2.455E+01 .            .            .   *        .            .
 1.200E-04  2.506E+01 .            .            .      *     .            .
 1.600E-04  2.546E+01 .            .            .        *   .            .
 2.000E-04  2.572E+01 .            .            .          * .            .
 2.400E-04  2.584E+01 .            .            .           *.            .
 2.800E-04  2.581E+01 .            .            .           *.            .
 3.200E-04  2.562E+01 .            .            .          * .            .
 3.600E-04  2.529E+01 .            .            .       *    .            .
 4.000E-04  2.483E+01 .            .            .    *       .            .
 4.400E-04  2.429E+01 .            .            . *          .            .
 4.800E-04  2.368E+01 .            .          * .            .            .
 5.200E-04  2.305E+01 .            .      *     .            .            .
 5.600E-04  2.243E+01 .            .  *         .            .            .
 6.000E-04  2.188E+01 .           *.            .            .            .
 6.400E-04  2.142E+01 .         *  .            .            .            .
 6.800E-04  2.108E+01 .       *    .            .            .            .
 7.200E-04  2.088E+01 .      *     .            .            .            .
 7.600E-04  2.083E+01 .     *      .            .            .            .
 8.000E-04  2.095E+01 .      *     .            .            .            .
 8.400E-04  2.122E+01 .        *   .            .            .            .
 8.800E-04  2.162E+01 .           *.            .            .            .
 9.200E-04  2.213E+01 .            .*           .            .            .
 9.600E-04  2.272E+01 .            .    *       .            .            .
 1.000E-03  2.334E+01 .            .        *   .            .            .
 1.040E-03  2.397E+01 .            .            *            .            .
 1.080E-03  2.455E+01 .            .            .   *        .            .
 1.120E-03  2.506E+01 .            .            .      *     .            .
 1.160E-03  2.546E+01 .            .            .        *   .            .
 1.200E-03  2.572E+01 .            .            .          * .            .
 1.240E-03  2.584E+01 .            .            .           *.            .
 1.280E-03  2.581E+01 .            .            .           *.            .
 1.320E-03  2.562E+01 .            .            .          * .            .
 1.360E-03  2.529E+01 .            .            .       *    .            .
 1.400E-03  2.483E+01 .            .            .    *       .            .
 1.440E-03  2.429E+01 .            .            . *          .            .
 1.480E-03  2.368E+01 .            .          * .            .            .
 1.520E-03  2.305E+01 .            .      *     .            .            .
 1.560E-03  2.244E+01 .            .  *         .            .            .
 1.600E-03  2.188E+01 .           *.            .            .            .
 1.640E-03  2.142E+01 .         *  .            .            .            .
 1.680E-03  2.108E+01 .       *    .            .            .            .
 1.720E-03  2.088E+01 .      *     .            .            .            .
 1.760E-03  2.083E+01 .     *      .            .            .            .
 1.800E-03  2.095E+01 .      *     .            .            .            .
 1.840E-03  2.122E+01 .        *   .            .            .            .
 1.880E-03  2.162E+01 .           *.            .            .            .
 1.920E-03  2.213E+01 .            .*           .            .            .
 1.960E-03  2.272E+01 .            .    *       .            .            .
 2.000E-03  2.334E+01 .            .        *   .            .            .
```

```
 TIME        V(7)

(*)---------    2.0000E+01  2.2000E+01  2.4000E+01  2.6000E+01  2.8000E+01
                      - - - - - - - - - - - - - - - - - - - - -
 2.040E-03  2.397E+01 .         .         *         .         .
 2.080E-03  2.455E+01 .         .         .  *      .         .
 2.120E-03  2.506E+01 .         .         .    *    .         .
 2.160E-03  2.546E+01 .         .         .      *  .         .
 2.200E-03  2.572E+01 .         .         .        *.         .
 2.240E-03  2.584E+01 .         .         .        *.         .
 2.280E-03  2.581E+01 .         .         .        *.         .
 2.320E-03  2.562E+01 .         .         .       * .         .
 2.360E-03  2.529E+01 .         .         .     *   .         .
 2.400E-03  2.483E+01 .         .         .   *     .         .
 2.440E-03  2.429E+01 .         .         .*        .         .
 2.480E-03  2.368E+01 .         .       * .         .         .
 2.520E-03  2.305E+01 .         .    *    .         .         .
 2.560E-03  2.243E+01 .         . *       .         .         .
 2.600E-03  2.188E+01 .        *.         .         .         .
 2.640E-03  2.142E+01 .      *  .         .         .         .
 2.680E-03  2.108E+01 .    *    .         .         .         .
 2.720E-03  2.088E+01 .   *     .         .         .         .
 2.760E-03  2.083E+01 .   *     .         .         .         .
 2.800E-03  2.095E+01 .    *    .         .         .         .
 2.840E-03  2.122E+01 .     *   .         .         .         .
 2.880E-03  2.162E+01 .       * .         .         .         .
 2.920E-03  2.213E+01 .         .*        .         .         .
 2.960E-03  2.272E+01 .         .   *     .         .         .
 3.000E-03  2.334E+01 .         .      *  .         .         .
 3.040E-03  2.397E+01 .         .         *         .         .
 3.080E-03  2.455E+01 .         .         .  *      .         .
 3.120E-03  2.506E+01 .         .         .    *    .         .
 3.160E-03  2.546E+01 .         .         .      *  .         .
 3.200E-03  2.572E+01 .         .         .        *.         .
 3.240E-03  2.584E+01 .         .         .        *.         .
 3.280E-03  2.581E+01 .         .         .        *.         .
 3.320E-03  2.562E+01 .         .         .       * .         .
 3.360E-03  2.529E+01 .         .         .     *   .         .
 3.400E-03  2.483E+01 .         .         .   *     .         .
 3.440E-03  2.429E+01 .         .         .*        .         .
 3.480E-03  2.368E+01 .         .       * .         .         .
 3.520E-03  2.305E+01 .         .    *    .         .         .
 3.560E-03  2.243E+01 .         . *       .         .         .
 3.600E-03  2.188E+01 .        *.         .         .         .
 3.640E-03  2.142E+01 .      *  .         .         .         .
 3.680E-03  2.108E+01 .    *    .         .         .         .
 3.720E-03  2.088E+01 .   *     .         .         .         .
 3.760E-03  2.083E+01 .   *     .         .         .         .
 3.800E-03  2.095E+01 .    *    .         .         .         .
 3.840E-03  2.122E+01 .     *   .         .         .         .
 3.880E-03  2.162E+01 .       * .         .         .         .
 3.920E-03  2.213E+01 .         .*        .         .         .
 3.960E-03  2.272E+01 .         .   *     .         .         .
 4.000E-03  2.334E+01 .         .      *  .         .         .
                      - - - - - - - - - - - - - - - - - - - - -
```

```
FOUR TRANSISTOR AMPLIFIER

 ****     FOURIER ANALYSIS                    TEMPERATURE =    0.000 DEG C

*************************************************************************************

FOURIER COMPONENTS OF TRANSIENT RESPONSE V(7)

 DC COMPONENT =    2.334581E+01

 HARMONIC   FREQUENCY    FOURIER     NORMALIZED    PHASE        NORMALIZED
    NO        (HZ)      COMPONENT    COMPONENT     (DEG)        PHASE (DEG)

     1     1.000E+03    2.514E+00    1.000E+00   -2.596E-01     0.000E+00
     2     2.000E+03    5.962E-03    2.372E-03    9.027E+01     9.053E+01
     3     3.000E+03    9.805E-04    3.900E-04    6.219E+00     6.478E+00
     4     4.000E+03    5.334E-05    2.122E-05   -7.882E+01    -7.856E+01
     5     5.000E+03    2.922E-05    1.162E-05    9.788E+01     9.814E+01
     6     6.000E+03    4.209E-05    1.674E-05    7.779E+01     7.805E+01
     7     7.000E+03    3.829E-05    1.523E-05    8.332E+01     8.358E+01
     8     8.000E+03    7.494E-05    2.981E-05    6.010E+01     6.036E+01
     9     9.000E+03    6.800E-05    2.705E-05    8.724E+01     8.750E+01

     TOTAL HARMONIC DISTORTION =   2.404185E-01 PERCENT
```

7.5.2 SPICE ANALYSIS DATA EXTRACTION

The SPICE analysis, as shown in Section 7.5.1, represents a large part of this text. This represents a relatively simple analysis. A more complex analysis could provide a serious amount of data to analyze. The analysis begins with an (element node table) followed by the BJT model parameters listing. These are a result of the .OPTIONS NODE command. The analysis continues with the circuit element summary, as a result of the .OPTIONS LIST command. The analysis then proceeds in the sequence shown in Table 7-1.

Table 7-1. Sequence of Analysis Events

Analysis Section	Initiating Commands
Temperature adjusted values	.TEMP
DC transfer curves - data	.DC/.PRINT DC
DC transfer curves - plot	.DC/.PLOT DC
Small signal bias solution	.OP
Operating point information	.TF
DC sensitivity analysis	.SENS
AC analysis - data	.AC/.PRINT AC/VIN
AC analysis - plot	.AC/.PLOT AC/VIN
Initial transient solution	.TRAN/VIN
Transient analysis - data	.TRAN/.PRINT TRAN/VIN
Transient analysis - plot	.TRAN/.PLOT TRAN/VIN
Fourier analysis	.FOUR

The analysis sequence is repeated three times. Each sequence is at one of the three specified temperatures. The first analysis is at 25° C, followed by 75° C, and 0° C. Each sequence begins by adjusting the temperature-sensitive values of all temperature-sensitive components. If temperature-sensitive resistors or other solid state devices had been specified, then they too would have been included. The values shown are only those values originally specified in the parameter information of the device.

The analysis proceeds to the DC portion. For the DC analysis portion, it was necessary to determine the input voltage range to the amplifier that provided the best linearity on the output. The linearity range, of the circuit, is relatively constant, but the center bias point varies considerably at each temperature. This can be seen in Figure 7-2.

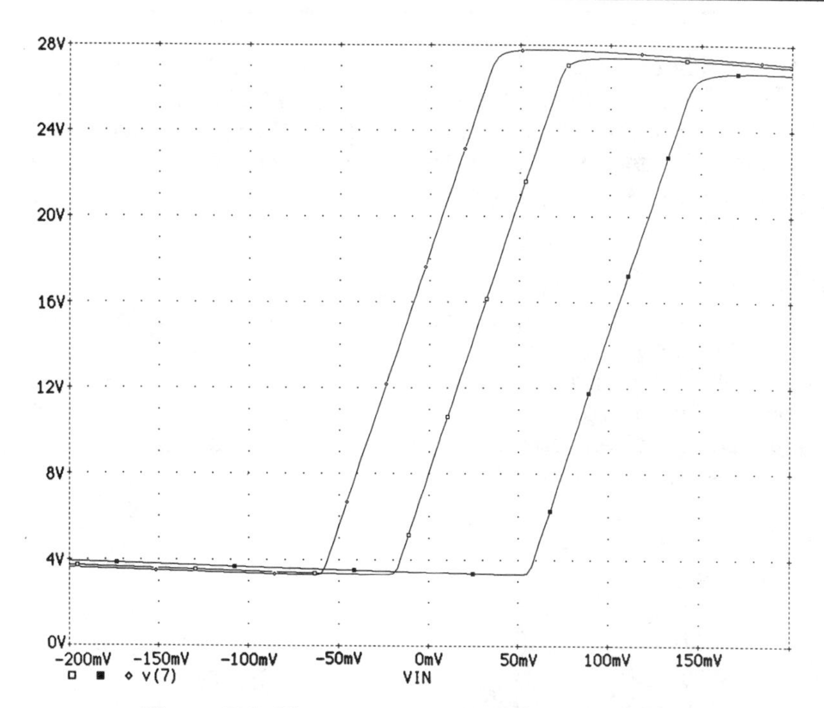

Figure 7-2. Three-temperature DC analysis plot.

The range is approximately 80 mV wide with the center bias point at 25° C, being approximately 27 mV. The bias point shifts to approximately 100 mV at 70° C and approximately -12 mV at 0° C. This indicates that the transistor has a -1.4 mV/° C temperature coefficient. The change in the center point is due to the VBE characteristics of the transistors. Since the VBE is a diode, as the temperature increases the VBE will decrease and the voltage applied to node 7 will shift. The observation that might be made, looking at Figure 7-2, is that the design is not be suitable for a directly coupled application over a wide temperature range due to the bias shift. The alternative solution might be to sacrifice high input resistance and put a biasing network in the input with a compensating diode to counteract the bias shift.

The small signal bias solution indicates that the voltage of node 7 changes from 8.0469 V at 25° C to 3.4186 V at 70° C. This represents a 4.62 V change on the output, but is misleading because the input bias point is at 0-V DC and at 70° C the amplifier is biased off (refer to Figure 7-2).

The power consumption increases directly with temperature, because of the falling output node voltage and the increasing voltage across R4. The result is slightly increased power consumption. The operating point information (.TF) shows that the ratio of V(7)/VIN, input resistance, and output resistance varies considerably over temperature. The reader is reminded that the circuit bias shift and the fixed 0-V DC input bias voltage will provide a misleading set of indications. The input resistance is approximately 144 MΩ. Note that at 70° C when the circuit is biased off, the input resistance approached 220 MΩ. This would be expected in a bias-off condition. The output resistance is approximately 20,000 Ω which is the value of R4. The value of 253.8 for the V(7) to VIN ratio is consistent with expected results. The value of -2.398 at 70° C appears in error. To explain this number, the reader is reminded that the amplifier is biased off at 70° C and this reading is consistent with an off condition. At 0° C, the value is 254.7, which agrees well with the 25° C value.

The DC analysis curves and data can be used to extract and compute differential gain by selecting a 10 mV region in the linear region of the curve. A review of the **.PRINT DC** data indicates shaded values, at each temperature, used for this calculation. Table 7-2 shows a summary of the computed differential gain values. The computed differential gain is in the range of 252 to 256. This computed range agrees well with the values computed by the .TF command. The values used for the gain calculation are shown in Table 7-2.

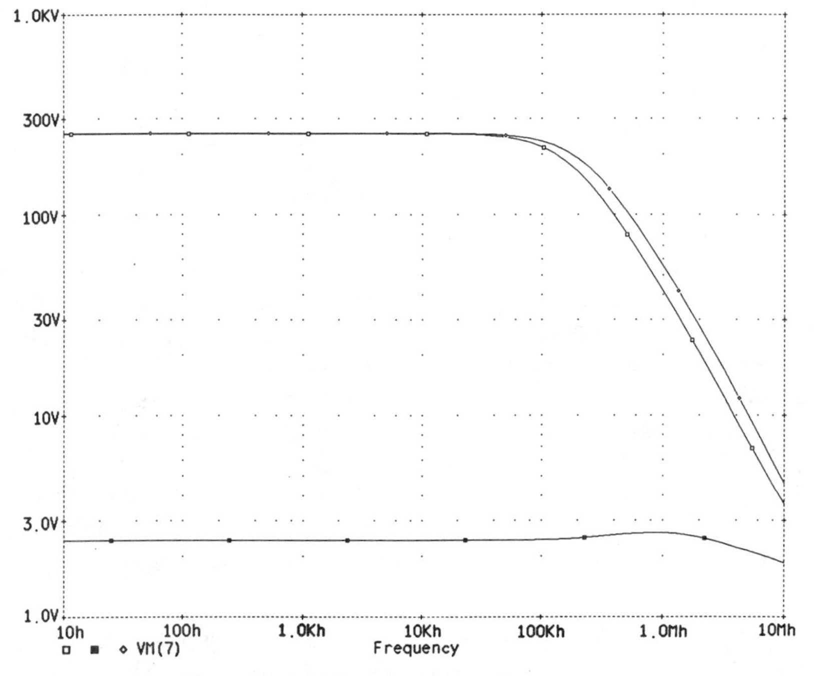

Figure 7-3. AC gain versus frequency.

The sensitivity analysis reveals that the most sensitive resistor, at all three temperatures, is R3. This sensitivity was based on a per unit change in the resistor. If the percentage change is to be evaluated, then resistors R2 and R3 are nearly equal in sensitivity, but they are opposite in their effects on output node 7. R2 increases the output voltage at a rate of approximately 2 V/ % change and R3 decreases the output at approximately the same rate.

The AC analysis indicates that at midband (1 KHz) of the analysis the gain is essentially independent of temperature. This assumes that the circuit is not biased off as in the 70°C case. The AC gain shown at 25°C and 0°C is in the approximately 254, which agrees with the computed differential gain. A combined plot of gain versus frequency, for all three temperatures is shown in Figure 7-3.

Some slight shift in the 3-dB points can be seen. This shift is minor and expected because of the effects of temperature on the solid state models. The analysis shows no lower 3-dB point due to the directly coupled design. The upper 3-dB point is in the range of 160 KHz to 200 KHz which, is impressive for an amplifier with a gain of 254. A plot of VM(7), VP(7), and VDB(7) is shown in Figure 7-4 and permits the reader to review the gain-phase relationship of the output in terms of circuit stability.

The transient analysis input waveform injected a 20-mV peak-to-peak 1 KHz sine wave. By measuring the peak-to-peak voltage at node 7, the gain at 1 KHz can be computed. The resulting output waveform at 25°C is shown in Figure 7-5. A calculation of gain, from the transient analysis data, compared to AC analysis predicted gain, reveals a high correlation as would be expected unless the input amplitude of the transient analysis source caused the output waveform to start clipping. This nonlinearity condition is not taken into account in the AC analysis.

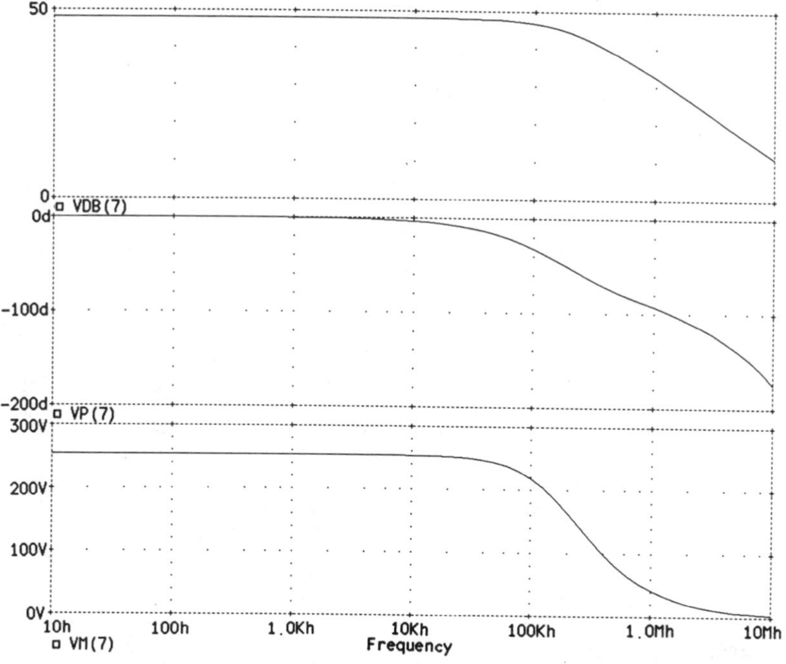

Figure 7-4. Plot of magnitude, phase, and decibel at node 7.

The final analysis is the Fourier analysis. The total harmonic distortion is consistent over the temperature range except, for the misleading result shown at 70°C caused by biasing. The distortion between 25°C and 0°C is because the bias point at 0°C is near the saturation point of the amplifier. This saturation point represents a nonlinear region of the amplifier and is subject to higher distortion.

Further analysis at various bias points and temperatures would be required to verify the relationship of bias and distortion.

The analysis data is summarized and shown in Table 7-2. The reader is encouraged to review the analysis data in great detail to gain a better understanding of how to utilize the data more effectively.

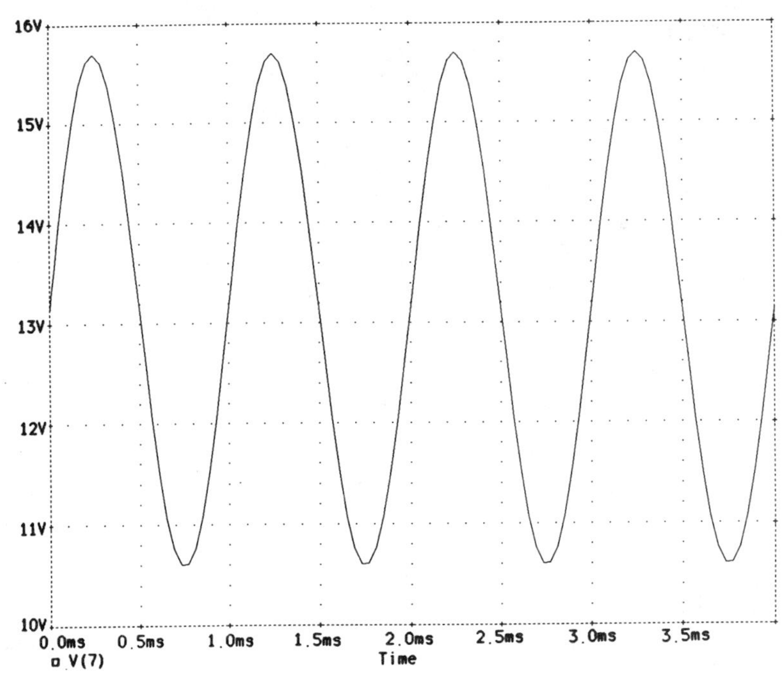

Figure 7-5. Waveform output at node 7 with a 20-mV P-P input sine wave.

Table 7-2. Analysis Data Summary.

Parameter	25°C	70°C	0°C
.DC			
Linearity range	-15 mV to 70 mV	57.5 mV to 142.5 mV	-55 mV to 32.5 mV
.OP			
V(7)	8.0469 V	3.4186 V	18.268 V
Power consumer (W)	238 mW	244 mW	224 mW
.TF			
V(7)/VIN (GAIN)	253.8	-2.398 *	254.7
Rin (OHMS)	144.2 MEG	219.7 MEG *	111.2 MEG
Rout (OHMS)	19.63 K	20.1 K	19.75 K

Parameter	25°C	70°C	0°C
.SENS			
Most sensitive	R3	R3	R3
.DC			
ΔV(7) for Δ in 10 mV	13.14 V - 10.59 V = 2.55 V	12.01 V - 9.45 V = 2.56 V	25.87 V - 23.35 V = 2.52 V
Differential gain	255	256	252
.AC			
Midband gain	253.8	2.398 *	254.7
Lower freq (HZ)	DC	DC	DC
Upper freq (HZ)	>160 KHz	>1 MHz *	>200 KHz
.TRAN			
P-P out (V)	15.68 V - 10.61 V = 5.07 V	3.395 V - 3.353 V = 0.042 V *	25.84 V - 20.83 V = 5.01 V
P-P in (V)	0.02	0.02	0.02
Gain	253.5	2.1 *	250.5
.FOUR			
THD in (%)	0.05365	2.376 *	0.2404

* = Misleading data due to input bias conditions.

7.6 SUMMARY

This section has been presented to demonstrate the combined analysis capability of SPICE. The amount of data that can be produced by and extracted from SPICE is nearly unlimited. The following represents a summary of the steps that should be taken in any SPICE analysis.

1. *Determine the problem.* The user must understand the problem and the critical information before an efficient analysis can be performed. If the user desires to analyze the DC transfer curve data associated with a DC circuit, no purpose is served by performing an AC or transient analysis. The information, although interesting, is not relevant to the problem.

2. *Define the environment.* If a circuit is to operate only at room temperature, what is gained by performing a temperature analysis? The only possible gain is the determination of the stability of the circuit and the margin of safety that exists. If a

circuit is stable at 27° C and oscillates at 40° C a concern might exist about the overall circuit stability with normal component tolerance variations.

3. *Model the circuit in sufficient detail.* If the circuit is not modeled in sufficient detail, the analysis will not accurately reflect circuit performance. Critical component parameters should be included to the extent that they are necessary for the analysis. If a low-frequency AC analysis is being performed on a RLC circuit, the time spent determining the inductor resistance and interwinding capacitance is not justified. If this is a high-frequency analysis, then the time is well spent.

4. *Use the SPICE commands efficiently.* SPICE offers the user a large variety of commands. The proper use of the SPICE program requires a good understanding of the commands and the capabilities of each. From experience, it can be stated that the better the commands are understood, the more efficient and meaningful the analysis can become.

5. *Keep the analysis to a minimum.* As a rule, the more focused the analysis becomes, the faster the analysis will run. This becomes significant when the circuit being analyzed is not yet proven. If you are modeling a circuit that requires a specific bias condition, the circuit should be optimized using the **.OP** capability before any DC, AC, or transient performance characteristics are evaluated. It makes sense to expand the analysis in a methodical fashion. This saves time and expense.

6. *Look for unexpected results.* The SPICE program is a tool, and like any tool requires proper use. If the user has no concept of the expected results, then the accuracy of the results can not be validated. If the user has a good idea of the expected results, then any inconsistency can be rapidly identified. Many times an analysis will produce a result that points to (1) a modeling problem or (2) a design problem. Good engineering practice requires that unexpected results be investigated, explained, and resolved. A failure to do this places the design and the integrity of the analysis in question.

APPENDIX A

SPICE MODELS

GENERAL PURPOSE LIBRARY

The following section is a collection of general purpose models which have been developed for general use over a number of years. The copying and distribution of these models is welcome and encouraged by the author.

DIODE MODELS

```
.MODEL D1N751   D(IS=880.5E-18 N=1 RS=.25 IKF=0 XTI=3 EG=1.11 CJO=175P M=.5516
+ VJ=.75 FC=.5 ISR=1.859N NR=2 BV=5.13 IBV=77.21M TT=1.443M)

.MODEL D1N752   D(IS=880.5E-18 N=1 RS=.25 IKF=0 XTI=3 EG=1.11 CJO=175P M=.5516
+ VJ=.75 FC=.5 ISR=1.859N NR=2 BV=5.641 IBV=.1203 TT=1.443M)

.MODEL D1N753   D(IS=880.5E-18 N=1 RS=.25 IKF=0 XTI=3 EG=1.11 CJO=175P M=.5516
+ VJ=.75 FC=.5 ISR=1.859N NR=2 BV=6.253 IBV=.1916 TT=1.443M)

.MODEL D1N754   D(IS=880.5E-18 N=1 RS=.25 IKF=0 XTI=3 EG=1.11 CJO=175P M=.5516
+ VJ=.75 FC=.5 ISR=1.859N NR=2 BV=6.863 IBV=.2723 TT=1.443M)

.MODEL D1N914 D (IS=1E-13 RS=16 CJO=2.2P TT=11.7NS BV=100 IBV=1E-13)

.MODEL D1N4934  D(IS=254P N=1.498 RS=6.847U IKF=19.23M XTI=3 EG=1.11 CJO=31.35P
+ M=.4379 VJ=.5581 FC=.5 ISR=51.42U NR=2 BV=100 IBV=100U TT=10.86N)

.MODEL D1N5401  D(IS=169.8F N=1.096 RS=10.11M IKF=0 XTI=3 EG=1.11 CJO=157.4P
+ M=1.63 VJ=15.05 FC=.5 ISR=18.17U NR=2 BV=99.98 IBV=258.6U TT=5.049U)
```

TRANSISTOR MODELS

```
.MODEL Q2N3904 NPN(IS=1.05E-15 ISE=4.12N NE=4 ISC=4.12N NC=4 BF=220 IKF=2E-1
+ VAF=80 CJC=4.32P CJE=5.27P RB=5 RE=.5 RC=1 TF=.617N TR=200N KF=1E-15 AF=1)

.MODEL Q2N3905 PNP(IS=1.05E-15 ISE=1.003NA NE=4 ISC=1.003NA NC=4 BF=220 IKF=1E-1
+ VAF=240 CJC=4.32P CJE=5.7P RB=3 RE=.5 RC=.2 TF=.335N TR=170N KF=4E-15 AF=1)
```

OPTICAL ISOLATOR MODEL

```
.SUBCKT OP4N25 1 2 4 5
* ANODE INPUT = 1; CATHODE INPUT = 2; EMITTER
* OUTPUT = 4; COLLECTOR OUTPUT = 5
VM 1 12 DC 0
D1 12 2 LED
.MODEL LED D(IS=25E-13 RS=0.75 CJO=35P N=2)
R1 10 11 450
C1 11 0 1000PF
H1 10 0 VM 3.33E-3
G1 5 6 11 0 1.0
Q1 5 6 4 QNPN
.MODEL QNPN NPN(IS=3.33E-11 NF=1.35 CJC=47.4P CJE=167P TF=.923N
+ TR=148N BF=150 BR=10 IKF=0.1 VAF=100)
.ENDS OP4N25
```

OPERATIONAL AMPLIFIER MODEL

```
.SUBCKT LM741 10 20 30 40 50
.MODEL DP D(IS=1E-14 N=2.118)
.MODEL DN D(IS=.971E-14 N=2.118)
.MODEL DV D(IS=6E-15 N=.1 XTI=.1)
.MODEL DI D(IS=1E-14 N=1)
VP 10 11 0
VN 20 21 0
DP 11 1 DP
DN 21 1 DN
FA 20 0 VN .276X
I1 1 0 140.5NA
C1 1 0 33.3FF IC=-.863V
FP 4 0 VP 162.2X
FN 0 4 VN 161.0X
GC 0 4 1 0 6.6N
RT 4 0 1
CT 4 0 80NF
G2 0 2 4 0 1X
R2 2 0 100K
CC 2 3 30PF
GO 3 0 2 0 161X
RD 3 0 60
DH 3 5 DV
DL 6 3 DV
VH 40 5 .7V
VL 6 50 2.2V
D1 3 9 DI
D2 9 3 DI
EX 9 0 POLY(2) 3 0 3 30 0 1 -1.64
RO 3 30 40
.ENDS LM741
```

The following library has been included herein with the written permission of MicroSim Corporation. PSPICE is a registered trademark of MicroSim Corporation.

```
* MICROSIM CORPORATION REDUCED LIBRARY

* COPYRIGHT 1990 BY MICROSIM CORPORATION
* THIS IS A REDUCED VERSION OF THE MASTER LIBRARY FOR MICROSIM'S STANDARD
* PARTS LIBRARIES. SOME COMPONENTS FROM EACH TYPE OF COMPONENT LIBRARY
* HAVE BEEN INCLUDED HERE.
*
* YOU ARE WELCOME TO MAKE AS MANY COPIES OF IT AS YOU FIND CONVENIENT.
*
* RELEASE DATE: JULY 1990
*
* IT TAKES TIME FOR PSPICE TO SCAN A LIBRARY FILE.  WHEN POSSIBLE, PSPICE
* CREATES AN INDEX FILE, CALLED <FILENAME>.IND, TO SPEED UP THE SEARCH
* PROCESS. THE INDEX FILE IS RE-CREATED WHENEVER PSPICE SENSES THAT IT MIGHT
* BE INVALID.

* THE FOLLOWING IS A SUMMARY OF PARTS IN THIS LIBRARY:
*
*          PART NAME          PART TYPE
*          ---------          ---------
*          Q2N2222A           NPN BIPOLAR TRANSISTOR
*          Q2N2907A           PNP BIPOLAR TRANSISTOR
*          Q2N3904            NPN BIPOLAR TRANSISTOR
*          Q2N3906            PNP BIPOLAR TRANSISTOR
*
*          D1N750             ZENER DIODE
*          MV2201             VOLTAGE VARIABLE CAPACITANCE DIODE
*          D1N4148            SWITCHING DIODE
*          MBD101             SWITCHING DIODE
*
*          J2N3819            N-CHANNEL JUNCTION FIELD EFFECT TRANSISTOR
*          J2N4393            N-CHANNEL JUNCTION FIELD EFFECT TRANSISTOR
*
*          LM324              LINEAR OPERATIONAL AMPLIFIER
*          UA741              LINEAR OPERATIONAL AMPLIFIER
*          LM111              VOLTAGE COMPARATOR
*
*          IRF150             N-TYPE POWER MOS FIELD EFFECT TRANSISTOR
*          IRF9140            P-TYPE POWER MOS FIELD EFFECT TRANSISTOR

*--------------------------------------------------------------------------
```

```
* LIBRARY OF BIPOLAR TRANSISTOR MODEL PARAMETERS
*
* THIS IS A REDUCED VERSION OF MICROSIM'S BIPOLAR TRANSISTOR MODEL LIBRARY.
* YOU ARE WELCOME TO MAKE AS MANY COPIES OF IT AS YOU FIND CONVENIENT.
*
.MODEL Q2N2222A NPN(IS=14.34F XTI=3 EG=1.11 VAF=74.03 BF=255.9 NE=1.307
+ ISE=14.34F IKF=.2847 XTB=1.5 BR=6.092 NC=2 ISC=0 IKR=0 RC=1
+ CJC=7.306P MJC=.3416 VJC=.75 FC=.5 CJE=22.01P MJE=.377 VJE=.75
+ TR=46.91N TF=411.1P ITF=.6 VTF=1.7 XTF=3 RB=10)
* NATIONAL       PID=19            CASE=TO18
* 88-09-07 BAM   CREATION

.MODEL Q2N2907A PNP(IS=650.6E-18 XTI=3 EG=1.11 VAF=115.7 BF=231.7 NE=1.829
+ ISE=54.81F IKF=1.079 XTB=1.5 BR=3.563 NC=2 ISC=0 IKR=0 RC=.715
+ CJC=14.76P MJC=.5383 VJC=.75 FC=.5 CJE=19.82P MJE=.3357 VJE=.75
+ TR=111.3N TF=603.7P ITF=.65 VTF=5 XTF=1.7 RB=10)
* NATIONAL       PID=63            CASE=TO18
* 88-09-09 BAM   CREATION

.MODEL Q2N3904  NPN(IS=6.734F XTI=3 EG=1.11 VAF=74.03 BF=416.4 NE=1.259
+ ISE=6.734F IKF=66.78M XTB=1.5 BR=.7371 NC=2 ISC=0 IKR=0 RC=1
+ CJC=3.638P MJC=.3085 VJC=.75 FC=.5 CJE=4.493P MJE=.2593 VJE=.75
+ TR=239.5N TF=301.2P ITF=.4 VTF=4 XTF=2 RB=10)
* NATIONAL       PID=23            CASE=TO92
* 88-09-08 BAM   CREATION

.MODEL Q2N3906  PNP(IS=1.41F XTI=3 EG=1.11 VAF=18.7 BF=180.7 NE=1.5 ISE=0
+ IKF=80M XTB=1.5 BR=4.977 NC=2 ISC=0 IKR=0 RC=2.5 CJC=9.728P
+ MJC=.5776 VJC=.75 FC=.5 CJE=8.063P MJE=.3677 VJE=.75 TR=33.42N
+ TF=179.3P ITF=.4 VTF=4 XTF=6 RB=10)
* NATIONAL       PID=66            CASE=TO92
* 88-09-09 BAM   CREATION
*
*------------------------------------------------------------------------
*
* LIBRARY OF DIODE MODEL PARAMETERS
*
* COPYRIGHT 1990 BY MICROSIM CORPORATION
* THIS IS A REDUCED VERSION OF MICROSIM'S DIODE MODEL LIBRARY.
* YOU ARE WELCOME TO MAKE AS MANY COPIES OF IT AS YOU FIND CONVENIENT.
*
*** ZENER DIODES ***
*
* "A" SUFFIX ZENERS HAVE THE SAME PARAMETERS (E.G., 1N750A HAS THE SAME
* PARAMETERS AS 1N750)
*

.MODEL D1N750   D(IS=880.5E-18 RS=.25 IKF=0 N=1 XTI=3 EG=1.11 CJO=175P M=.5516
+ VJ=.75 FC=.5 ISR=1.859N NR=2 BV=4.7 IBV=20.245M NBV=1.6989 IBVL=1.9556M NBVL=14.976
TBV1=-21.277U)
* MOTOROLA       PID=1N750         CASE=DO-35
* 89-9-18 GJG
* VZ = 4.7 @ 20MA, ZZ = 300 @ 1MA, ZZ = 12.5 @ 5MA, ZZ =2.6  @ 20MA

*** VOLTAGE-VARIABLE CAPACITANCE DIODES

.MODEL MV2201   D(IS=1.365P RS=1 IKF=0 N=1 XTI=3 EG=1.11 CJO=14.93P M=.4261
+ VJ=.75 FC=.5 ISR=16.02P NR=2 BV=25 IBV=10U)
* MOTOROLA       PID=MV2201        CASE=182-03
* 88-09-22 BAM   CREATION
```

```
*** SWITCHING DIODES ***

.MODEL D1N4148  D(IS=0.1P RS=16 CJO=2P TT=12N BV=100 IBV=0.1P)
* 85-??-??       ORIGINAL LIBRARY

.MODEL MBD101   D(IS=192.1P RS=.1 IKF=0 N=1 XTI=3 EG=1.11 CJO=893.8F M=98.29M
+ VJ=.75 FC=.5 ISR=16.91N NR=2 BV=5 IBV=10U)
* MOTOROLA       PID=MBD101       CASE=182-03
* 88-09-22 BAM   CREATION
*
*-----------------------------------------------------------------------------
*
* LIBRARY OF JUNCTION FIELD-EFFECT TRANSISTOR (JFET) MODEL PARAMETERS

* THIS IS A REDUCED VERSION OF MICROSIM'S JFET MODEL LIBRARY.
* YOU ARE WELCOME TO MAKE AS MANY COPIES OF IT AS YOU FIND CONVENIENT.

.MODEL J2N3819  NJF(BETA=1.304M BETATCE=-.5 RD=1 RS=1 LAMBDA=2.25M VTO=-3
+ VTOTC=-2.5M IS=33.57F ISR=322.4F N=1 NR=2 XTI=3 ALPHA=311.7
+ VK=243.6 CGD=1.6P M=.3622 PB=1 FC=.5 CGS=2.414P KF=9.882E-18
+ AF=1)
* NATIONAL       PID=50           CASE=TO92
* 88-08-01 RMN   BVMIN=25

.MODEL J2N4393  NJF(BETA=9.109M BETATCE=-.5 RD=1 RS=1 LAMBDA=6M VTO=-1.422
+ VTOTC=-2.5M IS=205.2F ISR=1.988P N=1 NR=2 XTI=3 ALPHA=20.98U
+ VK=123.7 CGD=4.57P M=.4069 PB=1 FC=.5 CGS=4.06P KF=123E-18 AF=1)
* NATIONAL       PID=51           CASE=TO18
* 88-07-13 BAM   BVMIN=40
*
*-----------------------------------------------------------------------------
*
* LIBRARY OF LINEAR IC DEFINITIONS

* THIS IS A REDUCED VERSION OF MICROSIM'S LINEAR SUBCIRCUIT LIBRARY.
* YOU ARE WELCOME TO MAKE AS MANY COPIES OF IT AS YOU FIND CONVENIENT.
*
* THE PARAMETERS IN THE OPAMP LIBRARY WERE DERIVED FROM THE DATA SHEETS
* FOR EACH PART.  THE MACROMODEL USED IS SIMILAR TO THE ONE DESCRIBED IN:
*
* MACROMODELING OF INTEGRATED CIRCUIT OPERATIONAL AMPLIFIERS
* BY GRAEME BOYLE, BARRY COHN, DONALD PEDERSON, AND JAMES SOLOMON
* IEEE JOURNAL OF SOLIE-STATE CIRCUITS, VOL. SC-9, NO. 6, DEC. 1974
*
* DIFFERENCES FROM THE REFERENCE (ABOVE) OCCUR IN THE OUTPUT LIMITING STAGE
* WHICH WAS MODIFIED TO REDUCE INTERNALLY GENERATED CURRENTS ASSOCIATED
* WITH OUTPUT VOLTAGE LIMITING, AS WELL AS SHORT-CIRCUIT CURRENT LIMITING.
*
* THE OPAMPS ARE MODELLED AT ROOM TEMPERATURE AND DO NOT TRACK CHANGES
* WITH TEMPERATURE.  THIS LIBRARY FILE CONTAINS MODELS FOR NOMINAL, NOT
* WORST CASE, DEVICES.
*----------------------------------------------------------------------------
```

```
* CONNECTIONS:   NON-INVERTING INPUT
*                | INVERTING INPUT
*                | | POSITIVE POWER SUPPLY
*                | | | NEGATIVE POWER SUPPLY
*                | | | | OUTPUT
*                | | | | |
.SUBCKT LM324    1 2 3 4 5
*
  C1   11 12 2.887E-12
  C2    6  7 30.00E-12
  DC    5 53 DX
  DE   54  5 DX
  DLP  90 91 DX
  DLN  92 90 DX
  DP    4  3 DX
  EGND 99  0 POLY(2) (3,0) (4,0) 0 .5 .5
  FB    7 99 POLY(5) VB VC VE VLP VLN 0 21.22E6 -20E6 20E6 20E6 -20E6
  GA    6  0 11 12 188.5E-6
  GCM   0  6 10 99 59.61E-9
  IEE   3 10 DC 15.09E-6
  HLIM 90  0 VLIM 1K
  Q1   11  2 13 QX
  Q2   12  1 14 QX
  R2    6  9 100.0E3
  RC1   4 11 5.305E3
  RC2   4 12 5.305E3
  RE1  13 10 1.845E3
  RE2  14 10 1.845E3
  REE  10 99 13.25E6
  RO1   8  5 50
  RO2   7 99 25
  RP    3  4 9.082E3
  VB    9  0 DC 0
  VC    3 53 DC 1.500
  VE   54  4 DC 0
  VLIM  7  8 DC 0
  VLP  91  0 DC 40
  VLN   0 92 DC 40
.MODEL DX D(IS=800.0E-18 RS=1)
.MODEL QX PNP(IS=800.0E-18 BF=166.7)
.ENDS
*-----------------------------------------------------------------
```

```
* CONNECTIONS:   NON-INVERTING INPUT
*                | INVERTING INPUT
*                | | POSITIVE POWER SUPPLY
*                | | | NEGATIVE POWER SUPPLY
*                | | | | OUTPUT
*                | | | | |
.SUBCKT UA741    1 2 3 4 5
*
  C1    11 12 8.661E-12
  C2     6  7 30.00E-12
  DC     5 53 DX
  DE    54  5 DX
  DLP   90 91 DX
  DLN   92 90 DX
  DP     4  3 DX
  EGND  99  0 POLY(2) (3,0) (4,0) 0 .5 .5
  FB     7 99 POLY(5) VB VC VE VLP VLN 0 10.61E6 -10E6 10E6 10E6 -10E6
  GA     6  0 11 12 188.5E-6
  GCM    0  6 10 99 5.961E-9
  IEE   10  4 DC 15.16E-6
  HLIM  90  0 VLIM 1K
  Q1    11  2 13 QX
  Q2    12  1 14 QX
  R2     6  9 100.0E3
  RC1    3 11 5.305E3
  RC2    3 12 5.305E3
  RE1   13 10 1.836E3
  RE2   14 10 1.836E3
  REE   10 99 13.19E6
  RO1    8  5 50
  RO2    7 99 100
  RP     3  4 18.16E3
  VB     9  0 DC 0
  VC     3 53 DC 1
  VE    54  4 DC 1
  VLIM   7  8 DC 0
  VLP   91  0 DC 40
  VLN    0 92 DC 40
.MODEL DX D(IS=800.0E-18 RS=1)
.MODEL QX NPN(IS=800.0E-18 BF=93.75)
.ENDS
*------------------------------------------------------------------------------
```

```
*** VOLTAGE COMPARATORS

* THE PARAMETERS IN THIS COMPARATOR LIBRARY WERE DERIVED FROM DATA SHEETS
* FOR EACH PARTS.  THE MACROMODEL USED WAS DEVELOPED BY MICROSIM
* CORPORATION, AND IS PRODUCED BY THE "PARTS" OPTION TO PSPICE.
*
* ALTHOUGH WE DO NOT USE IT, ANOTHER COMPARATOR MACRO MODEL IS
* DESCRIBED IN:
*
* AN INTEGRATED-CIRCUIT COMPARATOR MACROMODEL
* BY IAN GETREU, ANDREAS HADIWIDJAJA, AND JOHAN BRINCH
* IEEE JOURNAL OF SOLID-STATE CIRCUITS, VOL. SC-11, NO. 6, DEC. 1976
*
* THIS REFERENCE COVERS THE CONSIDERATIONS THAT GO INTO DUPLICATING THE
* BEHAVIOR OF VOLTAGE COMPARATORS.
*
* THE COMPARATORS ARE MODELLED AT ROOM TEMPERATURE.  THE MACRO MODEL
* DOES NOT TRACK CHANGES WITH TEMPERATURE.  THIS LIBRARY FILE CONTAINS
* MODELS FOR NOMINAL, NOT WORST CASE, DEVICES.
*
*-----------------------------------------------------------------------
* CONNECTIONS:   NON-INVERTING INPUT
*                | INVERTING INPUT
*                | | POSITIVE POWER SUPPLY
*                | | | NEGATIVE POWER SUPPLY
*                | | | | OPEN COLLECTOR OUTPUT
*                | | | | | OUTPUT GROUND
*                | | | | | |
.SUBCKT LM111    1 2 3 4 5 6
*
  F1    9  3 V1 1
  IEE   3  7 DC 100.0E-6
  VI1  21  1 DC .45
  VI2  22  2 DC .45
  Q1    9 21  7 QIN
  Q2    8 22  7 QIN
  Q3    9  8  4 QMO
  Q4    8  8  4 QMI
.MODEL QIN PNP(IS=800.0E-18 BF=833.3)
.MODEL QMI NPN(IS=800.0E-18 BF=1002)
.MODEL QMO NPN(IS=800.0E-18 BF=1000 CJC=1E-15 TR=118.8E-9)
  E1   10  6  9  4  1
  V1   10 11 DC 0
  Q5    5 11  6 QOC
.MODEL QOC NPN(IS=800.0E-18 BF=34.49E3 CJC=1E-15 TF=364.6E-12 TR=79.34E-9)
  DP    4  3 DX
  RP    3  4 6.122E3
.MODEL DX  D(IS=800.0E-18 RS=1)
*
.ENDS

*-----------------------------------------------------------------------
```

```
* LIBRARY OF MOSFET MODEL PARAMETERS (FOR "POWER" MOSFET DEVICES)
*
* THIS IS A REDUCED VERSION OF MICROSIM'S POWER MOSFET MODEL LIBRARY.
* YOU ARE WELCOME TO MAKE AS MANY COPIES OF IT AS YOU FIND CONVENIENT.
*
* THE "POWER" MOSFET DEVICE MODELS BENEFIT FROM RELATIVELY COMPLETE
* SPECIFICATION OF STATIC AND DYNAMIC CHARACTERISTICS BY THEIR
* MANUFACTURERS.  THE FOLLOWING EFFECTS ARE MODELED:
*
*        - DC TRANSFER CURVES IN FORWARD OPERATION,
*        - GATE DRIVE CHARACTERISTICS AND SWITCHING DELAY,
*        - "ON" RESISTANCE,
*        - REVERSE-MODE "BODY-DIODE" OPERATION.
*
* THE FACTORS NOT MODELED INCLUDE:
*
*        - MAXIMUM RATINGS (EG. HIGH-VOLTAGE BREAKDOWN),
*        - SAFE OPERATING AREA (EG. POWER DISSIPATION),
*        - LATCH-UP,
*        - NOISE.
*
* FOR HIGH-CURRENT SWITCHING APPLICATIONS, WE ADVISE THAT YOU INCLUDE
* SERIES INDUCTANCE ELEMENTS, FOR THE SOURCE AND DRAIN, IN YOUR CIRCUIT FILE.
* IN DOING SO, VOLTAGE SPIKES DUE TO DI/DT WILL BE MODELED.  ACCORDING TO THE
* 1985 INTERNATIONAL RECTIFIER DATABOOK, THE FOLLOWING CASE STYLES HAVE
* LEAD INDUCTANCE VALUES OF:
*
* TO-204 (MODIFIED TO-3) SOURCE = 12.5NH          DRAIN = 5.0NH
* TO-220                         SOURCE =  7.5NH          DRAIN = 3.5-4.5NH
* - - - - - - - - - - - - - - - - - - - - - - - - - - - - - - - - - - - - - -
*
.MODEL IRF150   NMOS(LEVEL=3 GAMMA=0 DELTA=0 ETA=0 THETA=0 KAPPA=0 VMAX=0 XJ=0
+ TOX=100N UO=600 PHI=.6 RS=1.624M KP=20.53U W=.3 L=2U VTO=2.831 RD=1.031M RDS=444.4K
+ CBD=3.229N PB=.8 MJ=.5 FC=.5 CGSO=9.027N CGDO=1.679N RG=13.89 IS=194E-18
+ N=1 TT=288N)
* INT'L RECTIFIER        PID=IRFC150      CASE=TO3
* 88-08-25 BAM   CREATION

.MODEL IRF9140   PMOS(LEVEL=3 GAMMA=0 DELTA=0 ETA=0 THETA=0 KAPPA=0 VMAX=0 XJ=0
+ TOX=100N UO=300 PHI=.6 RS=70.6M KP=10.15U W=1.9 L=2U VTO=-3.67 RD=60.66M RDS=444.4K
+ CBD=2.141N PB=.8 MJ=.5 FC=.5 CGSO=877.2P CGDO=369.3P RG=.811 IS=52.23E-18
+ N=2 TT=140N)
* INT'L RECTIFIER  PID=IRFC9140  CASE=TO3
* 88-08-25 BAM   CREATION
```

The following models have been graciously supplied and included herein with the permission of Texas Instruments.

```
* TEXAS INSTRUMENTS OPERATIONAL AMPLIFIER LIBRARY

* LF347 OPERATIONAL AMPLIFIER "MACROMODEL" SUBCIRCUIT
* CREATED USING PARTS RELEASE 4.01 ON 07/05/89 AT 13:19
* (REV N/A)
* CONNECTIONS:   NON-INVERTING INPUT
*                | INVERTING INPUT
*                | | POSITIVE POWER SUPPLY
*                | | | NEGATIVE POWER SUPPLY
*                | | | | OUTPUT
*                | | | | |
.SUBCKT LF347    1 2 3 4 5
*
  C1   11 12 3.498E-12
  C2    6  7 15.00E-12
  DC    5 53 DX
  DE   54  5 DX
  DLP  90 91 DX
  DLN  92 90 DX
  DP    4  3 DX
  EGND 99  0 POLY(2) (3,0) (4,0) 0 .5 .5
  FB    7 99 POLY(5) VB VC VE VLP VLN 0 14.15E6 -10E6 10E6 10E6 -10E6
  GA    6  0 11 12 282.8E-6
  GCM   0  6 10 99 1.590E-9
  ISS   3 10 DC 195.0E-6
  HLIM 90  0 VLIM 1K
  J1   11  2 10 JX
  J2   12  1 10 JX
  R2    6  9 100.0E3
  RD1   4 11 3.536E3
  RD2   4 12 3.536E3
  RO1   8  5 50
  RO2   7 99 25
  RP    3  4 15.00E3
  RSS  10 99 1.026E6
  VB    9  0 DC 0
  VC    3 53 DC 2.200
  VE   54  4 DC 2.200
  VLIM  7  8 DC 0
  VLP  91  0 DC 25
  VLN   0 92 DC 25
.MODEL DX D(IS=800.0E-18)
.MODEL JX PJF(IS=25.00E-12 BETA=235.1E-6 VTO=-1)
.ENDS
```

```
* LF351 OPERATIONAL AMPLIFIER "MACROMODEL" SUBCIRCUIT
* CREATED USING PARTS RELEASE 4.01 ON 07/05/89 AT 08:19
* (REV N/A)
* CONNECTIONS:   NON-INVERTING INPUT
*                | INVERTING INPUT
*                | | POSITIVE POWER SUPPLY
*                | | | NEGATIVE POWER SUPPLY
*                | | | | OUTPUT
*                | | | | |
.SUBCKT LF351    1 2 3 4 5
*
  C1    11 12 3.498E-12
  C2     6  7 15.00E-12
  DC     5 53 DX
  DE    54  5 DX
  DLP   90 91 DX
  DLN   92 90 DX
  DP     4  3 DX
  EGND  99  0 POLY(2) (3,0) (4,0) 0 .5 .5
  FB     7 99 POLY(5) VB VC VE VLP VLN 0 28.29E6 -30E6 30E6 30E6 -30E6
  GA     6  0 11 12 282.8E-6
  GCM    0  6 10 99 1.590E-9
  ISS    3 10 DC 195.0E-6
  HLIM  90  0 VLIM 1K
  J1    11  2 10 JX
  J2    12  1 10 JX
  R2     6  9 100.0E3
  RD1    4 11 3.536E3
  RD2    4 12 3.536E3
  RO1    8  5 50
  RO2    7 99 25
  RP     3  4 15.00E3
  RSS   10 99 1.026E6
  VB     9  0 DC 0
  VC     3 53 DC 2.200
  VE    54  4 DC 2.200
  VLIM   7  8 DC 0
  VLP   91  0 DC 30
  VLN    0 92 DC 30
.MODEL DX D(IS=800.0E-18)
.MODEL JX PJF(IS=12.50E-12 BETA=250.1E-6 VTO=-1)
.ENDS
```

```
* LF353 OPERATIONAL AMPLIFIER "MACROMODEL" SUBCIRCUIT
* CREATED USING PARTS RELEASE 4.01 ON 06/27/89 AT 08:19
* (REV N/A)
* CONNECTIONS:   NON-INVERTING INPUT
*                | INVERTING INPUT
*                | | POSITIVE POWER SUPPLY
*                | | | NEGATIVE POWER SUPPLY
*                | | | | OUTPUT
*                | | | | |
.SUBCKT LF353    1 2 3 4 5
*
  C1    11 12 3.498E-12
  C2     6  7 15.00E-12
  DC     5 53 DX
  DE    54  5 DX
  DLP   90 91 DX
  DLN   92 90 DX
  DP     4  3 DX
  EGND  99  0 POLY(2) (3,0) (4,0) 0 .5 .5
  FB     7 99 POLY(5) VB VC VE VLP VLN 0 28.29E6 -30E6 30E6 30E6 -30E6
  GA     6  0 11 12 282.8E-6
  GCM    0  6 10 99 1.590E-9
  ISS    3 10 DC 195.0E-6
  HLIM  90  0 VLIM 1K
  J1    11  2 10 JX
  J2    12  1 10 JX
  R2     6  9 100.0E3
  RD1    4 11 3.536E3
  RD2    4 12 3.536E3
  RO1    8  5 50
  RO2    7 99 11.62
  RP     3  4 15.00E3
  RSS   10 99 1.026E6
  VB     9  0 DC 0
  VC     3 53 DC 2.200
  VE    54  4 DC 2.200
  VLIM   7  8 DC 0
  VLP   91  0 DC 30
  VLN    0 92 DC 30
.MODEL DX D(IS=800.0E-18)
.MODEL JX PJF(IS=12.50E-12 BETA=250.1E-6 VTO=-1)
.ENDS
```

```
* LF411C OPERATIONAL AMPLIFIER "MACROMODEL" SUBCIRCUIT
* CREATED USING PARTS RELEASE 4.01 ON 06/27/89 AT 08:19
* (REV N/A)
* CONNECTIONS:   NON-INVERTING INPUT
*                | INVERTING INPUT
*                | | POSITIVE POWER SUPPLY
*                | | | NEGATIVE POWER SUPPLY
*                | | | | OUTPUT
*                | | | | |
.SUBCKT LF411C   1 2 3 4 5
*
  C1    11 12 3.498E-12
  C2     6  7 15.00E-12
  DC     5 53 DX
  DE    54  5 DX
  DLP   90 91 DX
  DLN   92 90 DX
  DP     4  3 DX
  EGND  99  0 POLY(2) (3,0) (4,0) 0 .5 .5
  FB     7 99 POLY(5) VB VC VE VLP VLN 0 28.29E6 -30E6 30E6 30E6 -30E6
  GA     6  0 11 12 282.8E-6
  GCM    0  6 10 99 1.590E-9
  ISS    3 10 DC 195.0E-6
  HLIM  90  0 VLIM 1K
  J1    11  2 10 JX
  J2    12  1 10 JX
  R2     6  9 100.0E3
  RD1    4 11 3.536E3
  RD2    4 12 3.536E3
  RO1    8  5 50
  RO2    7 99 25
  RP     3  4 15.00E3
  RSS   10 99 1.026E6
  VB     9  0 DC 0
  VC     3 53 DC 2.200
  VE    54  4 DC 2.200
  VLIM   7  8 DC 0
  VLP   91  0 DC 30
  VLN    0 92 DC 30
.MODEL DX D(IS=800.0E-18)
.MODEL JX PJF(IS=12.50E-12 BETA=250.1E-6 VTO=-1)
.ENDS
```

```
* LF412C OPERATIONAL AMPLIFIER "MACROMODEL" SUBCIRCUIT
* CREATED USING PARTS RELEASE 4.01 ON 06/27/89 AT 08:19
* (REV N/A)
* CONNECTIONS:   NON-INVERTING INPUT
*                | INVERTING INPUT
*                | | POSITIVE POWER SUPPLY
*                | | | NEGATIVE POWER SUPPLY
*                | | | | OUTPUT
*                | | | | |
.SUBCKT LF412C   1 2 3 4 5
*
  C1    11 12 3.498E-12
  C2     6  7 15.00E-12
  DC     5 53 DX
  DE    54  5 DX
  DLP   90 91 DX
  DLN   92 90 DX
  DP     4  3 DX
  EGND  99  0 POLY(2) (3,0) (4,0) 0 .5 .5
  FB     7 99 POLY(5) VB VC VE VLP VLN 0 28.29E6 -30E6 30E6 30E6 -30E6
  GA     6  0 11 12 282.8E-6
  GCM    0  6 10 99 1.590E-9
  ISS    3 10 DC 195.0E-6
  HLIM  90  0 VLIM 1K
  J1    11  2 10 JX
  J2    12  1 10 JX
  R2     6  9 100.0E3
  RD1    4 11 3.536E3
  RD2    4 12 3.536E3
  RO1    8  5 50
  RO2    7 99 25
  RP     3  4 15.00E3
  RSS   10 99 1.026E6
  VB     9  0 DC 0
  VC     3 53 DC 2.200
  VE    54  4 DC 2.200
  VLIM   7  8 DC 0
  VLP   91  0 DC 30
  VLN    0 92 DC 30
.MODEL DX D(IS=800.0E-18)
.MODEL JX PJF(IS=12.50E-12 BETA=250.1E-6 VTO=-1)
.ENDS
```

```
* LM301A OPERATIONAL AMPLIFIER "MACROMODEL" SUBCIRCUIT
* CREATED USING PARTS RELEASE 4.01 ON 09/01/89 AT 13:14
* (REV N/A)
* CONNECTIONS:   NON-INVERTING INPUT
*                | INVERTING INPUT
*                | | POSITIVE POWER SUPPLY
*                | | | NEGATIVE POWER SUPPLY
*                | | | | OUTPUT
*                | | | | |  COMPENSATION
*                | | | | | / \
.SUBCKT LM301A   1 2 3 4 5 6 7
*
  C1   11 12 7.977E-12
  DC    5 53 DX
  DE   54  5 DX
  DLP  90 91 DX
  DLN  92 90 DX
  DP    4  3 DX
  EGND 99  0 POLY(2) (3,0) (4,0) 0 .5 .5
  FB    7 99 POLY(5) VB VC VE VLP VLN 0 42.44E6 -40E6 40E6 40E6 -40E6
  GA    6  0 11 12 188.5E-6
  GCM   0  6 10 99 3.352E-9
  IEE  10  4 DC 15.14E-6
  HLIM 90  0 VLIM 1K
  Q1   11  2 13 QX
  Q2   12  1 14 QX
  R2    6  9 100.0E3
  RC1   3 11 5.305E3
  RC2   3 12 5.305E3
  RE1  13 10 1.839E3
  RE2  14 10 1.839E3
  REE  10 99 13.21E6
  RO1   8  5 50
  RO2   7 99 25
  RP    3  4 16.81E3
  VB    9  0 DC 0
  VC    3 53 DC 2.600
  VE   54  4 DC 2.600
  VLIM  7  8 DC 0
  VLP  91  0 DC 25
  VLN   0 92 DC 25
.MODEL DX D(IS=800.0E-18)
.MODEL QX NPN(IS=800.0E-18 BF=107.1)
.ENDS
```

```
* LM307 OPERATIONAL AMPLIFIER "MACROMODEL" SUBCIRCUIT
* CREATED USING PARTS RELEASE 4.01 ON 09/01/89 AT 10:21
* (REV N/A)
* CONNECTIONS:   NON-INVERTING INPUT
*                | INVERTING INPUT
*                | | POSITIVE POWER SUPPLY
*                | | | NEGATIVE POWER SUPPLY
*                | | | | OUTPUT
*                | | | | |
.SUBCKT LM307    1 2 3 4 5
*
  C1    11 12 8.887E-12
  C2     6  7 30.00E-12
  DC     5 53 DX
  DE    54  5 DX
  DLP   90 91 DX
  DLN   92 90 DX
  DP     4  3 DX
  EGND 99   0 POLY(2) (3,0) (4,0) 0 .5 .5
  FB     7 99 POLY(5) VB VC VE VLP VLN 0 42.44E6 -40E6 40E6 40E6 -40E6
  GA     6  0 11 12 188.5E-6
  GCM    0  6 10 99 3.352E-9
  IEE   10  4 DC 15.14E-6
  HLIM 90   0 VLIM 1K
  Q1    11  2 13 QX
  Q2    12  1 14 QX
  R2     6  9 100.0E3
  RC1    3 11 5.305E3
  RC2    3 12 5.305E3
  RE1   13 10 1.839E3
  RE2   14 10 1.839E3
  REE   10 99 13.21E6
  RO1    8  5 50
  RO2    7 99 25
  RP     3  4 16.81E3
  VB     9  0 DC 0
  VC     3 53 DC 2.600
  VE    54  4 DC 2.600
  VLIM   7  8 DC 0
  VLP   91  0 DC 25
  VLN    0 92 DC 25
.MODEL DX D(IS=800.0E-18)
.MODEL QX NPN(IS=800.0E-18 BF=107.1)
.ENDS
```

```
* LM308 OPERATIONAL AMPLIFIER "MACROMODEL" SUBCIRCUIT
* CREATED USING PARTS RELEASE 4.01 ON 08/30/89 AT 09:24
* (REV N/A)
* CONNECTIONS:   NON-INVERTING INPUT
*                | INVERTING INPUT
*                | | POSITIVE POWER SUPPLY
*                | | | NEGATIVE POWER SUPPLY
*                | | | | OUTPUT
*                | | | | |  COMPENSATION
*                | | | | | / \
.SUBCKT LM308    1 2 3 4 5 6 7
*
  C1   11 12 6.887E-12
  DC    5 53 DX
  DE   54  5 DX
  DLP  90 91 DX
  DLN  92 90 DX
  DP    4  3 DX
  EGND 99  0 POLY(2) (3,0) (4,0) 0 .5 .5
  FB    7 99 POLY(5) VB VC VE VLP VLN 0 19.59E6 -20E6 20E6 20E6 -20E6
  GA    6  0 11 12 122.5E-6
  GCM   0  6 10 99 6.891E-9
  IEE  10  4 DC 6.003E-6
  HLIM 90  0 VLIM 1K
  Q1   11  2 13 QX
  Q2   12  1 14 QX
  R2    6  9 100.0E3
  RC1   3 11 8.161E3
  RC2   3 12 8.161E3
  RE1  13 10 -460.3
  RE2  14 10 -460.3
  REE  10 99 33.32E6
  RO1   8  5 125
  RO2   7 99 125
  RP    3  4 102.0E3
  VB    9  0 DC 0
  VC    3 53 DC 2.600
  VE   54  4 DC 2.600
  VLIM  7  8 DC 0
  VLP  91  0 DC 6
  VLN   0 92 DC 6
.MODEL DX D(IS=800.0E-18)
.MODEL QX NPN(IS=800.0E-18 BF=2.000E3)
.ENDS
```

```
* LM318 OPERATIONAL AMPLIFIER "MACROMODEL" SUBCIRCUIT
* CREATED USING PARTS RELEASE 4.01 ON 09/08/89 AT 08:27
* (REV N/A)
* CONNECTIONS:   NON-INVERTING INPUT
*                | INVERTING INPUT
*                | | POSITIVE POWER SUPPLY
*                | | | NEGATIVE POWER SUPPLY
*                | | | | OUTPUT
*                | | | | |
.SUBCKT LM318    1 2 3 4 5
*
  C1    11 12 8.50E-12
  C2     6  7 25.00E-12
  DC     5 53 DX
  DE    54  5 DX
  DLP   90 91 DX
  DLN   92 90 DX
  DP     4  3 DX
  EGND  99  0 POLY(2) (3,0) (4,0) 0 .5 .5
  FB     7 99 POLY(5) VB VC VE VLP VLN 0 1.697E6 -2E6 2E6 2E6 -2E6
  GA     6  0 11 12 2.474E-3
  GCM    0  6 10 99 13.25E-9
  IEE   10  4 DC 1.750E-3
  HLIM  90  0 VLIM 1K
  Q1    11  2 13 QX
  Q2    12  1 14 QX
  R2     6  9 100.0E3
  RC1    3 11 424.4
  RC2    3 12 424.4
  RE1   13 10 394.7
  RE2   14 10 394.7
  REE   10 99 114.3E3
  RO1    8  5 50
  RO2    7 99 50
  RP     3  4 9.231E3
  VB     9  0 DC 0
  VC     3 53 DC 2.700
  VE    54  4 DC 2.700
  VLIM   7  8 DC 0
  VLP   91  0 DC 21
  VLN    0 92 DC 21
.MODEL DX D(IS=800.0E-18)
.MODEL QX NPN(IS=800.0E-18 BF=5.833E3)
.ENDS
```

```
* LM324 OPERATIONAL AMPLIFIER "MACROMODEL" SUBCIRCUIT
* CREATED USING PARTS RELEASE 4.01 ON 09/08/89 AT 10:54
* (REV N/A)
* CONNECTIONS:   NON-INVERTING INPUT
*                | INVERTING INPUT
*                | | POSITIVE POWER SUPPLY
*                | | | NEGATIVE POWER SUPPLY
*                | | | | OUTPUT
*                | | | | |
.SUBCKT LM324    1 2 3 4 5
*
  C1    11 12 5.544E-12
  C2     6  7 20.00E-12
  DC     5 53 DX
  DE    54  5 DX
  DLP   90 91 DX
  DLN   92 90 DX
  DP     4  3 DX
  EGND  99  0 POLY(2) (3,0) (4,0) 0 .5 .5
  FB     7 99 POLY(5) VB VC VE VLP VLN 0 15.91E6 -20E6 20E6 20E6 -20E6
  GA     6  0 11 12 125.7E-6
  GCM    0  6 10 99 7.067E-9
  IEE    3 10 DC 10.04E-6
  HLIM  90  0 VLIM 1K
  Q1    11  2 13 QX
  Q2    12  1 14 QX
  R2     6  9 100.0E3
  RC1    4 11 7.957E3
  RC2    4 12 7.957E3
  RE1   13 10 2.773E3
  RE2   14 10 2.773E3
  REE   10 99 19.92E6
  RO1    8  5 50
  RO2    7 99 50
  RP     3  4 30.31E3
  VB     9  0 DC 0
  VC     3 53 DC 2.100
  VE    54  4 DC .6
  VLIM   7  8 DC 0
  VLP   91  0 DC 40
  VLN    0 92 DC 40
.MODEL DX D(IS=800.0E-18)
.MODEL QX PNP(IS=800.0E-18 BF=250)
.ENDS
```

```
* LM348 OPERATIONAL AMPLIFIER "MACROMODEL" SUBCIRCUIT
* CREATED USING PARTS RELEASE 4.01 ON 09/14/89 AT 15:56
* (REV N/A)
* CONNECTIONS:   NON-INVERTING INPUT
*                | INVERTING INPUT
*                | | POSITIVE POWER SUPPLY
*                | | | NEGATIVE POWER SUPPLY
*                | | | | OUTPUT
*                | | | | |
.SUBCKT LM348    1 2 3 4 5
*
  C1    11 12 9.461E-12
  C2     6  7 30.00E-12
  DC     5 53 DX
  DE    54  5 DX
  DLP   90 91 DX
  DLN   92 90 DX
  DP     4  3 DX
  EGND 99  0 POLY(2) (3,0) (4,0) 0 .5 .5
  FB     7 99 POLY(5) VB VC VE VLP VLN 0 4.715E6 -5E6 5E6 5E6 -5E6
  GA    6  0 11 12 256.2E-6
  GCM   0  6 10 99 4.023E-9
  IEE  10  4 DC 15.06E-6
  HLIM 90  0 VLIM 1K
  Q1    11  2 13 QX
  Q2    12  1 14 QX
  R2     6  9 100.0E3
  RC1    3 11 4.420E3
  RC2    3 12 4.420E3
  RE1   13 10 968
  RE2   14 10 968
  REE   10 99 13.28E6
  RO1    8  5 150
  RO2    7 99 150
  RP     3  4 51.28E3
  VB     9  0 DC 0
  VC     3 53 DC 3.600
  VE    54  4 DC 3.600
  VLIM   7  8 DC 0
  VLP   91  0 DC 25
  VLN    0 92 DC 25
.MODEL DX D(IS=800.0E-18)
.MODEL QX NPN(IS=800.0E-18 BF=250)
.ENDS
```

```
* LM358 OPERATIONAL AMPLIFIER "MACROMODEL" SUBCIRCUIT
* CREATED USING PARTS RELEASE 4.01 ON 09/08/89 AT 10:54
* (REV N/A)
* CONNECTIONS:   NON-INVERTING INPUT
*                | INVERTING INPUT
*                | | POSITIVE POWER SUPPLY
*                | | | NEGATIVE POWER SUPPLY
*                | | | | OUTPUT
*                | | | | |
.SUBCKT LM358    1 2 3 4 5
*
  C1   11 12 5.544E-12
  C2    6  7 20.00E-12
  DC    5 53 DX
  DE   54  5 DX
  DLP  90 91 DX
  DLN  92 90 DX
  DP    4  3 DX
  EGND 99  0 POLY(2) (3,0) (4,0) 0 .5 .5
  FB    7 99 POLY(5) VB VC VE VLP VLN 0 15.91E6 -20E6 20E6 20E6 -20E6
  GA    6  0 11 12 125.7E-6
  GCM   0  6 10 99 7.067E-9
  IEE   3 10 DC 10.04E-6
  HLIM 90  0 VLIM 1K
  Q1   11  2 13 QX
  Q2   12  1 14 QX
  R2    6  9 100.0E3
  RC1   4 11 7.957E3
  RC2   4 12 7.957E3
  RE1  13 10 2.773E3
  RE2  14 10 2.773E3
  REE  10 99 19.92E6
  RO1   8  5 50
  RO2   7 99 50
  RP    3  4 30.31E3
  VB    9  0 DC 0
  VC    3 53 DC 2.100
  VE   54  4 DC .6
  VLIM  7  8 DC 0
  VLP  91  0 DC 40
  VLN   0 92 DC 40
.MODEL DX D(IS=800.0E-18)
.MODEL QX PNP(IS=800.0E-18 BF=250)
.ENDS
```

```
* MC1458 OPERATIONAL AMPLIFIER "MACROMODEL" SUBCIRCUIT
* CREATED USING PARTS RELEASE 4.01 ON 07/05/89 AT 09:09
* (REV N/A)
* CONNECTIONS:   NON-INVERTING INPUT
*                | INVERTING INPUT
*                | | POSITIVE POWER SUPPLY
*                | | | NEGATIVE POWER SUPPLY
*                | | | | OUTPUT
*                | | | | |
.SUBCKT MC1458   1 2 3 4 5
*
  C1    11 12 4.664E-12
  C2     6  7 20.00E-12
  DC     5 53 DX
  DE    54  5 DX
  DLP   90 91 DX
  DLN   92 90 DX
  DP     4  3 DX
  EGND  99  0 POLY(2) (3,0) (4,0) 0 .5 .5
  FB     7 99 POLY(5) VB VC VE VLP VLN 0 10.61E6 -10E6 10E6 10E6 -10E6
  GA     6  0 11 12 137.7E-6
  GCM    0  6 10 99 2.574E-9
  IEE   10  4 DC 10.16E-6
  HLIM  90  0 VLIM 1K
  Q1    11  2 13 QX
  Q2    12  1 14 QX
  R2     6  9 100.0E3
  RC1    3 11 7.957E3
  RC2    3 12 7.957E3
  RE1   13 10 2.740E3
  RE2   14 10 2.740E3
  REE   10 99 19.69E6
  RO1    8  5 150
  RO2    7 99 150
  RP     3  4 18.11E3
  VB     9  0 DC 0
  VC     3 53 DC 2.600
  VE    54  4 DC 2.600
  VLIM   7  8 DC 0
  VLP   91  0 DC 25
  VLN    0 92 DC 25
.MODEL DX D(IS=800.0E-18)
.MODEL QX NPN(IS=800.0E-18 BF=62.50)
.ENDS
```

```
* MC3403 OPERATIONAL AMPLIFIER "MACROMODEL" SUBCIRCUIT
* CREATED USING PARTS RELEASE 4.01 ON 09/08/89 AT 13:59
* (REV N/A)
* CONNECTIONS:   NON-INVERTING INPUT
*                | INVERTING INPUT
*                | | POSITIVE POWER SUPPLY
*                | | | NEGATIVE POWER SUPPLY
*                | | | | OUTPUT
*                | | | | |
.SUBCKT MC3403   1 2 3 4 5
*
  C1   11 12 7.544E-12
  C2    6  7 20.00E-12
  DC    5 53 DX
  DE   54  5 DX
  DLP  90 91 DX
  DLN  92 90 DX
  DP    4  3 DX
  EGND 99  0 POLY(2) (3,0) (4,0) 0 .5 .5
  FB    7 99 POLY(5) VB VC VE VLP VLN 0 42.44E6 -40E6 40E6 40E6 -40E6
  GA    6  0 11 12 130.7E-6
  GCM   0  6 10 99 2.235E-9
  IEE   3 10 DC 12.40E-6
  HLIM 90  0 VLIM 1K
  Q1   11  2 13 QX
  Q2   12  1 14 QX
  R2    6  9 100.0E3
  RC1   4 11 7.957E3
  RC2   4 12 7.957E3
  RE1  13 10 3.529E3
  RE2  14 10 3.529E3
  REE  10 99 16.13E6
  RO1   8  5 37.50
  RO2   7 99 37.50
  RP    3  4 43.62E3
  VB    9  0 DC 0
  VC    3 53 DC 2.600
  VE   54  4 DC 2.600
  VLIM  7  8 DC 0
  VLP  91  0 DC 30
  VLN   0 92 DC 30
.MODEL DX D(IS=800.0E-18)
.MODEL QX PNP(IS=800.0E-18 BF=30)
.ENDS
```

```
* NE5534 OPERATIONAL AMPLIFIER "MACROMODEL" SUBCIRCUIT
* CREATED USING PARTS RELEASE 4.01 ON 04/10/89 AT 12:41
* (REV N/A)
* CONNECTIONS:   NON-INVERTING INPUT
*                | INVERTING INPUT
*                | | POSITIVE POWER SUPPLY
*                | | | NEGATIVE POWER SUPPLY
*                | | | | OUTPUT
*                | | | | |  COMPENSATION
*                | | | | | / \
.SUBCKT NE5534   1 2 3 4 5 6 7
*
  C1    11 12 7.703E-12
  DC     5 53 DX
  DE    54  5 DX
  DLP   90 91 DX
  DLN   92 90 DX
  DP     4  3 DX
  EGND 99  0 POLY(2) (3,0) (4,0) 0 .5 .5
  FB     7 99 POLY(5) VB VC VE VLP VLN 0 2.893E6 -3E6 3E6 3E6 -3E6
  GA     6  0 11 12 1.382E-3
  GCM    0  6 10 99 13.82E-9
  IEE   10  4 DC 133.0E-6
  HLIM 90  0 VLIM 1K
  Q1    11  2 13 QX
  Q2    12  1 14 QX
  R2     6  9 100.0E3
  RC1    3 11 723.3
  RC2    3 12 723.3
  RE1   13 10 329
  RE2   14 10 329
  REE   10 99 1.504E6
  RO1    8  5 50
  RO2    7 99 25
  RP     3  4 7.757E3
  VB     9  0 DC 0
  VC     3 53 DC 2.700
  VE    54  4 DC 2.700
  VLIM   7  8 DC 0
  VLP   91  0 DC 38
  VLN    0 92 DC 38
.MODEL DX D(IS=800.0E-18)
.MODEL QX NPN(IS=800.0E-18 BF=132)
.ENDS
```

```
* TL022C OPERATIONAL AMPLIFIER "MACROMODEL" SUBCIRCUIT
* CREATED USING PARTS RELEASE 4.01 ON 09/14/89 AT 10:48
* (REV N/A)
* CONNECTIONS:   NON-INVERTING INPUT
*                | INVERTING INPUT
*                | | POSITIVE POWER SUPPLY
*                | | | NEGATIVE POWER SUPPLY
*                | | | | OUTPUT
*                | | | | |
.SUBCKT TL022C   1 2 3 4 5
*
  C1   11 12 3.498E-12
  C2    6  7 15.00E-12
  DC    5 53 DX
  DE   54  5 DX
  DLP  90 91 DX
  DLN  92 90 DX
  DP    4  3 DX
  EGND 99  0 POLY(2) (3,0) (4,0) 0 .5 .5
  FB    7 99 POLY(5) VB VC VE VLP VLN 0 1.697E6 -2E6 2E6 2E6 -2E6
  GA    6  0 11 12 47.13E-6
  GCM   0  6 10 99 6.657E-9
  IEE   3 10 DC 7.700E-6
  HLIM 90  0 VLIM 1K
  Q1   11  2 13 QX
  Q2   12  1 14 QX
  R2    6  9 100.0E3
  RC1   4 11 21.22E3
  RC2   4 12 21.22E3
  RE1  13 10 13.95E3
  RE2  14 10 13.95E3
  REE  10 99 25.97E6
  RO1   8  5 125
  RO2   7 99 125
  RP    3  4 245.1E3
  VB    9  0 DC 0
  VC    3 53 DC 2.600
  VE   54  4 DC 2.600
  VLIM  7  8 DC 0
  VLP  91  0 DC 6
  VLN   0 92 DC 6
.MODEL DX D(IS=800.0E-18)
.MODEL QX PNP(IS=800.0E-18 BF=37.50)
.ENDS
```

```
* TL031 OPERATIONAL AMPLIFIER "MACROMODEL" SUBCIRCUIT
* CREATED USING PARTS RELEASE 4.01 ON 06/08/89 AT 12:57
* (REV N/A)
* CONNECTIONS:   NON-INVERTING INPUT
*                | INVERTING INPUT
*                | | POSITIVE POWER SUPPLY
*                | | | NEGATIVE POWER SUPPLY
*                | | | | OUTPUT
*                | | | | |
.SUBCKT TL031    1 2 3 4 5
*
  C1    11 12 3.498E-12
  C2     6  7 15.00E-12
  CSS   10 99 11.38E-12
  DC     5 53 DX
  DE    54  5 DX
  DLP   90 91 DX
  DLN   92 90 DX
  DP     4  3 DX
  EGND  99  0 POLY(2) (3,0) (4,0) 0 .5 .5
  FB     7 99 POLY(5) VB VC VE VLP VLN 0 936.5E3 -900E3 900E3 900E3 -900E3
  GA     6  0 11 12 113.1E-6
  GCM    0  6 10 99 2.257E-9
  ISS    3 10 DC 76.50E-6
  HLIM  90  0 VLIM 1K
  J1    11  2 10 JX
  J2    12  1 10 JX
  R2     6  9 100.0E3
  RD1    4 11 8.841E3
  RD2    4 12 8.841E3
  RO1    8  5 135
  RO2    7 99 135
  RP     3  4 138.5E3
  RSS   10 99 2.614E6
  VB     9  0 DC 0
  VC     3 53 DC 1.700
  VE    54  4 DC 1.800
  VLIM   7  8 DC 0
  VLP   91  0 DC 8
  VLN    0 92 DC 8
.MODEL DX D(IS=800.0E-18)
.MODEL JX PJF(IS=1.000E-12 BETA=140E-6 VTO=-1)
.ENDS
```

```
* TL032 OPERATIONAL AMPLIFIER "MACROMODEL" SUBCIRCUIT
* CREATED USING PARTS RELEASE 4.01 ON 06/08/89 AT 12:57
* (REV N/A)
* CONNECTIONS:   NON-INVERTING INPUT
*                | INVERTING INPUT
*                | | POSITIVE POWER SUPPLY
*                | | | NEGATIVE POWER SUPPLY
*                | | | | OUTPUT
*                | | | | |
.SUBCKT TL032    1 2 3 4 5
*
  C1   11 12 3.498E-12
  C2    6  7 15.00E-12
  CSS  10 99 11.38E-12
  DC    5 53 DX
  DE   54  5 DX
  DLP  90 91 DX
  DLN  92 90 DX
  DP    4  3 DX
  EGND 99  0 POLY(2) (3,0) (4,0) 0 .5 .5
  FB    7 99 POLY(5) VB VC VE VLP VLN 0 936.5E3 -900E3 900E3 900E3 -900E3
  GA   6  0 11 12 113.1E-6
  GCM   0  6 10 99 2.257E-9
  ISS   3 10 DC 76.50E-6
  HLIM 90  0 VLIM 1K
  J1   11  2 10 JX
  J2   12  1 10 JX
  R2    6  9 100.0E3
  RD1   4 11 8.841E3
  RD2   4 12 8.841E3
  RO1   8  5 135
  RO2   7 99 135
  RP    3  4 138.5E3
  RSS  10 99 2.614E6
  VB    9  0 DC 0
  VC    3 53 DC 1.700
  VE   54  4 DC 1.800
  VLIM  7  8 DC 0
  VLP  91  0 DC 8
  VLN   0 92 DC 8
.MODEL DX D(IS=800.0E-18)
.MODEL JX PJF(IS=1.000E-12 BETA=140E-6 VTO=-1)
.ENDS
```

```
* TL034 OPERATIONAL AMPLIFIER "MACROMODEL" SUBCIRCUIT
* CREATED USING PARTS RELEASE 4.01 ON 06/08/89 AT 12:57
* (REV N/A)
* CONNECTIONS:   NON-INVERTING INPUT
*                | INVERTING INPUT
*                | | POSITIVE POWER SUPPLY
*                | | | NEGATIVE POWER SUPPLY
*                | | | | OUTPUT
*                | | | | |
.SUBCKT TL034    1 2 3 4 5
*
  C1    11 12 3.498E-12
  C2     6  7 15.00E-12
  CSS   10 99 11.38E-12
  DC     5 53 DX
  DE    54  5 DX
  DLP   90 91 DX
  DLN   92 90 DX
  DP     4  3 DX
  EGND  99  0 POLY(2) (3,0) (4,0) 0 .5 .5
  FB     7 99 POLY(5) VB VC VE VLP VLN 0 936.5E3 -900E3 900E3 900E3 -900E3
  GA     6  0 11 12 113.1E-6
  GCM    0  6 10 99 2.257E-9
  ISS    3 10 DC 76.50E-6
  HLIM  90  0 VLIM 1K
  J1    11  2 10 JX
  J2    12  1 10 JX
  R2     6  9 100.0E3
  RD1    4 11 8.841E3
  RD2    4 12 8.841E3
  RO1    8  5 135
  RO2    7 99 135
  RP     3  4 138.5E3
  RSS   10 99 2.614E6
  VB     9  0 DC 0
  VC     3 53 DC 1.700
  VE    54  4 DC 1.800
  VLIM   7  8 DC 0
  VLP   91  0 DC 8
  VLN    0 92 DC 8
.MODEL DX D(IS=800.0E-18)
.MODEL JX PJF(IS=1.000E-12 BETA=140E-6 VTO=-1)
.ENDS
```

```
* TL044C OPERATIONAL AMPLIFIER "MACROMODEL" SUBCIRCUIT
* CREATED USING PARTS RELEASE 4.01 ON 09/14/89 AT 10:48
* (REV N/A)
* CONNECTIONS:   NON-INVERTING INPUT
*                | INVERTING INPUT
*                | | POSITIVE POWER SUPPLY
*                | | | NEGATIVE POWER SUPPLY
*                | | | | OUTPUT
*                | | | | |
.SUBCKT TL044C   1 2 3 4 5
*
  C1   11 12 3.498E-12
  C2    6  7 15.00E-12
  DC    5 53 DX
  DE   54  5 DX
  DLP  90 91 DX
  DLN  92 90 DX
  DP    4  3 DX
  EGND 99  0 POLY(2) (3,0) (4,0) 0 .5 .5
  FB    7 99 POLY(5) VB VC VE VLP VLN 0 1.697E6 -2E6 2E6 2E6 -2E6
  GA    6  0 11 12 47.13E-6
  GCM   0  6 10 99 6.657E-9
  IEE   3 10 DC 7.700E-6
  HLIM 90  0 VLIM 1K
  Q1   11  2 13 QX
  Q2   12  1 14 QX
  R2    6  9 100.0E3
  RC1   4 11 21.22E3
  RC2   4 12 21.22E3
  RE1  13 10 13.95E3
  RE2  14 10 13.95E3
  REE  10 99 25.97E6
  RO1   8  5 125
  RO2   7 99 125
  RP    3  4 245.1E3
  VB    9  0 DC 0
  VC    3 53 DC 2.600
  VE   54  4 DC 2.600
  VLIM  7  8 DC 0
  VLP  91  0 DC 6
  VLN   0 92 DC 6
.MODEL DX D(IS=800.0E-18)
.MODEL QX PNP(IS=800.0E-18 BF=37.50)
.ENDS
```

```
*****************************************************************************
* TL051 OPERATIONAL AMPLIFIER "MACROMODEL" SUBCIRCUIT
* CREATED USING PARTS RELEASE 4.01 ON 04/12/89 AT 09:57
* (REV N/A)
* CONNECTIONS:   NON-INVERTING INPUT
*                | INVERTING INPUT
*                | | POSITIVE POWER SUPPLY
*                | | | NEGATIVE POWER SUPPLY
*                | | | | OUTPUT
*                | | | | |
.SUBCKT TL051    1 2 3 4 5
*
  C1    11 12 3.988E-12
  C2     6  7 15.00E-12
  DC     5 53 DX
  DE    54  5 DX
  DLP   90 91 DX
  DLN   92 90 DX
  DP     4  3 DX
  EGND 99   0 POLY(2) (3,0) (4,0) 0 .5 .5
  FB     7 99 POLY(5) VB VC VE VLP VLN 0 2.875E6 -3E6 3E6 3E6 -3E6
  GA    6   0 11 12 292.2E-6
  GCM   0   6 10 99 6.542E-9
  ISS   3  10 DC 300.0E-6
  HLIM 90   0 VLIM 1K
  J1    11  2 10 JX
  J2    12  1 10 JX
  R2    6   9 100.0E3
  RD1    4 11 3.422E3
  RD2    4 12 3.422E3
  RO1    8  5 125
  RO2    7 99 125
  RP     3  4 11.11E3
  RSS   10 99 666.7E3
  VB     9  0 DC 0
  VC     3 53 DC 3
  VE    54  4 DC 3.700
  VLIM   7  8 DC 0
  VLP   91  0 DC 28
  VLN    0 92 DC 28
.MODEL DX D(IS=800.0E-18)
.MODEL JX PJF(IS=15.00E-12 BETA=185.2E-6 VTO=-1)
.ENDS
```

```
********************************************************************************
* TL052 OPERATIONAL AMPLIFIER "MACROMODEL" SUBCIRCUIT
* CREATED USING PARTS RELEASE 4.01 ON 04/12/89 AT 09:57
* (REV N/A)
* CONNECTIONS:   NON-INVERTING INPUT
*                | INVERTING INPUT
*                | | POSITIVE POWER SUPPLY
*                | | | NEGATIVE POWER SUPPLY
*                | | | | OUTPUT
*                | | | | |
.SUBCKT TL052    1 2 3 4 5
*
  C1    11 12 3.988E-12
  C2     6  7 15.00E-12
  DC     5 53 DX
  DE    54  5 DX
  DLP   90 91 DX
  DLN   92 90 DX
  DP     4  3 DX
  EGND  99  0 POLY(2) (3,0) (4,0) 0 .5 .5
  FB     7 99 POLY(5) VB VC VE VLP VLN 0 2.875E6 -3E6 3E6 3E6 -3E6
  GA     6  0 11 12 292.2E-6
  GCM    0  6 10 99 6.542E-9
  ISS    3 10 DC 300.0E-6
  HLIM  90  0 VLIM 1K
  J1    11  2 10 JX
  J2    12  1 10 JX
  R2     6  9 100.0E3
  RD1    4 11 3.422E3
  RD2    4 12 3.422E3
  RO1    8  5 125
  RO2    7 99 125
  RP     3  4 11.11E3
  RSS   10 99 666.7E3
  VB     9  0 DC 0
  VC     3 53 DC 3
  VE    54  4 DC 3.700
  VLIM   7  8 DC 0
  VLP   91  0 DC 28
  VLN    0 92 DC 28
.MODEL DX D(IS=800.0E-18)
.MODEL JX PJF(IS=15.00E-12 BETA=185.2E-6 VTO=-1)
.ENDS
```

```
* TL054 OPERATIONAL AMPLIFIER "MACROMODEL" SUBCIRCUIT
* CREATED USING PARTS RELEASE 4.01 ON 04/12/89 AT 09:57
* (REV N/A)
* CONNECTIONS:   NON-INVERTING INPUT
*                | INVERTING INPUT
*                | | POSITIVE POWER SUPPLY
*                | | | NEGATIVE POWER SUPPLY
*                | | | | OUTPUT
*                | | | | |
.SUBCKT TL054    1 2 3 4 5
*
  C1    11 12 3.988E-12
  C2     6  7 15.00E-12
  DC     5 53 DX
  DE    54  5 DX
  DLP   90 91 DX
  DLN   92 90 DX
  DP     4  3 DX
  EGND 99   0 POLY(2) (3,0) (4,0) 0 .5 .5
  FB     7 99 POLY(5) VB VC VE VLP VLN 0 2.875E6 -3E6 3E6 3E6 -3E6
  GA    6   0 11 12 292.2E-6
  GCM   0   6 10 99 6.542E-9
  ISS   3  10 DC 300.0E-6
  HLIM 90   0 VLIM 1K
  J1    11  2 10 JX
  J2    12  1 10 JX
  R2    6   9 100.0E3
  RD1    4 11 3.422E3
  RD2    4 12 3.422E3
  RO1    8  5 125
  RO2    7 99 125
  RP     3  4 11.11E3
  RSS   10 99 666.7E3
  VB     9  0 DC 0
  VC     3 53 DC 3
  VE    54  4 DC 3.700
  VLIM   7  8 DC 0
  VLP   91  0 DC 28
  VLN    0 92 DC 28
.MODEL DX D(IS=800.0E-18)
.MODEL JX PJF(IS=15.00E-12 BETA=185.2E-6 VTO=-1)
.ENDS
```

```
* TL060 OPERATIONAL AMPLIFIER "MACROMODEL" SUBCIRCUIT
* CREATED USING PARTS RELEASE 4.01 ON 09/22/89 AT 13:08
* (REV N/A)
* CONNECTIONS:   NON-INVERTING INPUT
*                | INVERTING INPUT
*                | | POSITIVE POWER SUPPLY
*                | | | NEGATIVE POWER SUPPLY
*                | | | | OUTPUT
*                | | | | |  COMPENSATION
*                | | | | | / \
.SUBCKT TL060    1 2 3 4 5 6 7
*
  C1   11 12 2.332E-12
  DC    5 53 DX
  DE   54  5 DX
  DLP  90 91 DX
  DLN  92 90 DX
  DP    4  3 DX
  EGND 99  0 POLY(2) (3,0) (4,0) 0 .5 .5
  FB    7 99 POLY(5) VB VC VE VLP VLN 0 477.4E3 -500E3 500E3 500E3 -500E3
  GA    6  0 11 12 62.84E-6
  GCM   0  6 10 99 2.178E-9
  ISS   3 10 DC 35.00E-6
  HLIM 90  0 VLIM 1K
  J1   11  2 10 JX
  J2   12  1 10 JX
  R2    6  9 100.0E3
  RD1   4 11 15.91E3
  RD2   4 12 15.91E3
  RO1   8  5 200
  RO2   7 99 200
  RP    3  4 150.0E3
  RSS  10 99 5.714E6
  VB    9  0 DC 0
  VC    3 53 DC 2.130
  VE   54  4 DC 2.130
  VLIM  7  8 DC 0
  VLP  91  0 DC 15
  VLN   0 92 DC 15
.MODEL DX D(IS=800.0E-18)
.MODEL JX PJF(IS=15.00E-12 BETA=64E-6 VTO=-1)
.ENDS
```

```
* TL061 OPERATIONAL AMPLIFIER "MACROMODEL" SUBCIRCUIT
* CREATED USING PARTS RELEASE 4.01 ON 06/28/89 AT 10:42
* (REV N/A)
* CONNECTIONS:   NON-INVERTING INPUT
*                | INVERTING INPUT
*                | | POSITIVE POWER SUPPLY
*                | | | NEGATIVE POWER SUPPLY
*                | | | | OUTPUT
*                | | | | |
.SUBCKT TL061    1 2 3 4 5
*
  C1    11 12 3.498E-12
  C2     6  7 15.00E-12
  DC     5 53 DX
  DE    54  5 DX
  DLP   90 91 DX
  DLN   92 90 DX
  DP     4  3 DX
  EGND 99   0 POLY(2) (3,0) (4,0) 0 .5 .5
  FB     7 99 POLY(5) VB VC VE VLP VLN 0 318.3E3 -300E3 300E3 300E3 -300E3
  GA     6  0 11 12 94.26E-6
  GCM    0  6 10 99 1.607E-9
  ISS    3 10 DC 52.50E-6
  HLIM 90   0 VLIM 1K
  J1    11  2 10 JX
  J2    12  1 10 JX
  R2     6  9 100.0E3
  RD1    4 11 10.61E3
  RD2    4 12 10.61E3
  RO1    8  5 200
  RO2    7 99 200
  RP     3  4 150.0E3
  RSS   10 99 3.810E6
  VB     9  0 DC 0
  VC     3 53 DC 2.200
  VE    54  4 DC 2.200
  VLIM   7  8 DC 0
  VLP   91  0 DC 15
  VLN    0 92 DC 15
.MODEL DX D(IS=800.0E-18)
.MODEL JX PJF(IS=15.00E-12 BETA=100.5E-6 VTO=-1)
.ENDS
```

```
* TL062 OPERATIONAL AMPLIFIER "MACROMODEL" SUBCIRCUIT
* CREATED USING PARTS RELEASE 4.01 ON 06/28/89 AT 10:42
* (REV N/A)
* CONNECTIONS:   NON-INVERTING INPUT
*                | INVERTING INPUT
*                | | POSITIVE POWER SUPPLY
*                | | | NEGATIVE POWER SUPPLY
*                | | | | OUTPUT
*                | | | | |
.SUBCKT TL062    1 2 3 4 5
*
  C1   11 12 3.498E-12
  C2    6  7 15.00E-12
  DC    5 53 DX
  DE   54  5 DX
  DLP  90 91 DX
  DLN  92 90 DX
  DP    4  3 DX
  EGND 99  0 POLY(2) (3,0) (4,0) 0 .5 .5
  FB    7 99 POLY(5) VB VC VE VLP VLN 0 318.3E3 -300E3 300E3 300E3 -300E3
  GA    6  0 11 12 94.26E-6
  GCM   0  6 10 99 1.607E-9
  ISS   3 10 DC 52.50E-6
  HLIM 90  0 VLIM 1K
  J1   11  2 10 JX
  J2   12  1 10 JX
  R2    6  9 100.0E3
  RD1   4 11 10.61E3
  RD2   4 12 10.61E3
  RO1   8  5 200
  RO2   7 99 200
  RP    3  4 150.0E3
  RSS  10 99 3.810E6
  VB    9  0 DC 0
  VC    3 53 DC 2.200
  VE   54  4 DC 2.200
  VLIM  7  8 DC 0
  VLP  91  0 DC 15
  VLN   0 92 DC 15
.MODEL DX D(IS=800.0E-18)
.MODEL JX PJF(IS=15.00E-12 BETA=100.5E-6 VTO=-1)
.ENDS
```

```
* TL064 OPERATIONAL AMPLIFIER "MACROMODEL" SUBCIRCUIT
* CREATED USING PARTS RELEASE 4.01 ON 06/28/89 AT 10:42
* (REV N/A)
* CONNECTIONS:   NON-INVERTING INPUT
*                | INVERTING INPUT
*                | | POSITIVE POWER SUPPLY
*                | | | NEGATIVE POWER SUPPLY
*                | | | | OUTPUT
*                | | | | |
.SUBCKT TL064    1 2 3 4 5
*
  C1    11 12 3.498E-12
  C2     6  7 15.00E-12
  DC     5 53 DX
  DE    54  5 DX
  DLP   90 91 DX
  DLN   92 90 DX
  DP     4  3 DX
  EGND  99  0 POLY(2) (3,0) (4,0) 0 .5 .5
  FB     7 99 POLY(5) VB VC VE VLP VLN 0 318.3E3 -300E3 300E3 300E3 -300E3
  GA     6  0 11 12 94.26E-6
  GCM    0  6 10 99 1.607E-9
  ISS    3 10 DC 52.50E-6
  HLIM  90  0 VLIM 1K
  J1    11  2 10 JX
  J2    12  1 10 JX
  R2     6  9 100.0E3
  RD1    4 11 10.61E3
  RD2    4 12 10.61E3
  RO1    8  5 200
  RO2    7 99 200
  RP     3  4 150.0E3
  RSS   10 99 3.810E6
  VB     9  0 DC 0
  VC     3 53 DC 2.200
  VE    54  4 DC 2.200
  VLIM   7  8 DC 0
  VLP   91  0 DC 15
  VLN    0 92 DC 15
.MODEL DX D(IS=800.0E-18)
.MODEL JX PJF(IS=15.00E-12 BETA=100.5E-6 VTO=-1)
.ENDS
```

```
* TL066 OPERATIONAL AMPLIFIER "MACROMODEL" SUBCIRCUIT
* CREATED USING PARTS RELEASE 4.01 ON 09/15/89 AT 09:28
* (REV N/A)
* CONNECTIONS:   NON-INVERTING INPUT
*                | INVERTING INPUT
*                | | POSITIVE POWER SUPPLY
*                | | | NEGATIVE POWER SUPPLY
*                | | | | OUTPUT
*                | | | | |
.SUBCKT TL066    1 2 3 4 5
*
  C1   11 12 3.498E-12
  C2    6  7 15.00E-12
  DC    5 53 DX
  DE   54  5 DX
  DLP  90 91 DX
  DLN  92 90 DX
  DP    4  3 DX
  EGND 99  0 POLY(2) (3,0) (4,0) 0 .5 .5
  FB    7 99 POLY(5) VB VC VE VLP VLN 0 318.3E3 -300E3 300E3 300E3 -300E3
  GA    6  0 11 12 94.26E-6
  GCM   0  6 10 99 1.171E-8
  ISS   3 10 DC 52.50E-6
  HLIM 90  0 VLIM 1K
  J1   11  2 10 JX
  J2   12  1 10 JX
  R2    6  9 100.0E3
  RD1   4 11 10.61E3
  RD2   4 12 10.61E3
  RO1   8  5 200
  RO2   7 99 200
  RP    3  4 150.0E3
  RSS  10 99 3.810E6
  VB    9  0 DC 0
  VC    3 53 DC 2.200
  VE   54  4 DC 2.200
  VLIM  7  8 DC 0
  VLP  91  0 DC 15
  VLN   0 92 DC 15
.MODEL DX D(IS=800.0E-18)
.MODEL JX PJF(IS=15.00E-12 BETA=100.5E-6 VTO=-1)
.ENDS
```

```
* TL070 OPERATIONAL AMPLIFIER "MACROMODEL" SUBCIRCUIT
* CREATED USING PARTS RELEASE 4.01 ON 09/22/89 AT 09:52
* (REV N/A)
* CONNECTIONS:   NON-INVERTING INPUT
*                | INVERTING INPUT
*                | | POSITIVE POWER SUPPLY
*                | | | NEGATIVE POWER SUPPLY
*                | | | | OUTPUT
*                | | | | |  COMPENSATION
*                | | | | | / \
.SUBCKT TL070    1 2 3 4 5 6 7
*
  C1    11 12 5.197E-12
  DC     5 53 DX
  DE    54  5 DX
  DLP   90 91 DX
  DLN   92 90 DX
  DP     4  3 DX
  EGND  99  0 POLY(2) (3,0) (4,0) 0 .5 .5
  FB     7 99 POLY(5) VB VC VE VLP VLN 0 3.929E6 -4E6 4E6 4E6 -4E6
  GA     6  0 11 12 361.3E-6
  GCM    0  6 10 99 1.908E-9
  ISS    3 10 DC 234.0E-6
  HLIM  90  0 VLIM 1K
  J1    11  2 10 JX
  J2    12  1 10 JX
  R2     6  9 100.0E3
  RD1    4 11 2.947E3
  RD2    4 12 2.947E3
  RO1    8  5 150
  RO2    7 99 150
  RP     3  4 21.43E3
  RSS   10 99 854.7E3
  VB     9  0 DC 0
  VC     3 53 DC 2.180
  VE    54  4 DC 2.180
  VLIM   7  8 DC 0
  VLP   91  0 DC 25
  VLN    0 92 DC 25
.MODEL DX D(IS=800.0E-18)
.MODEL JX PJF(IS=32.50E-12 BETA=311E-6 VTO=-1)
.ENDS
```

```
* TL071 OPERATIONAL AMPLIFIER "MACROMODEL" SUBCIRCUIT
* CREATED USING PARTS RELEASE 4.01 ON 06/16/89 AT 13:08
* (REV N/A)
* CONNECTIONS:   NON-INVERTING INPUT
*                | INVERTING INPUT
*                | | POSITIVE POWER SUPPLY
*                | | | NEGATIVE POWER SUPPLY
*                | | | | OUTPUT
*                | | | | |
.SUBCKT TL071    1 2 3 4 5
*
  C1   11 12 3.498E-12
  C2    6  7 15.00E-12
  DC    5 53 DX
  DE   54  5 DX
  DLP  90 91 DX
  DLN  92 90 DX
  DP    4  3 DX
  EGND 99  0 POLY(2) (3,0) (4,0) 0 .5 .5
  FB    7 99 POLY(5) VB VC VE VLP VLN 0 4.715E6 -5E6 5E6 5E6 -5E6
  GA    6  0 11 12 282.8E-6
  GCM   0  6 10 99 8.942E-9
  ISS   3 10 DC 195.0E-6
  HLIM 90  0 VLIM 1K
  J1   11  2 10 JX
  J2   12  1 10 JX
  R2    6  9 100.0E3
  RD1   4 11 3.536E3
  RD2   4 12 3.536E3
  RO1   8  5 150
  RO2   7 99 150
  RP    3  4 2.143E3
  RSS  10 99 1.026E6
  VB    9  0 DC 0
  VC    3 53 DC 2.200
  VE   54  4 DC 2.200
  VLIM  7  8 DC 0
  VLP  91  0 DC 25
  VLN   0 92 DC 25
.MODEL DX D(IS=800.0E-18)
.MODEL JX PJF(IS=15.00E-12 BETA=270.1E-6 VTO=-1)
.ENDS
```

```
* TL072 OPERATIONAL AMPLIFIER "MACROMODEL" SUBCIRCUIT
* CREATED USING PARTS RELEASE 4.01 ON 06/16/89 AT 13:08
* (REV N/A)
* CONNECTIONS:   NON-INVERTING INPUT
*                | INVERTING INPUT
*                | | POSITIVE POWER SUPPLY
*                | | | NEGATIVE POWER SUPPLY
*                | | | | OUTPUT
*                | | | | |
.SUBCKT TL072    1 2 3 4 5
*
  C1    11 12 3.498E-12
  C2     6  7 15.00E-12
  DC     5 53 DX
  DE    54  5 DX
  DLP   90 91 DX
  DLN   92 90 DX
  DP     4  3 DX
  EGND  99  0 POLY(2) (3,0) (4,0) 0 .5 .5
  FB     7 99 POLY(5) VB VC VE VLP VLN 0 4.715E6 -5E6 5E6 5E6 -5E6
  GA     6  0 11 12 282.8E-6
  GCM    0  6 10 99 8.942E-9
  ISS    3 10 DC 195.0E-6
  HLIM  90  0 VLIM 1K
  J1    11  2 10 JX
  J2    12  1 10 JX
  R2     6  9 100.0E3
  RD1    4 11 3.536E3
  RD2    4 12 3.536E3
  RO1    8  5 150
  RO2    7 99 150
  RP     3  4 2.143E3
  RSS   10 99 1.026E6
  VB     9  0 DC 0
  VC     3 53 DC 2.200
  VE    54  4 DC 2.200
  VLIM   7  8 DC 0
  VLP   91  0 DC 25
  VLN    0 92 DC 25
.MODEL DX D(IS=800.0E-18)
.MODEL JX PJF(IS=15.00E-12 BETA=270.1E-6 VTO=-1)
.ENDS
```

```
* TL074 OPERATIONAL AMPLIFIER "MACROMODEL" SUBCIRCUIT
* CREATED USING PARTS RELEASE 4.01 ON 06/16/89 AT 13:08
* (REV N/A)
* CONNECTIONS:   NON-INVERTING INPUT
*                | INVERTING INPUT
*                | | POSITIVE POWER SUPPLY
*                | | | NEGATIVE POWER SUPPLY
*                | | | | OUTPUT
*                | | | | |
.SUBCKT TL074    1 2 3 4 5
*
  C1    11 12 3.498E-12
  C2     6  7 15.00E-12
  DC     5 53 DX
  DE    54  5 DX
  DLP   90 91 DX
  DLN   92 90 DX
  DP     4  3 DX
  EGND  99  0 POLY(2) (3,0) (4,0) 0 .5 .5
  FB     7 99 POLY(5) VB VC VE VLP VLN 0 4.715E6 -5E6 5E6 5E6 -5E6
  GA     6  0 11 12 282.8E-6
  GCM    0  6 10 99 8.942E-9
  ISS    3 10 DC 195.0E-6
  HLIM  90  0 VLIM 1K
  J1    11  2 10 JX
  J2    12  1 10 JX
  R2     6  9 100.0E3
  RD1    4 11 3.536E3
  RD2    4 12 3.536E3
  RO1    8  5 150
  RO2    7 99 150
  RP     3  4 2.143E3
  RSS   10 99 1.026E6
  VB     9  0 DC 0
  VC     3 53 DC 2.200
  VE    54  4 DC 2.200
  VLIM   7  8 DC 0
  VLP   91  0 DC 25
  VLN    0 92 DC 25
.MODEL DX D(IS=800.0E-18)
.MODEL JX PJF(IS=15.00E-12 BETA=270.1E-6 VTO=-1)
.ENDS
```

```
* TL075 OPERATIONAL AMPLIFIER "MACROMODEL" SUBCIRCUIT
* CREATED USING PARTS RELEASE 4.01 ON 06/16/89 AT 13:08
* (REV N/A)
* CONNECTIONS:   NON-INVERTING INPUT
*                | INVERTING INPUT
*                | | POSITIVE POWER SUPPLY
*                | | | NEGATIVE POWER SUPPLY
*                | | | | OUTPUT
*                | | | | |
.SUBCKT TL075    1 2 3 4 5
*
  C1    11 12 3.498E-12
  C2     6  7 15.00E-12
  DC     5 53 DX
  DE    54  5 DX
  DLP   90 91 DX
  DLN   92 90 DX
  DP     4  3 DX
  EGND  99  0 POLY(2) (3,0) (4,0) 0 .5 .5
  FB     7 99 POLY(5) VB VC VE VLP VLN 0 4.715E6 -5E6 5E6 5E6 -5E6
  GA     6  0 11 12 282.8E-6
  GCM    0  6 10 99 8.942E-9
  ISS    3 10 DC 195.0E-6
  HLIM  90  0 VLIM 1K
  J1    11  2 10 JX
  J2    12  1 10 JX
  R2     6  9 100.0E3
  RD1    4 11 3.536E3
  RD2    4 12 3.536E3
  RO1    8  5 150
  RO2    7 99 150
  RP     3  4 2.143E3
  RSS   10 99 1.026E6
  VB     9  0 DC 0
  VC     3 53 DC 2.200
  VE    54  4 DC 2.200
  VLIM   7  8 DC 0
  VLP   91  0 DC 25
  VLN    0 92 DC 25
.MODEL DX D(IS=800.0E-18)
.MODEL JX PJF(IS=15.00E-12 BETA=270.1E-6 VTO=-1)
.ENDS
```

```
* TL080 OPERATIONAL AMPLIFIER "MACROMODEL" SUBCIRCUIT
* CREATED USING PARTS RELEASE 4.01 ON 09/15/89 AT 12:46
* (REV N/A)
* CONNECTIONS:   NON-INVERTING INPUT
*                | INVERTING INPUT
*                | | POSITIVE POWER SUPPLY
*                | | | NEGATIVE POWER SUPPLY
*                | | | | OUTPUT
*                | | | | |  COMPENSATION
*                | | | | | / \
.SUBCKT TL080    1 2 3 4 5 6 7
*
  C1   11 12 5.197E-12
  DC    5 53 DX
  DE   54  5 DX
  DLP  90 91 DX
  DLN  92 90 DX
  DP    4  3 DX
  EGND 99  0 POLY(2) (3,0) (4,0) 0 .5 .5
  FB    7 99 POLY(5) VB VC VE VLP VLN 0 3.803E6 -4E6 4E6 4E6 -4E6
  GA   6  0 11 12 377.6E-6
  GCM   0  6 10 99 9.882E-9
  ISS   3 10 DC 234.0E-6
  HLIM 90  0 VLIM 1K
  J1   11  2 10 JX
  J2   12  1 10 JX
  R2    6  9 100.0E3
  RD1   4 11 2.852E3
  RD2   4 12 2.852E3
  RO1   8  5 150
  RO2   7 99 150
  RP    3  4 21.43E3
  RSS  10 99 854.7E3
  VB    9  0 DC 0
  VC    3 53 DC 2.200
  VE   54  4 DC 2.200
  VLIM  7  8 DC 0
  VLP  91  0 DC 25
  VLN   0 92 DC 25
.MODEL DX D(IS=800.0E-18)
.MODEL JX PJF(IS=15.00E-12 BETA=332E-6 VTO=-1)
.ENDS
```

```
* TL081 OPERATIONAL AMPLIFIER "MACROMODEL" SUBCIRCUIT
* CREATED USING PARTS RELEASE 4.01 ON 06/16/89 AT 13:08
* (REV N/A)
* CONNECTIONS:   NON-INVERTING INPUT
*                | INVERTING INPUT
*                | | POSITIVE POWER SUPPLY
*                | | | NEGATIVE POWER SUPPLY
*                | | | | OUTPUT
*                | | | | |
.SUBCKT TL081    1 2 3 4 5
*
  C1    11 12 3.498E-12
  C2     6  7 15.00E-12
  DC     5 53 DX
  DE    54  5 DX
  DLP   90 91 DX
  DLN   92 90 DX
  DP     4  3 DX
  EGND 99  0 POLY(2) (3,0) (4,0) 0 .5 .5
  FB     7 99 POLY(5) VB VC VE VLP VLN 0 4.715E6 -5E6 5E6 5E6 -5E6
  GA     6  0 11 12 282.8E-6
  GCM    0  6 10 99 8.942E-9
  ISS    3 10 DC 195.0E-6
  HLIM 90  0 VLIM 1K
  J1    11  2 10 JX
  J2    12  1 10 JX
  R2     6  9 100.0E3
  RD1    4 11 3.536E3
  RD2    4 12 3.536E3
  RO1    8  5 150
  RO2    7 99 150
  RP     3  4 2.143E3
  RSS   10 99 1.026E6
  VB     9  0 DC 0
  VC     3 53 DC 2.200
  VE    54  4 DC 2.200
  VLIM   7  8 DC 0
  VLP   91  0 DC 25
  VLN    0 92 DC 25
.MODEL DX D(IS=800.0E-18)
.MODEL JX PJF(IS=15.00E-12 BETA=270.1E-6 VTO=-1)
.ENDS
```

```
* TL082 OPERATIONAL AMPLIFIER "MACROMODEL" SUBCIRCUIT
* CREATED USING PARTS RELEASE 4.01 ON 06/16/89 AT 13:08
* (REV N/A)
* CONNECTIONS:   NON-INVERTING INPUT
*                | INVERTING INPUT
*                | | POSITIVE POWER SUPPLY
*                | | | NEGATIVE POWER SUPPLY
*                | | | | OUTPUT
*                | | | | |
.SUBCKT TL082    1 2 3 4 5
*
  C1   11 12 3.498E-12
  C2    6  7 15.00E-12
  DC    5 53 DX
  DE   54  5 DX
  DLP  90 91 DX
  DLN  92 90 DX
  DP    4  3 DX
  EGND 99  0 POLY(2) (3,0) (4,0) 0 .5 .5
  FB    7 99 POLY(5) VB VC VE VLP VLN 0 4.715E6 -5E6 5E6 5E6 -5E6
  GA    6  0 11 12 282.8E-6
  GCM   0  6 10 99 8.942E-9
  ISS   3 10 DC 195.0E-6
  HLIM 90  0 VLIM 1K
  J1   11  2 10 JX
  J2   12  1 10 JX
  R2    6  9 100.0E3
  RD1   4 11 3.536E3
  RD2   4 12 3.536E3
  RO1   8  5 150
  RO2   7 99 150
  RP    3  4 2.143E3
  RSS  10 99 1.026E6
  VB    9  0 DC 0
  VC    3 53 DC 2.200
  VE   54  4 DC 2.200
  VLIM  7  8 DC 0
  VLP  91  0 DC 25
  VLN   0 92 DC 25
.MODEL DX D(IS=800.0E-18)
.MODEL JX PJF(IS=15.00E-12 BETA=270.1E-6 VTO=-1)
.ENDS
```

```
* TL083 OPERATIONAL AMPLIFIER "MACROMODEL" SUBCIRCUIT
* CREATED USING PARTS RELEASE 4.01 ON 06/16/89 AT 13:08
* (REV N/A)
* CONNECTIONS:   NON-INVERTING INPUT
*                | INVERTING INPUT
*                | | POSITIVE POWER SUPPLY
*                | | | NEGATIVE POWER SUPPLY
*                | | | | OUTPUT
*                | | | | |
.SUBCKT TL083    1 2 3 4 5
*
  C1    11 12 3.498E-12
  C2     6  7 15.00E-12
  DC     5 53 DX
  DE    54  5 DX
  DLP   90 91 DX
  DLN   92 90 DX
  DP     4  3 DX
  EGND  99  0 POLY(2) (3,0) (4,0) 0 .5 .5
  FB     7 99 POLY(5) VB VC VE VLP VLN 0 4.715E6 -5E6 5E6 5E6 -5E6
  GA     6  0 11 12 282.8E-6
  GCM    0  6 10 99 8.942E-9
  ISS    3 10 DC 195.0E-6
  HLIM  90  0 VLIM 1K
  J1    11  2 10 JX
  J2    12  1 10 JX
  R2     6  9 100.0E3
  RD1    4 11 3.536E3
  RD2    4 12 3.536E3
  RO1    8  5 150
  RO2    7 99 150
  RP     3  4 2.143E3
  RSS   10 99 1.026E6
  VB     9  0 DC 0
  VC     3 53 DC 2.200
  VE    54  4 DC 2.200
  VLIM   7  8 DC 0
  VLP   91  0 DC 25
  VLN    0 92 DC 25
.MODEL DX D(IS=800.0E-18)
.MODEL JX PJF(IS=15.00E-12 BETA=270.1E-6 VTO=-1)
.ENDS
```

```
* TL084 OPERATIONAL AMPLIFIER "MACROMODEL" SUBCIRCUIT
* CREATED USING PARTS RELEASE 4.01 ON 06/16/89 AT 13:08
* (REV N/A)
* CONNECTIONS:   NON-INVERTING INPUT
*                | INVERTING INPUT
*                | | POSITIVE POWER SUPPLY
*                | | | NEGATIVE POWER SUPPLY
*                | | | | OUTPUT
*                | | | | |
.SUBCKT TL084    1 2 3 4 5
*
  C1    11 12 3.498E-12
  C2     6  7 15.00E-12
  DC     5 53 DX
  DE    54  5 DX
  DLP   90 91 DX
  DLN   92 90 DX
  DP     4  3 DX
  EGND  99  0 POLY(2) (3,0) (4,0) 0 .5 .5
  FB     7 99 POLY(5) VB VC VE VLP VLN 0 4.715E6 -5E6 5E6 5E6 -5E6
  GA     6  0 11 12 282.8E-6
  GCM    0  6 10 99 8.942E-9
  ISS    3 10 DC 195.0E-6
  HLIM  90  0 VLIM 1K
  J1    11  2 10 JX
  J2    12  1 10 JX
  R2     6  9 100.0E3
  RD1    4 11 3.536E3
  RD2    4 12 3.536E3
  RO1    8  5 150
  RO2    7 99 150
  RP     3  4 2.143E3
  RSS   10 99 1.026E6
  VB     9  0 DC 0
  VC     3 53 DC 2.200
  VE    54  4 DC 2.200
  VLIM   7  8 DC 0
  VLP   91  0 DC 25
  VLN    0 92 DC 25
.MODEL DX D(IS=800.0E-18)
.MODEL JX PJF(IS=15.00E-12 BETA=270.1E-6 VTO=-1)
.ENDS
```

```
* TL085 OPERATIONAL AMPLIFIER "MACROMODEL" SUBCIRCUIT
* CREATED USING PARTS RELEASE 4.01 ON 06/16/89 AT 13:08
* (REV N/A)
* CONNECTIONS:   NON-INVERTING INPUT
*                | INVERTING INPUT
*                | | POSITIVE POWER SUPPLY
*                | | | NEGATIVE POWER SUPPLY
*                | | | | OUTPUT
*                | | | | |
.SUBCKT TL085    1 2 3 4 5
*
  C1   11 12 3.498E-12
  C2    6  7 15.00E-12
  DC    5 53 DX
  DE   54  5 DX
  DLP  90 91 DX
  DLN  92 90 DX
  DP    4  3 DX
  EGND 99  0 POLY(2) (3,0) (4,0) 0 .5 .5
  FB    7 99 POLY(5) VB VC VE VLP VLN 0 4.715E6 -5E6 5E6 5E6 -5E6
  GA    6  0 11 12 282.8E-6
  GCM   0  6 10 99 8.942E-9
  ISS   3 10 DC 195.0E-6
  HLIM 90  0 VLIM 1K
  J1   11  2 10 JX
  J2   12  1 10 JX
  R2    6  9 100.0E3
  RD1   4 11 3.536E3
  RD2   4 12 3.536E3
  RO1   8  5 150
  RO2   7 99 150
  RP    3  4 2.143E3
  RSS  10 99 1.026E6
  VB    9  0 DC 0
  VC    3 53 DC 2.200
  VE   54  4 DC 2.200
  VLIM  7  8 DC 0
  VLP  91  0 DC 25
  VLN   0 92 DC 25
.MODEL DX D(IS=800.0E-18)
.MODEL JX PJF(IS=15.00E-12 BETA=270.1E-6 VTO=-1)
.ENDS
```

```
* TL087 OPERATIONAL AMPLIFIER "MACROMODEL" SUBCIRCUIT
* CREATED USING PARTS RELEASE 4.01 ON 06/28/89 AT 14:04
* (REV N/A)
* CONNECTIONS:   NON-INVERTING INPUT
*                | INVERTING INPUT
*                | | POSITIVE POWER SUPPLY
*                | | | NEGATIVE POWER SUPPLY
*                | | | | OUTPUT
*                | | | | |
.SUBCKT TL087    1 2 3 4 5
*
  C1   11 12 3.887E-12
  C2   6  7 12.00E-12
  DC    5 53 DX
  DE   54  5 DX
  DLP  90 91 DX
  DLN  92 90 DX
  DP    4  3 DX
  EGND 99  0 POLY(2) (3,0) (4,0) 0 .5 .5
  FB    7 99 POLY(5) VB VC VE VLP VLN 0 6.189E6 -6E6 6E6 6E6 -6E6
  GA    6  0 11 12 282.8E-6
  GCM   0  6 10 99 3.560E-9
  ISS   3 10 DC 270.0E-6
  HLIM 90  0 VLIM 1K
  J1   11  2 10 JX
  J2   12  1 10 JX
  R2    6  9 100.0E3
  RD1   4 11 3.536E3
  RD2   4 12 3.536E3
  RO1   8  5 60
  RO2   7 99 60
  RP    3  4 11.54E3
  RSS  10 99 740.7E3
  VB    9  0 DC 0
  VC    3 53 DC 2.200
  VE   54  4 DC 2.200
  VLIM  7  8 DC 0
  VLP  91  0 DC 30
  VLN   0 92 DC 30
.MODEL DX D(IS=800.0E-18)
.MODEL JX PJF(IS=15.00E-12 BETA=165.3E-6 VTO=-1)
.ENDS
```

```
* TL088 OPERATIONAL AMPLIFIER "MACROMODEL" SUBCIRCUIT
* CREATED USING PARTS RELEASE 4.01 ON 06/28/89 AT 14:04
* (REV N/A)
* CONNECTIONS:   NON-INVERTING INPUT
*                | INVERTING INPUT
*                | | POSITIVE POWER SUPPLY
*                | | | NEGATIVE POWER SUPPLY
*                | | | | OUTPUT
*                | | | | |
.SUBCKT TL088    1 2 3 4 5
*
  C1    11 12 3.887E-12
  C2    6  7 12.00E-12
  DC     5 53 DX
  DE    54  5 DX
  DLP   90 91 DX
  DLN   92 90 DX
  DP     4  3 DX
  EGND  99  0 POLY(2) (3,0) (4,0) 0 .5 .5
  FB     7 99 POLY(5) VB VC VE VLP VLN 0 6.189E6 -6E6 6E6 6E6 -6E6
  GA     6  0 11 12 282.8E-6
  GCM    0  6 10 99 3.560E-9
  ISS    3 10 DC 270.0E-6
  HLIM  90  0 VLIM 1K
  J1    11  2 10 JX
  J2    12  1 10 JX
  R2     6  9 100.0E3
  RD1    4 11 3.536E3
  RD2    4 12 3.536E3
  RO1    8  5 60
  RO2    7 99 60
  RP     3  4 11.54E3
  RSS   10 99 740.7E3
  VB     9  0 DC 0
  VC     3 53 DC 2.200
  VE    54  4 DC 2.200
  VLIM   7  8 DC 0
  VLP   91  0 DC 30
  VLN    0 92 DC 30
.MODEL DX D(IS=800.0E-18)
.MODEL JX PJF(IS=15.00E-12 BETA=165.3E-6 VTO=-1)
.ENDS
```

```
* TL321 OPERATIONAL AMPLIFIER "MACROMODEL" SUBCIRCUIT
* CREATED USING PARTS RELEASE 4.01 ON 09/14/89 AT 13:31
* (REV N/A)
* CONNECTIONS:   NON-INVERTING INPUT
*                | INVERTING INPUT
*                | | POSITIVE POWER SUPPLY
*                | | | NEGATIVE POWER SUPPLY
*                | | | | OUTPUT
*                | | | | |
.SUBCKT TL321    1 2 3 4 5
*
  C1    11 12 4.664E-12
  C2     6  7 20.00E-12
  DC     5 53 DX
  DE    54  5 DX
  DLP   90 91 DX
  DLN   92 90 DX
  DP     4  3 DX
  EGND  99  0 POLY(2) (3,0) (4,0) 0 .5 .5
  FB     7 99 POLY(5) VB VC VE VLP VLN 0 15.91E6 -20E6 20E6 20E6 -20E6
  GA     6  0 11 12 132.7E-6
  GCM    0  6 10 99 3.974E-9
  IEE    3 10 DC 10.09E-6
  HLIM  90  0 VLIM 1K
  Q1    11  2 13 QX
  Q2    12  1 14 QX
  R2     6  9 100.0E3
  RC1    4 11 7.957E3
  RC2    4 12 7.957E3
  RE1   13 10 2.759E3
  RE2   14 10 2.759E3
  REE   10 99 19.82E6
  RO1    8  5 50
  RO2    7 99 50
  RP     3  4 15.08E3
  VB     9  0 DC 0
  VC     3 53 DC 2.600
  VE    54  4 DC .6
  VLIM   7  8 DC 0
  VLP   91  0 DC 40
  VLN    0 92 DC 40
.MODEL DX D(IS=800.0E-18)
.MODEL QX PNP(IS=800.0E-18 BF=111.1)
.ENDS
```

```
* TL322 OPERATIONAL AMPLIFIER "MACROMODEL" SUBCIRCUIT
* CREATED USING PARTS RELEASE 4.01 ON 09/08/89 AT 13:59
* (REV N/A)
* CONNECTIONS:   NON-INVERTING INPUT
*                | INVERTING INPUT
*                | | POSITIVE POWER SUPPLY
*                | | | NEGATIVE POWER SUPPLY
*                | | | | OUTPUT
*                | | | | |
.SUBCKT TL322    1 2 3 4 5
*
  C1    11 12 7.544E-12
  C2     6  7 20.00E-12
  DC     5 53 DX
  DE    54  5 DX
  DLP   90 91 DX
  DLN   92 90 DX
  DP     4  3 DX
  EGND 99   0 POLY(2) (3,0) (4,0) 0 .5 .5
  FB     7 99 POLY(5) VB VC VE VLP VLN 0 42.44E6 -40E6 40E6 40E6 -40E6
  GA    6   0 11 12 130.7E-6
  GCM    0  6 10 99 2.235E-9
  IEE    3 10 DC 12.40E-6
  HLIM 90   0 VLIM 1K
  Q1    11  2 13 QX
  Q2    12  1 14 QX
  R2     6  9 100.0E3
  RC1    4 11 7.957E3
  RC2    4 12 7.957E3
  RE1   13 10 3.529E3
  RE2   14 10 3.529E3
  REE   10 99 16.13E6
  RO1    8  5 37.50
  RO2    7 99 37.50
  RP     3  4 43.62E3
  VB     9  0 DC 0
  VC    3  53 DC 2.600
  VE    54  4 DC 2.600
  VLIM   7  8 DC 0
  VLP   91  0 DC 30
  VLN    0 92 DC 30
.MODEL DX D(IS=800.0E-18)
.MODEL QX PNP(IS=800.0E-18 BF=30)
.ENDS
```

```
*
* UA741 OPERATIONAL AMPLIFIER "MACROMODEL" SUBCIRCUIT
* CREATED USING PARTS RELEASE 4.01 ON 07/05/89 AT 09:09
* (REV N/A)
* CONNECTIONS:   NON-INVERTING INPUT
*                | INVERTING INPUT
*                | | POSITIVE POWER SUPPLY
*                | | | NEGATIVE POWER SUPPLY
*                | | | | OUTPUT
*                | | | | |
.SUBCKT UA741    1 2 3 4 5
*
  C1   11 12 4.664E-12
  C2    6  7 20.00E-12
  DC    5 53 DX
  DE   54  5 DX
  DLP  90 91 DX
  DLN  92 90 DX
  DP    4  3 DX
  EGND 99  0 POLY(2) (3,0) (4,0) 0 .5 .5
  FB    7 99 POLY(5) VB VC VE VLP VLN 0 10.61E6 -10E6 10E6 10E6 -10E6
  GA   6  0 11 12 137.7E-6
  GCM   0  6 10 99 2.574E-9
  IEE  10  4 DC 10.16E-6
  HLIM 90  0 VLIM 1K
  Q1   11  2 13 QX
  Q2   12  1 14 QX
  R2    6  9 100.0E3
  RC1   3 11 7.957E3
  RC2   3 12 7.957E3
  RE1  13 10 2.740E3
  RE2  14 10 2.740E3
  REE  10 99 19.69E6
  RO1   8  5 150
  RO2   7 99 150
  RP    3  4 18.11E3
  VB    9  0 DC 0
  VC   3 53 DC 2.600
  VE   54  4 DC 2.600
  VLIM  7  8 DC 0
  VLP  91  0 DC 25
  VLN   0 92 DC 25
.MODEL DX D(IS=800.0E-18)
.MODEL QX NPN(IS=800.0E-18 BF=62.50)
.ENDS
```

```
* UA747 OPERATIONAL AMPLIFIER "MACROMODEL" SUBCIRCUIT
* CREATED USING PARTS RELEASE 4.01 ON 07/05/89 AT 09:09
* (REV N/A)
* CONNECTIONS:   NON-INVERTING INPUT
*                | INVERTING INPUT
*                | | POSITIVE POWER SUPPLY
*                | | | NEGATIVE POWER SUPPLY
*                | | | | OUTPUT
*                | | | | |
.SUBCKT UA747    1 2 3 4 5
*
  C1    11 12 4.664E-12
  C2     6  7 20.00E-12
  DC     5 53 DX
  DE    54  5 DX
  DLP   90 91 DX
  DLN   92 90 DX
  DP     4  3 DX
  EGND  99  0 POLY(2) (3,0) (4,0) 0 .5 .5
  FB     7 99 POLY(5) VB VC VE VLP VLN 0 10.61E6 -10E6 10E6 10E6 -10E6
  GA    6  0 11 12 137.7E-6
  GCM    0  6 10 99 2.574E-9
  IEE   10  4 DC 10.16E-6
  HLIM 90   0 VLIM 1K
  Q1    11  2 13 QX
  Q2    12  1 14 QX
  R2     6  9 100.0E3
  RC1    3 11 7.957E3
  RC2    3 12 7.957E3
  RE1   13 10 2.740E3
  RE2   14 10 2.740E3
  REE   10 99 19.69E6
  RO1    8  5 150
  RO2    7 99 150
  RP     3  4 18.11E3
  VB     9  0 DC 0
  VC    3 53 DC 2.600
  VE    54  4 DC 2.600
  VLIM   7  8 DC 0
  VLP   91  0 DC 25
  VLN    0 92 DC 25
.MODEL DX D(IS=800.0E-18)
.MODEL QX NPN(IS=800.0E-18 BF=62.50)
.ENDS
```

```
* UA748 OPERATIONAL AMPLIFIER "MACROMODEL" SUBCIRCUIT
* CREATED USING PARTS RELEASE 4.01 ON 09/01/89 AT 13:14
* (REV N/A)
* CONNECTIONS:   NON-INVERTING INPUT
*                | INVERTING INPUT
*                | | POSITIVE POWER SUPPLY
*                | | | NEGATIVE POWER SUPPLY
*                | | | | OUTPUT
*                | | | | |  COMPENSATION
*                | | | | | / \
.SUBCKT UA748    1 2 3 4 5 6 7
*
  C1    11 12 7.977E-12
  DC     5 53 DX
  DE    54  5 DX
  DLP   90 91 DX
  DLN   92 90 DX
  DP     4  3 DX
  EGND  99  0 POLY(2) (3,0) (4,0) 0 .5 .5
  FB     7 99 POLY(5) VB VC VE VLP VLN 0 42.44E6 -40E6 40E6 40E6 -40E6
  GA     6  0 11 12 188.5E-6
  GCM    0  6 10 99 3.352E-9
  IEE   10  4 DC 15.14E-6
  HLIM  90  0 VLIM 1K
  Q1    11  2 13 QX
  Q2    12  1 14 QX
  R2     6  9 100.0E3
  RC1    3 11 5.305E3
  RC2    3 12 5.305E3
  RE1   13 10 1.839E3
  RE2   14 10 1.839E3
  REE   10 99 13.21E6
  RO1    8  5 50
  RO2    7 99 25
  RP     3  4 16.81E3
  VB     9  0 DC 0
  VC     3 53 DC 2.600
  VE    54  4 DC 2.600
  VLIM   7  8 DC 0
  VLP   91  0 DC 25
  VLN    0 92 DC 25
.MODEL DX D(IS=800.0E-18)
.MODEL QX NPN(IS=800.0E-18 BF=107.1)
.ENDS
```

APPENDIX B

SPICE COMMAND AND DEVICE DESCRIPTION SUMMARY

Component/command category	Basic form	Statement parameters/format	Chapter
BASIC COMPONENTS			
Capacitor, voltage dependent	C	CXXXXXXX N+ N- POLY C0 C1 C2 ... [IC = INCOND]	2.2.4
Capacitor	C	CXXXXXXX N+ N- CVALUE [IC=IN-COND]	2.2.2
Current controlled current source, dependent	F	FXXXXXXX N+ N- VNAM VALUE	5.1.3
Current source, independent	I	IXXXXXXX N+ N- [DC] value [AC value [ACPHASE]] [PULSE I1 I2 [td [tr [tf [pw [per]]]]]] or [SIN IO IA [freq [td [kd]]]] or [EXP I1 I2 [td1 [t1 [td2 [t2]]]]] or [PWL T1 I1 T2 I2 ... TN IN] or [SFFM IO IA FREQ [mdi [fs]]]	2.2.7, 4.4, 6.1
Current controlled voltage source, dependent	H	HXXXXXXX N+ N- VNAM VALUE	5.1.4
Inductor, current dependent	L	LYYYYYYY N+ N- POLY L0 L1 L2 ... [IC = INCOND]	2.2.4
Inductor	L	LXXXXXXX N+ N- LVALUE [IC=IN-COND]	2.2.3

Component/command category	Basic form	Statement parameters/format	Chapter
Resistor	R	RXXXXXXX N1 N2 VALUE [TC = TC1-[,TC2]]	2.2.1
Subcircuit calls	X	XYYYYYYY N1 [N2 N3 ...] SUBNAM	5.2.2
Transformer	K	KXXXXXXX LXXXXXXX LYYYYYYY MC	2.2.5
Transmission line	T	TXXXXXXX N1 N2 N3 N4 Z0=VALUE [TD=VALUE] [F=FREQ [NL=NRMLEN]] [IC=V1,I1,V2,I2]	7.3.3
Voltage controlled voltage source, dependent	E	EXXXXXXX N+ N- NC+ NC- VALUE	5.1.2
Voltage controlled current source, dependent	G	GXXXXXXX N+ N- NC+ NC- VALUE	5.1.1
Voltage source, independent	V	VXXXXXXX N+ N- [DC] value [AC value [ACPHASE]] [PULSE V1 V2 [td [tr [tf [pw [per]]]]]] or [SIN VO VA [freq [td [kd]]]] or [EXP V1 V2 [td1 [t1 [td2 [t2]-]]]] or [PWL T1 V1 T2 V2 ... TN VN] or [SFFM VO VA FREQ [mdi [fs]]]	2.2.6, 4.4, 6.1
NON-ANALYSIS IMPLEMENTING COMMANDS			
Operating point statement	.OP	.OP	2.3.1
Options statement	.OPTIONS	.OPTIONS OPT1 OPT2 ... (or OPT=OPT-VAL ...)	2.4
Sensitivity analysis statement	.SENS	.SENS OV1 [OV2 ...]	4.2
Temperature analysis statement	.TEMP	.TEMP T1 [T2] [T3] [TN]	7.1
Transfer function statement	.TF	.TF OUTPUT-NODE INPUT-SOURCE	2.3.2
DC ANALYSIS COMMANDS			
DC sweep range description statement	.DC	.DC SRC1 VALSTRT1 VALSTOP1 VAL-INCR1 [SRC2 START2 STOP2 INCR2]	4.7
Plot command for dc analysis	.PLOT DC	.PLOT DC OV1 [OV2 ... OV8]	4.5, 4.6
Print command for dc analysis	.PRINT DC	.PRINT DC OV1 [OV2 ... OV8]	4.5

AC ANALYSIS COMMANDS

Component/command category	Basic form	Statement parameters/format	Chapter
AC sweep range description statement	.AC	.AC DEC NP FS FE .AC OCT NP FS FE .AC LIN NP FS FE	3.5.5
Distortion analysis statement	.DISTO	.DISTO RLOAD [INTER [SWK2 [REFPWR [SPW2]]]]	7.3.2
Noise analysis statement	.NOISE	.NOISE V(N1,[N2]) INSRC SINT	7.3.1
Plot command for ac analysis	.PLOT AC	.PLOT AC OV1 [OV2 ... OV8]	3.3.7, 3.4
Print command for ac analysis	.PRINT AC	.PRINT AC OV1 [OV2 ... OV8]	3.3.7
TRANSIENT ANALYSIS COMMANDS			
Fourier analysis statement	.FOUR	.FOUR FREQ OV1 [0V2 0V3 ...]	6.2
Initial conditions statement	.IC	.IC V(NODNUM)=VAL V(NODNUM)=VAL ...	6.3.2
Plot command for transient analysis	.PLOT TRAN	.PLOT TRAN V(X) I(VX)	6.3.3
Print command for transient analysis	.PRINT TRAN	.PRINT TRAN V(X) I(VX)	6.3.3
Transient analysis description statement	.TRAN	.TRAN TSTEP TSTOP [TSTART [TMAX]] [UIC]	6.3.1
SOLID STATE DEVICE DESIGNATORS/MODEL COMMANDS			
Diode model parameter description	.MODEL MNAME D	.MODEL MNAME D [PNAME1=PVAL1 PNAME2=PVAL2 ...]	4.3.1
Diode circuit interconnect statement	D	DXXXXXXX NA NC MNAME [AREA] [OFF] [IC=VD]	4.3.1
JFET model parameter description	.MODEL MNAME NJF .MODEL MNAME PJF	.MODEL MNAME NJF [PNAME1=PVAL1 PNAME2=PVAL2 ...] .MODEL MNAME PJF [PNAME1=PVAL1 PNAME2=PVAL2 ...]	4.3.4
JFET circuit interconnect statement	J	JXXXXXXX ND NG NS MNAME [AREA] [OFF] [IC=VDS,VGS]	4.3.4
MOSFET model parameter description	.MODEL MNAME NMOS .MODEL MNAME PMOS	.MODEL MNAME NMOS [PNAME1-=PVAL1 PNAME2=PVAL2 ...] .MODEL MNAME PMOS [PNAME1-=PVAL1 PNAME2=PVAL2 ...]	4.3.5

Component/command category	Basic form	Statement parameters/format	Chapter
MOSFET circuit interconnect statement	M	MXXXXXXX ND NG NS NB MNAME [L=VAL] [W=VAL] [AD=VAL] [AS=VAL] [PD=VAL] [PS=VAL] [NRD=VAL] [NRS=VAL] [OFF] [IC=VDS,VGS,VBS]	4.3.5
Transistor model parameter description	.MODEL MNAME NPN	.MODEL MNAME NPN [PNAME1=PVAL1 PNAME2=PVAL2 ...]	4.3.3
	.MODEL MNAME PNP	.MODEL MNAME PNP [PNAME1=PVAL1 PNAME2=PVAL2 ...]	
Transistor circuit interconnect statement	Q	QXXXXXXX NC NB NE [NS] MNAME [AREA] [OFF] [IC=VBE,VCE]	4.3.3

SUB-CIRCUIT COMMANDS

Component/command category	Basic form	Statement parameters/format	Chapter
Subcircuit definitions statement	.SUBCKT	.SUBCKT SUBNAM N1 [N2 N3 ...] ******* SUB-CIRCUIT DATA ******* .ENDS [SUBNAM]	5.2.1

COMMAND AND SPICE DEVICE DESCRIPTION SUMMARY SORTED BY BASIC FORM

Basic form	Statement parameters/format	Component/command category	Chapter
.AC	.AC DEC NP FS FE .AC OCT NP FS FE .AC LIN NP FS FE	AC sweep range description statement	3.5.5
.DC	.DC SRC1 VALSTRT1 VALSTOP1 VAL-INCR1 [SRC2 START2 STOP2 INCR2]	DC sweep range description statement	4.7
.DISTO	.DISTO RLOAD [INTER [SWK2 [REFPWR [SPW2]]]]	Distortion analysis statement	7.3.2
.FOUR	.FOUR FREQ OV1 [0V2 0V3 ...]	Fourier analysis statement	6.2
.IC	.IC V(NODNUM)=VAL V(NODNUM)=VAL ...	Initial conditions statement	6.3.2
.MODEL MNAME D	.MODEL MNAME D [PNAME1=PVAL1 PNAME2=PVAL2 ...]	Diode model parameter description	4.3.1
.MODEL MNAME NJF .MODEL MNAME PJF	.MODEL MNAME NJF [PNAME1=PVAL1 PNAME2=PVAL2 ...] .MODEL MNAME PJF [PNAME1=PVAL1 PNAME2=PVAL2 ...]	JFET model parameter description	4.3.4
.MODEL MNAME NPN .MODEL MNAME PNP	.MODEL MNAME NPN [PNAME1=PVAL1 PNAME2=PVAL2 ...] .MODEL MNAME PNP [PNAME1=PVAL1 PNAME2=PVAL2 ...]	Transistor model parameter description	4.3.3
.MODEL MNAME NMOS .MODEL MNAME PMOS	.MODEL MNAME NMOS [PNAME1-=PVAL1 PNAME2=PVAL2 ...] .MODEL MNAME PMOS [PNAME1-=PVAL1 PNAME2=PVAL2 ...]	MOSFET model parameter description	4.3.5
.NOISE	.NOISE V(N1,[N2]) INSRC SINT	Noise analysis statement	7.3.1
.OP	.OP	Operating point statement	2.3.1
.OPTIONS	.OPTIONS OPT1 OPT2 ... (or OPT-=OPTVAL ...)	Options statement	2.4

Basic form	Statement parameters/format	Component/command category	Chapter
.PLOT TRAN	.PLOT TRAN V(X) I(VX)	Plot command for transient analysis	6.3.3
.PLOT AC	.PLOT AC OV1 [OV2 ... OV8]	Plot command for ac analysis	3.3.7, 3.4
.PLOT DC	.PLOT DC OV1 [OV2 ... OV8]	Plot command for dc analysis	4.5, 4.6
.PRINT DC	.PRINT DC OV1 [OV2 ... OV8]	Print command for dc analysis	4.5
.PRINT AC	.PRINT AC OV1 [OV2 ... OV8]	Print command for ac analysis	3.3.7
.PRINT TRAN	.PRINT TRAN V(X) I(VX)	Print command for transient analysis	6.3.3
.SENS	.SENS OV1 [OV2 ...]	Sensitivity analysis statement	4.2
.SUBCKT	.SUBCKT SUBNAM N1 [N2 N3 ...] ** SUBCIRCUIT DATA** .ENDS [SUBNAM]	Subcircuit definitions statement	5.2.1
.TEMP	.TEMP T1 [T2] [T3] [TN]	Temperature analysis statement	7.1
.TF	.TF OUTPUT-NODE INPUT-SOURCE	Transfer function statement	2.3.2
.TRAN	.TRAN TSTEP TSTOP [TSTART [TMAX]] [UIC]	Transient analysis description statement	6.3.1
C	CXXXXXXX N+ N- CVALUE [IC=INCOND]	Capacitor	2.2.2
C	CXXXXXXX N+ N- POLY C0 C1 C2 ... [IC = INCOND]	Capacitor, voltage dependent	2.2.4
D	DXXXXXXX NA NC MNAME [AREA] [OFF] [IC=VD]	Diode circuit interconnect statement	4.3.1
E	EXXXXXXX N+ N- NC+ NC- VALUE	Voltage controlled voltage source, dependent	5.1.2
F	FXXXXXXX N+ N- VNAM VALUE	Current controlled current source, dependent	5.1.3
G	GXXXXXXX N+ N- NC+ NC- VALUE	Voltage controlled current source, dependent	5.1.1
H	HXXXXXXX N+ N- VNAM VALUE	Current controlled voltage source, dependent	5.1.4

Basic form	Statement parameters/format	Component/command category	Chapter
I	IXXXXXXX N+ N- [DC] value [ACMAG [ACPHASE]] [PULSE I1 I2 [td [tr [tf [pw [per]]]]]] or [SIN IO IA [freq [td [kd]]]] or [EXP I1 I2 [td1 [t1 [td2 [t2]]]]] or [PWL T1 I1 T2 I2 ... TN IN] or [SFFM IO IA FREQ [mdi [fs]]]	Current source, independent	2.2.7, 4.4, 6.1
J	JXXXXXXX ND NG NS MNAME [AREA] [OFF] [IC=VDS,VGS]	JFET circuit interconnect statement	4.3.4
K	KXXXXXXX LXXXXXXX LYYYYYYY MC	Transformer	2.2.5
L	LXXXXXXX N+ N- LVALUE [IC=INCOND]	Inductor	2.2.3
L	LYYYYYYY N+ N- POLY L0 L1 L2 ... [IC = INCOND]	Inductor, current dependent	2.2.4
M	MXXXXXXX ND NG NS NB MNAME [L=VAL] [W=VAL] [AD=VAL] [AS=VAL] [PD=VAL] [PS=VAL] [NRD=VAL] [NRS=VAL] [OFF] [IC=VDS,VGS,VBS]	MOSFET circuit interconnect statement	4.3.5
Q	QXXXXXXX NC NB NE [NS] MNAME [AREA] [OFF] [IC=VBE,VCE]	Transistor circuit interconnect statement	4.3.3
R	RXXXXXXX N1 N2 VALUE [TC = TC1 [,TC2]]	Resistor	2.2.1
T	TXXXXXXX N1 N2 N3 N4 Z0=VALUE [TD=VALUE] [F=FREQ [NL=NRMLEN]] [IC=V1,I1,V2,I2]	Transmission line	7.3.3
V	VXXXXXXX N+ N- [DC] value [ACMAG [ACPHASE]] [PULSE V1 V2 [td [tr [tf [pw [per]]]]]] or [SIN VO VA [freq [td [kd]]]] or [EXP V1 V2 [td1 [t1 [td2 [t2]]]]] or [PWL T1 V1 T2 V2 ... TN VN] or [SFFM VO VA FREQ [mdi [fs]]]	Voltage source, independent	2.2.6, 4.4, 6.1
X	XYYYYYYY N1 [N2 N3 ...] SUBNAM	Subcircuit calls	5.2.2

INDEX